HEAVY METALS IN THE ENVIRONMENT
ENVIRONMENT
Microorganisms and Bioremediation

CRC Press
Taylor & Francis Group
6000 Broken Sound Parkway NW, Suite 300
Boca Raton, FL 33487-2742

First issued in paperback 2020

© 2018 by Taylor & Francis Group, LLC
CRC Press is an imprint of Taylor & Francis Group, an Informa business

No claim to original U.S. Government works

ISBN-13: 978-1-138-03580-5 (hbk)
ISBN-13: 978-0-367-78157-6 (pbk)

Library of Congress Cataloging-in-Publication Data

Names: Donati, Edgardo R., editor.
Title: Heavy metals in the environment : microorganisms and bioremediation / editor, Edgardo R. Donati, National University of La Plata, La Plata, Argentina.
Description: Boca Raton, FL : CRC Press, 2018. | "A science publishers book." | Includes bibliographical references and index.
Identifiers: LCCN 2017044396 | ISBN 9781138035805 (hardback)
Subjects: LCSH: Heavy metals--Environmental aspects. | Bioremediation.
Classification: LCC TD196.M4 H4295 2018 | DDC 628.4/2--dc23
LC record available at https://lccn.loc.gov/2017044396

Visit the Taylor & Francis Web site at
http://www.taylorandfrancis.com

and the CRC Press Web site at
http://www.crcpress.com

PREFACE

Heavy metals and metalloids are released into the environment from anthropogenic activities without rare or null concern or government regulation in many countries. Sources of such pollutants are industrial effluents, municipal waste treatment plants, landfill leaching and mining activities, among others. Since heavy metals cannot be degraded, they persist and are accumulated over time, increasing the human exposure and causing serious negative environmental consequences. Although numerous physicochemical technologies have been developed in the last decades to remediate sites contaminated with heavy metal(loid)s, most are expensive and/or inefficient at low metal concentrations and large polluted areas. Hence, many new treatments have emerged in the last few decades.

In bioremediation processes, biological technologies are used to remediate contaminated environments. These processes offer high specificity in the removal of some particular heavy metal(loid)s of interest while also offering operational flexibility. In the 17 chapters written by experts in the field, this book deals with different approaches to the topic.

The first part comprises some aspects of the interaction between microbial communities and microorganisms and heavy metal(loid)s, including mechanisms of resistance to such pollutants. In the second part, different strategies for bioremediation are described: biosorption and bioaccumulation, bioprecipitation, biosolubilization, and also phytoremediation. The third part elucidates particular bioremediation cases for some of the most relevant heavy metal(loid)s: arsenic, cadmium, cobalt, copper, lead, and mercury. The last part comprises three chapters with field applications including an application using wetlands. These 17 chapters configure a comprehensive understanding of this area including some novel and interesting approaches to the topic.

I would like to acknowledge the efforts of all the contributors for bringing the book to fruition. The continued assistance of the Editorial Department of CRC Press is also highly appreciated.

Edgardo R. Donati

CONTENTS

PART I
INTERACTION METAL(LOID)S: MICROORGANISMS

CHAPTER 1

Microbial Communities and the Interaction with Heavy Metals and Metalloids
Impact and Adaptation

María Alejandra Lima, María Sofía Urbieta and
Edgardo R. Donati*

1. Introduction

1.1 Heavy metals and their influence on microbial diversity

The millenary activity of extracting heavy metals (HM) is nowadays considered to be one of the main economic activities of the world with México, Chile, China and Australia being some of the countries with higher productions. Due to their particular properties, HM were and still are, used in many industries and production activities. Consequently, HM consumption has increased at a higher rate than population which has led to the generation and accumulation of tons of wastes. The inappropriate disposal of HM wastes has caused their mobility into the environment causing serious alterations in the ecosystems and triggering health problems in populations close to the polluted areas. The mobility of HM and metalloids into the environment also increases their distribution (especially in water systems) and bioavailability. The mobility and bioavailability of these pollutants are controlled by many chemical and biochemical processes such as precipitation/dissolution, sorption/desorption, complexation/dissociation and oxidation/reduction. The increase in HM and metalloids concentration in the environment directly affects microbial diversity, community structure and metabolic activity. It also

CINDEFI (CCT La Plata-CONICET, UNLP), Facultad de Ciencias Exactas, Universidad Nacional de La Plata, Calle 47 y 115, (1900) La Plata, Argentina.
* Corresponding author: msurbieta@biol.unlp.edu.ar

causes the disappearance of susceptible species and the increase of tolerant species and/or the increase of the expression of resistance mechanisms (Zampieri et al., 2016).

1.2 Study of the impact of HM in microbial communities

The impact of heavy metals in the diversity of microbial communities has been studied for several years. The methods used have evolved with the passage of time. Culture dependent techniques are not quite suitable to study microbial diversity because only a small fraction of the microorganisms present in a sample can be successfully recovered in culture, even more if the sample contains any sort of contaminant such as heavy metals. Traditional molecular ecology techniques have been extensively used to characterize the microbial diversity in environments contaminated with heavy metals, either natural or from anthropogenic origin, such as mine tailings or industrial areas. Among the most commonly used techniques is DGGE (denaturating gradient gel electrophoresis) that involves the amplification of a short fragment of the 16S rRNA gene (aprox. 500 bp) of all the members of a microbial community that then are electrophoretically separated by sequence differences by the denaturizing effect of urea and formamide. In this way the band pattern of each sample reflexes the complexity of the microbial community. Moreover, the bands can be cut from the gel, the DNA eluded, re-amplified and sequenced to know the identity of the microorganisms present. An important advantage of DGGE is that it allows for a rapid comparison among different samples, for example the microbial community in a sample after and before being exposed to heavy metals. Another strategy that is used is the amplification of the whole 16S rRNA gene, cloning to construct a sequence library and sequencing the clones. The success of covering the entire microbial diversity of a sample depends on the number of clones sequenced, which is why it is recommended to do a DGGE run initially to assess the potential number of species present. Among the most modern approaches, the meta-omics techniques allows for a much deeper study in several aspects of the community behavior.

Metagenomics, the massive sequencing of the microbial community DNA, can be applied to a specific gene, for example the 16S rRNA, or to the entire genetic material. 16S rRNA metagenomics studies produce millions of sequences, which are longer with the advances of sequencing technologies, that can be used in robust statistics analysis of microbial community structure and diversity and, by comparison of different samples, the changes suffered by any stress source.

Metatranscriptomics deals with the massive sequencing of the whole RNA of a microbial community and gives information on the genes that are expressed in certain conditions and time. In the same way metaproteomics reveals all the proteins produced by the community. Both, metatranscriptomics and metaproteomics, allow the comparison of RNA or proteins, respectively, in two different conditions, which is fundamental to understand the metabolic changes in a microbial community impacted by heavy metals.

A completely different approach was taken by Wang and co-worker (2010) who used microcalorimetry to assess and compare the toxic effect of heavy metals, such as As, Cu, Cd, Cr, Co, Pb, and Zn, on soil microbial activities and community. The samples of soil were supplemented with glucose, ammonium sulfate and each heavy metal. The metabolic activity was estimated by calculating certain thermodynamic parameters associated with growth rate, such as microbial growth rate constant, total heat evolution, metabolic enthalpy and mass specific heat rate. According to the authors these parameters can act as indicators of changes occurred in soil due to their exposure to heavy metals, and allow them to propose a ranking of toxicity (Cr > Pb > As > Co > Zn > Cd > Cu).

2. Heavy metals' resistance mechanisms

Metals and metalloids can be classified into three categories according to their biological roles and effects on microorganisms (Prabhakaran et al., 2016):

- *Essential metals* (Na, Ca, K, Mn, Mg, V, Fe, Cu, Co, Mo, Ni, Zn, and W): have recognized biological role, however, their ions become toxic if the concentration increases.
- *Toxic metals* (Ag, Sn, Cd, Au, Ti, Hg, Pb, Al and metalloids Ge, Sb, As, and Se): do not have a biological role and can interfere in cellular processes.
- *Non-essential* (Rb, Sr, Cs, and T): do not have a defined biological role and have no toxicity.

Essential metals function as co-factors for enzymes and stabilizers of protein structures and bacterial cell walls, and help maintaining osmotic balance. For instance, Fe, Cu, and Ni are involved in redox processes and Mg and Zn stabilize various enzymes and DNA through electrostatic forces. Also, Fe, Mg, Ni, and Co are a part of molecular complexes with a wide array of functions. Toxicity occurs through the displacement of essential metals from their native binding sites or through ligand interactions. Toxic metals bind with higher affinity to thiol groups and oxygen binding sites than essential metals do. Toxicity results from alterations in the conformational structure of nucleic acids and proteins and due to interference with oxidative phosphorylation and osmotic balance (Bruins et al., 2000).

Microorganisms are the oldest member of the living system and possess high adaptability strength to thrive in adverse conditions. They primarily respond to the changes in the environment by altering their genetic system or by transferring the genetic elements among other mechanisms for maintaining the structure and function of ecosystem (Das et al., 2016).

Bacteria can use a set of direct or indirect mechanism in order to avoid the toxic effect of HM; for example:

Metal expulsion by a permeability barrier. Before entering the cell some HM can be retained by alterations in the cell wall or outer membrane to protect metal-sensitive internal components. Certain periplasmic proteins that are able to bind to specific metals have been described in different species. Giner-Lamia et al. (2015) studied a Cu-binding protein named CopM that is encoded in the *CopMRS* operon from the cyanobacterium *Synechocystis* sp. PCC 6803; the expression of the operon is activated by the product of CopMR when Cu ions are in the media, CopM bind Cu(I) in the periplasm and export it outside the cell avoiding the intracellular Cu accumulation (Fig. 1a). Similar proteins involved in Cu resistance were found in other microorganisms: CusF in *E. coli* (Franke et al., 2003), CueP in *Salmonella typhimurium* (Pontel and Soncini, 2009) and CopK in *C. metallidurans* (Mergeay et al., 2003). Also, the production of exopolysaccharides (EPS) by some microorganisms allows the retention of metal ions outside the cell. The presence of many functional anionic groups in EPS, such as carboxyl, phosphoryl, sulfhydryl, phenolic and hydroxyl groups, function as metal-binding sites with the possibility of cation exchange (Fig. 1b), physical sorption, complexation and/or precipitation of HM (More et al., 2014). Deschatre et al. (2015) studied an EPS produced by a marine

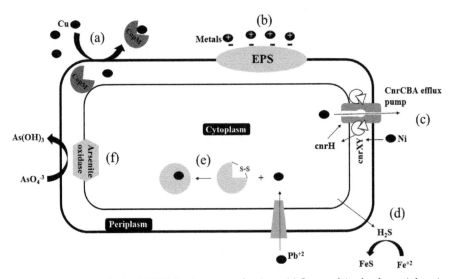

Figure 1. Overview of microbial HM resistance mechanisms. (a) Cu expulsion by the periplasmic Cu-binding protein CopM. (b) HM cations retention by anionic groups of EPS. (c) Ni expulsion from the cell by a specific efflux pump. (d) Extracellular Fe precipitation as FeS by SRB. (e) Intracellular Pb sequestration by the Pb-binding protein pbrD. (f) Oxidation of arsenite to a less toxic form by the arsenite oxidase.

bacteria which showed the maximum sorption capacities to be as high as 400 mg/g EPS and 256 mg/g EPS for Cu (II) and Ag (I), respectively, which are highly promising for bioremediation of these metals.

Active transport for expulsion of the metal from the cell. Microorganisms use active transport mechanisms to export toxic metals from their cytoplasm. These mechanisms can be encoded in the chromosome or on plasmids. Non-essential metals normally enter the cell through normal nutrient transport systems but are rapidly exported. These efflux systems, that can be ATP dependent or not, are highly specific for the cation or anion they export (Bruins et al., 2000). Ni resistance is generally mediated by active transport; Grass et al. (2000) reported the mechanisms of the operon *cnrYHXCBAT* founded in *Cupriavidus metallidurans* which encodes the highly efficient CnrCBA efflux pump that is activated by the sigma factor cnrH and the membrane-bound proteins cnrY and cnrX at μM concentrations of Ni. The mechanism is illustrated in Fig. 1c, when periplasmic Ni binds to cnrX, the anti-sigma factor cnrXY release cnrH which indirectly initiates the transcription of the CnrCBA efflux pump, necessary to expel Ni from the cell. In the case of Cu, one of the main systems of expulsion mediated by transporters, is that codified by the operon *copYABZ*, where *copA* and *copB* are P-type ATPases, also able to transport other metals such as Ag, Zn, and Cd. Besaury et al. (2013) reported high abundances of the ATPase CopA were found in Cu-contaminated Chilean marine sediments.

Extracellular metal sequestration. Microorganisms produce and expel metabolites such as phosphate, glutathione, oxalate, sulfur, among others that can bind metal and the complex metal-metabolite cannot pass through the cell membrane, thus decreasing or avoiding the absorption and intracellular harm. Sulfate reducing bacteria (SRB) that generate hydrogen sulfide through dissimilatory sulfate reduction are a great example of this mechanism; sulfide reacts with certain metals such as Cu, Zn, or Fe (Fig. 1d) forming insoluble precipitates thus decreasing the concentration of bioavailable metal. This mechanism could be critical for certain species. For instance when an acidophilic metal tolerant *Desulfosporosinus* was isolated from an abandoned gold mine highly contaminated with arsenic, the strain could not grow at very low concentrations of As even though the major tolerance mechanisms were present in its genome. The authors propose that As, could be precipitated as sulfides by SRB; in fact two different arsenic sulfides were found in the sediments samples of the tailing, thus, lowering the bioavailable concentration of As in the environment (Mardanov et al., 2016).

Intracellular metal sequestration. Bacteria are capable to complex metals and accumulate them in the cytoplasm. Complexants include glutathione and bacterial metallothioneins, also, Pb and Hg may accumulate in polyphosphate inclusions (Guillan, 2016). Jarosławiecka and Piotrowska-Seget (2014) described the mechanisms of Pb resistance in *C. metallidurans* CH34; this species has the *pbrABCD* operon that encoded a group of proteins involved in Pb resistance. In the presence of $Pb^{+2}pbrR$, a regulatory protein, induced the expression of *pbr*, one of the product of this transcription is *pbrD*, an intracellular Pb binding protein in which the binding-site is rich in cysteine residues and also bears a large number of proline and serine residues that sequester Pb reducing its toxic effects (Fig. 1e). Similarly the *mer* operon is induced by the presence of Hg^{+2}; *merR* is the activator of the expression and *merC* is the protein involved in the transport and accumulation of Hg. Some bacteria have the ppk gene that encodes polyphosphate kinase involved in polyphosphate biosynthesis. These negatively charged orthophosphate polymers are capable of binding Hg ions among others (Das et al., 2016).

Detoxification/Transformation to a less toxic form. It includes processes such as oxidation or reduction, methylation or volatilization. Detoxification of As is one of the most clear examples; As(V) is less toxic and less mobile than As(III) and thus some microorganisms use oxidation as a resistance strategy. The ArsC protein is involved in the oxidation of As(III) to As(V) for the purpose of detoxification only. But some chemolithoautotrops have the *aioAB* genes that codify for an arsenite oxidase that couples oxygen reduction to arsenite oxidation during fixation of CO_2 (Yamamura and Amachi, 2014) (Fig. 1f). Lesser known is the *arsM* gene that codifies for an arsenite S-adenosylmethionine methyltransferase that transforms inorganic As in its methylated less toxic form but is not specific for arsenic. In addition to this, biovolatilization occurs in microorganisms capable of removing Hg in a process mediated by a mercury ion reductase coded in the *merA* gene; this enzyme reduces Hg^{+2} to the volatile form Hg^0 which diffuses out of the cell.

The analysis of nine sedimentary samples from three sites in Xiawan port of Xiangjiang river (China) with different HM concentrations (Cu, Pb, Zn, As, Cd, Ni, Hg, Cr, and S) demonstrated that the site with higher contamination had more functional genes involving metal resistance (Jie et al., 2016). Therefore it would be possible to consider microbial composition, community structure and their functional genes as indicators of HM and metalloid pollution.

3. Adaptation: Are microorganisms resistant or tolerant to heavy metals?

The so far known mechanisms to avoid HM poisoning are varied and not all of them are specific to every metal. The knowledge of the specific genes involved in the active stress response to a certain metal allows to "predict" the mechanisms that can be used to survive in contaminated environments and the possibility to use them as resistance markers. However, there are other factors that need to be considered such as physicochemical conditions of the environment and possible lateral gene transfer (LGT) which allow the propagation of resistance to HM in bacteria.

Acidophiles (microorganisms that grow at pH values less than 3) are an interesting example to analyze resistance and tolerance. At their optimal pH growth, many toxic metals are more soluble than at neutral pH and exist as toxic free ions. Consequently, acidophiles are often described as highly resistant to metals and possess more efficient active resistance systems than neutrophiles. However, acidophiles in general do not appear to have more metal resistance genes than neutrophiles, and their growth in high metal concentrations could be partially explained by an intrinsic tolerance due to the environment in which they live (Dopson et al., 2014). Some of these passive systems include:

- Complexation of metal cations by sulfate ions, which are very abundant in natural or anthropogenic acidic environments. This way the concentration of free ions that can enter the cytoplasm is significantly lower than the total concentration of the metal.

- Chemiosmotic barrier produced by the internal positive transmembrane potential characteristic of acidophiles, which prevents the free passage of H^+ and metal cations into the cytoplasm.

- Competition of metal cations with H^+ for metal-binding sites on the cell surface.

- Formation of a biofilm that acts as a barrier to toxic metals where they can be sequestered, immobilized, mineralized and/or precipitated. Biofilms decrease metals bioavailability and retard their diffusion reducing cell exposure and increasing bacterial tolerance (Koechler et al., 2015).

However, the active systems involve proteins and enzymes codified in metal resistance genes, such as efflux pumps dependent on ATP or H^+, enzymes that convert metals into less toxic species or metal-binding protein implicated in intracellular sequestration.

LGT (lateral gene transfer) is one of the main mechanisms that microbial communities use to adapt to stress conditions and changes in the environment. It has also been suggested as the main force pushing microbial genome evolution (Hemme et al., 2016). Metagenomic studies evince that HM resistance genes have been laterally transferred within communities. For example, a study done on the microbial community of groundwater contaminated with HM showed that *Rhodanobacter* was the predominant species and that it has a recombinational hot spot in which numerous metal resistance genes were subjected to LGT and/or duplication, particularly those coding for $Co^{2+}/Zn^{2+}/Cd^{2+}$ efflux pumps and mercuric resistance.

Based on the results, the authors propose that the acquisition of genes critical for survival, growth, and reproduction via LGT is the most rapid and effective way to enable microorganisms in microbial communities to adapt to the abrupt environmental stresses (Hemme et al., 2016).

In conclusion, microorganisms may have tolerance and resistance mechanisms that allow them to develop in habitats with high concentrations of HM. The tolerance can be related to passive or intrinsic mechanisms as a consequence of their metabolisms and/or the environment in which they develop. However, resistance mechanisms are associated with specific genes that a lot of times are triggered by the accumulation of certain metals. In addition, some of these capabilities can be acquired through LGT. The combinations of these mechanisms have a crucial role in the evolution of the microbial communities in response to stress conditions.

4. Study case: Adaptation to increasing concentrations of arsenite and arsenate by microbial consortia obtained from an environmental sample with low As content

In this section we describe the process of adaptation to arsenic species of three different microbial consortia from a sample with low concentration of As.

The sample was collected from Copahue geothermal area, a natural environment located in the Northwest corner of Cordillera de los Andes in Neuquén province (Argentina), specifically from "Salto del Agrio", a point with a temperature of 15.5°C, pH 3.66, Eh 368.8 mV, dissolved arsenic less than 0.02 mg/L and dissolved sulphate 0.03 mg/L. In order to obtain heterotrophic, autotrophic and anaerobic consortia, sediment samples were enriched in Luria Bertani (LB) medium, Mackintosh basal salt solution (MAC) (Mackintosh, 1978) at pH 3 with 10 g/L of sterile sulfur powder or using Posgate B medium (Posgate, 1984) at pH 7, respectively. Microbial consortia obtained were supplemented with increasing concentrations of As(III) as $NaAsO_2$ and As(V) as $Na_2HAsO_4 \cdot 7\,H_2O$, starting from concentrations of 5 mM and 25 mM respectively. Salts were directly added to the culture media, except for anaerobic cultures where stock solutions (62 mM $NaAsO_2$ and 500 mM $Na_2HAsO_4 \cdot 7\,H_2O$) were prepared with distilled water and filtered with 0.22 µm membrane. Growth of cultures was monitored periodically by measuring absorbance at 600 nm (heterotrophic cultures) or recounting cells in suspension in a Neubauer chamber (autotrophic and anaerobic cultures). Cultures were used as inoculum (10%) in fresh media with a higher concentration of As(III) or As(V) when OD^{600} was higher than 0.6 or the number of cells was higher than 1×10^7 per mL. A limit of expected growth time was established: up to 30 days for heterotrophic and autotrophic cultures or 40 days for anaerobic cultures. Cultures with the same concentrations of arsenic species and without inoculum were set as sterile controls. Then, the influence of arsenic on microbial growth was evaluated using the heterotrophic and autotrophic cultures that achieved growth at higher concentration of arsenic species. These cultures were used as inoculum (10%) in the respective fresh media with and without arsenic. Growth was evaluated every 24 hours. All cultures were performed in duplicate.

Table 1 shows the maximum concentration of As(III) and As(V) tolerated by each consortia. These proved to be able to adapt to relatively high concentrations of both As species. All microbial consortia exhibited more tolerance to As(V), probably due to their lower toxicity, with respect to As(III). Also, Fig. 2 presents the growth profile of higher As(III) (Het(As^{+3} 20 mM)) and As(V) (Het(As^{+5} 450 mM)) tolerant heterotrophic consortia compared with their controls without arsenic added, called Het(As^{+3} 0 mM) and Het(As^{+5} 0 mM), respectively. The maximum OD^{600} recorded for Het(As^{+5} 450 mM) was reached on the 4th day of incubation while its control (Het(As^{+5} 0 mM)) reached that on the 3rd day; these data indicate that the growth of the microbial consortium was scarcely altered with the addition of 33.75 g/L (450 mM) of As(V). However, for Het(As^{+3} 20 mM) the maximum OD^{600} was reached at the 8th day of incubation and its control (Het(As^{+3} 0 mM)) achieved that on the 2nd day, i.e., the increase of As(III) concentration in the culture meant a delayed growth. Aditionally, Fig. 3 presents the growth profile of

Table 1. Maximum concentration of arsenic species tolerated.

Consortium	As^{+3} tolerated (mM)	As^{+5} tolerated (mM)
Heterotrophic	20	450
Autotrophic	15	150
Anaerobic	–	50

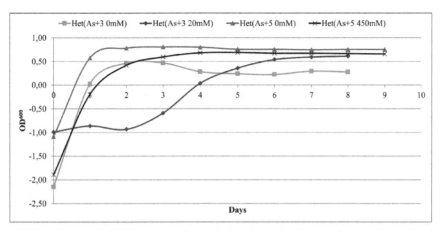

Figure 2. Growth profile of heterotrophic As(III) and As(V) tolerant cultures compared with their controls without As added.

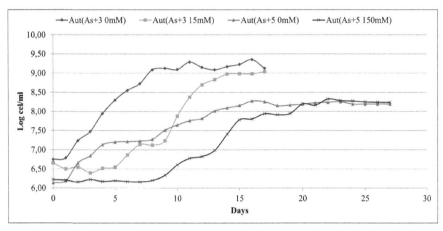

Figure 3. Growth profile of autotrophic As(III) and As(V) tolerant cultures compared with their controls without As added.

higher As(III) (Aut(As^{+3} 15 mM)) and As(V) (Aut(As^{+5} 150 mM)) tolerant autotrophic consortia compared with their controls without arsenic added, called Aut(As^{+3} 0 mM) and Aut(As^{+5} 0 mM) respectively. The maximum cell concentration per mL for Aut(As^{+5} 150 mM) was reached after 22 days of incubation while its control (Aut(As^{+5} 0 mM)) reached that after 16 days; in the case of Aut(As^{+3} 15 mM) the maximum concentration of cells per mL was reached after 19 days of incubation while in its control (Aut(As^{+3} 0 mM)) it was reached after 16 days.

These figures show that the cultures with arsenic added had longer lag phase than the controls without the addition of arsenic. However, the final concentration of biomass was similar in both cases.

Although, the anaerobic consortium failed to grow in As(III) concentrations tested, however, it showed considerable growth in As(V); after 40 days of incubation 1.48 x 10^8 cells/mL (with an initial of 9.79 x 10^5 cells/mL) were counted in the culture with the addition of 50 mM As(V). Moreover, a brownish precipitate was observed in the inoculated cultures, unlike the sterile controls.

In summary, the results presented here demonstrate the adaptive capacity of microorganisms present in the sediments of Lower Río Agrio, which have almost undetectable soluble As concentration, at increased concentrations of As(III) and As(V). The three consortia obtained, particularly the autotrophs, managed to grow at relatively high concentration of As salts. Even though there are many reports of different acidophilic autotrophic species that are able to tolerate high concentrations of As, most assays have been done using Fe(II) in the media which may reduce the bioavailability of As since the couple Fe(II)/Fe(III) has extremely similar redox potential to the couple As(III)/As(V).

5. Conclusion

In this chapter we have briefly reviewed the tools that microorganisms posses to interact/tolerate/resist the presence of heavy metals in their growth environment and the strategies to study such microbial tools and to follow changes in diversity and structure of microbial communities affected by heavy metals. Many studies performed by culturing or classical molecular ecology techniques (DGGE, cloning and sequencing of the 16S RNAr gene) show that microbial diversity diminishes, drastically in same cases, when microbial communities are exposed to heavy metals. However, recent works done using high throughput strategies such as metagenomic sequencing reveal that microbial diversity does not seem to be affected by heavy metals contaminations although the more resistant species are favored. As we could witness throughout the chapter, the world of microbial-heavy metals interactions is quite complex and its study and correct interpretation of the information obtained demand all our effort and imagination.

References

Besaury L., Bodilis J., Delgas F., Andrade S., De la Iglesia R., Ouddane B., Quillet L. Abundance and diversity of copper resistance genes cusA and copA in microbial communities in relation to the impact of copper on Chilean marine sediments. Mar. Pollut. Bull. 67, 16–25, 2013.

Bruins M.R., Kapil S., Oehme F.W. Microbial resistance to metals in the environment. Ecotoxicol. Environ. Saf. 45, 198–207, 2000.

Das S., Dash H.R., Chakraborty J. Genetic basis and importance of metal resistant genes in bacteria for bioremediation of contaminated environments with toxic metal pollutants. Appl. Microbiol. Biotechnol. 100, 2967–2984, 2016.

Deschatre M., Ghillebaert F., Guezennec J., Simon-Colin C. Study of biosorption of copper and silver by marine bacterial exopolysaccharides. WIT Trans. Ecol. Environ. 196, 549–559, 2015.

Dopson M., Ossandon F.J., Lövgren L., Holmes D.S. Metal resistance or tolerance? Acidophiles confront high metal loads via both abiotic and biotic mechanisms. Front. Microbiol. 5, 157–161, 2014.

Franke S., Grass G., Rensing C., Nies D.H. Molecular analysis of the copper-transporting efflux system CusCFBA of *Escherichia coli*. J. Bacteriol. 185, 3804–3812, 2003.

Giner–Lamia J., López–Maury L., Florencio F.J. CopM is a novel copper-binding protein involved in copper resistance in *Synechocystis* sp. PCC 6803. Microbiology Open 4, 167–185, 2015.

Grass G., Große C., Nies D.H. Regulation of the cnr cobalt and nickel resistance determinant from *Ralstonia* sp. strain CH34. J. Bacteriol. 182, 1390–1398, 2000.

Guillan D.C. Metal resistance systems in cultivated bacteria: are they found in complex communities? Curr. Opin. Biotechnol. 38, 123–130, 2016.

Hemme C.L., Green S.J., Rishishwar L., Prakash O., Pettenato A., Chakraborty R., Deutschbauer A.M., Van Nostrand J.D., Wu L., He Z., Jordan I.K., Hazen T.C., Arkin A.P., Kostka J.E., Zhou J. Lateral gene transfer in a heavy metal–contaminated–groundwater microbial community. MBio 7, e02234–15, 2016.

Jarosławiecka A., Piotrowska-Seget Z. Lead resistance in microorganisms. Microbiology 160, 12–25, 2014.

Jie S., Li M., Gan M., Zhu J., Yin H., Liu X. Microbial functional genes enriched in the Xiangjiang River sediments with heavy metal contamination. BMC Microbiol. 16, 179–192, 2016.

Koechler S., Farasin J., Cleiss-Arnold J., Arséne-Ploetze F. Toxic metal resistance in biofilms: diversity of microbial responses and their evolution. Res. Microbiol. 166, 764–773, 2015.

Mackintosh M. Nitrogen fixation by *Thiobacillus ferrooxidans*. J. Gen. Microbiol. 105, 215–218, 1978.

Mardanov A.V., Panova I.A., Beletsky A.V., Avakyan M.R., Kadnikov V.V., Antsiferov D.V., Banks D., Frank Y.A., Pimenov N.V., Ravin N.V., Karnachuk O.V. Genomic insights into a new acidophilic, copper-resistant *Desulfosporosinus* isolate from the oxidized tailings area of an abandoned gold mine. FEMS Microbiol. Ecol. 92, fiw111, 2016.

Mergeay M., Monchy S., Vallaeys T., Auquier V., Benotmane A., Bertin P., Taghavi S., Dunn J., van der Lelie D., Wattiez R. *Ralstonia metallidurans*, a bacterium specifically adapted to toxic metals: towards a catalogue of metal-responsive genes. FEMS Microbiol. Rev. 27, 385–410, 2003.

More T.T., Yadav J.S.S., Yan S., Tyagi R.D., Surampalli R.Y. Extracellular polymeric substances of bacteria and their potential environmental applications. J. Environ. Manage. 144, 1–25, 2014.

Pontel L.B., Soncini F.C. Alternative periplasmic copper-resistance mechanisms in Gram negative bacteria. Mol. Microbiol. 73, 212–225, 2009.

Postgate J. The Sulphate-Reducing Bacteria. Cambridge University Press, Cambridge, 1984.

Prabhakaran P., Ashraf M.A., Aqma W.S. Microbial stress response to heavy metal in the environment. RSC Adv. 6, 109862–109877, 2016.

Yamamura S., Amachi S. Microbiology of inorganic arsenic: from metabolism to bioremediation. J. Biosci. Bioeng. 118, 1–9, 2014.

Wang F., Yao J., Si Y., Chen H., Russel M., Chen K., Qian Y., Zaray G., Bramanti E. Short-time effect of heavy metals upon microbial community activity. J. Hazard. Mater. 173, 510–516, 2010.

Zampieri B.D.B., Bartelochi Pinto A., Schultz L., De Oliveira M.A., Fernandes Cardoso de Oliveira A.J. Diversity and distribution of heavy metal-resistant bacteria in polluted sediments of the Araça Bay, São Sebastião (SP), and the relationship between heavy metals and organic matter concentrations. Microb. Ecol. 72, 582–594, 2016.

CHAPTER 2

Mechanisms of Bacterial Heavy Metal Resistance and Homeostasis
An Overview

Pallavee Srivastava and *Meenal Kowshik**

1. Introduction

Heavy metals are naturally occurring elements that have high atomic weights and a density that is at least 5 times greater than that of water (Tchounwou et al., 2012). These include the transition metals and the metalloids such as arsenic (As), tellurium (Te), antimony (Sb), and germanium (Ge). The transition metals have incompletely filled d-orbitals that allow the formation of complexes. Thus, these metals play an integral role in the life processes of microorganisms as 'trace elements' (Nies, 1999). Although metals such as iron (Fe), nickel (Ni), copper (Cu), manganese (Mn), zinc (Zn), and vanadium (V) are relevant for the biochemical processes of bacterial cells, however, their concentrations are tightly controlled as they exert toxicity at higher concentrations. Trace metals act as micronutrients and are used for redox processes, to stabilize molecules through electrostatic interactions, as components of various enzymes, and for regulation of osmotic pressure (Bruins et al., 2000). However, metals such as mercury (Hg), cadmium (Cd), lead (Pb), and silver (Ag) have no physiological function and are extremely toxic to microorganisms. Majority of the bacteria are exposed to these metals as they are ubiquitously present since the origin of cellular life. Bacteria have

Department of Biological Sciences, Birla Institute of Technology and Science-Pilani, K K Birla Goa Campus, Zuarinagar, Goa, India, 403726.
* Corresponding author: meenal@goa.bits-pilani.ac.in

therefore evolved specific genes and transport systems for uptake of essential metals and efflux of toxic non-essential metals (Silver, 1998).

Biologically relevant metals are tightly regulated within the bacterial cell through comprehensive regulatory and protein-coding machinery devoted to maintaining 'homeostasis. Homeostasis is the maintenance of an optimal bioavailable concentration, mediated by the balancing of metal uptake and intracellular trafficking with efflux/storage processes so that the "right" metal is inserted into the "right" macromolecule at the appropriate time (Waldron and Robinson, 2009). This is brought about by the formation of specific protein-metal coordination complexes that facilitate uptake, intracellular trafficking within various compartments, storage for efficient incorporation and ultimately efflux if the metal is present in excess. The process entails various membrane spanning metal transporters, metal sensing riboswitches, metallochaperones and specific protein-protein complexes for metal-ligand exchange. Each component of this machinery is usually selective for its target metal ion(s) and tightly regulated so that in the presence of excess metal ions, activation of the detoxification systems results in a decrease in uptake and a concomitant increase in efflux. However, in the case of metal ion depletion, an increase in uptake with a simultaneous decrease of efflux is observed (Outten and O'Halloran, 2001).

Non-essential metals that are toxic result in the development of various metal resistance and tolerance mechanisms in bacteria. On first contact with the toxic biologically non-essential metal, bacteria try to prevent the entry of the metal ion either by creating a permeability barrier which offers limited protection or through extracellular secretion of biomolecules that sequester metals and prevent metal influx. The metal ions that do enter the bacterial cells may be (i) actively transported out of the cell; (ii) rendered inactive via intracellular chelation preventing the exposure of the intracellular sensitive components; or (iii) converted to non-toxic form via enzymatic detoxification. Bacteria may adapt to the presence of noxious metals by either lowering the sensitivity of the cellular components through mutation or increasing the production of certain cellular components to prevent downstream inactivation. A combination of two or more of these mechanisms enables bacteria to overcome the toxicity of non-essential metals (Bruins et al., 2000; Nies, 1999). The forthcoming sections will provide an overview of the molecular mechanisms involved in metal homeostasis and resistance in bacteria.

2. Metal homeostasis in bacteria

Bacteria are exposed to varying concentrations of metals in their surroundings. In an event of depletion or accumulation of metals to toxic levels, bacteria activate cellular responses to maintain metal homeostasis. This is achieved primarily through metal-sensing/metalloregulatory proteins that control transport and storage of target metal ion(s). The allosteric binding of target metal ions to these proteins brings about a conformational change in the DNA-binding domains that eventually results in transcriptional repression/ derepression/activation of the downstream genes. These regulatory metal sensing proteins control the expression of various transporters, metallochaperones, and intermediary protein complexes involved in influx or efflux of the metal(s) for maintaining homeostasis (Ma et al., 2009). Recent studies have also implicated the role of RNA-mediated regulatory response involving riboswitches in metal ion homeostasis (Furukawa et al., 2015; Ramesh and Winkler, 2010). The metal ion(s) present in the vicinity of the bacterium is transported across the impermeable membrane through the metal transporters in a directional fashion. These transporters are integral membrane proteins embedded in the inner or plasma membrane of the organism and include ATP binding cassette (ABC) and Nramp transporters. ABC transporters have been identified and characterized for nearly all biologically relevant metal ions, while Nramps have been identified only as Mn and Fe transporters (Ma et al., 2009). Once transported across the membrane, specialized proteins designated as 'metallochaperones' associate with the metal in such a way that it can be readily transferred to an appropriate acceptor protein within a cellular compartment such as the periplasm and cytosol. The transfer of the metal to the acceptor proteins is essentially an intermolecular metal-ligand exchange brought about by the formation of transient protein-protein complexes (Tottey et al., 2005). In an event of toxic build-up of metal within the bacterial cell, the organism exports the toxicant out of the cell through efflux pumps that include cation diffusion facilitators (CDFs), P-type ATPases, and tripartite RND (resistance-nodulation-cell division) transporters (Kolaj-Robin et al., 2015; Argüello et al., 2007). Figure 1 illustrates the common mechanisms involved in maintaining essential metal ion homeostasis in bacteria.

The metalloregulatory proteins that collectively maintain metal homeostasis within the bacterial cell function as transcriptional regulators of genes that encode membrane bound transporters involved in metal influx/efflux and to a lesser extent genes that encode intracellular chelators, metallothioneins, siderophores, and bacterioferritins (Andrew et al., 2003; Barkay et al., 2003). Interestingly, the regulators belonging to different sensor families may regulate the expressions of genes with identical functions in different organisms, and are usually present as a part of metal-sensing operon (Fig. 2) or regulon in tandem with the corresponding transporter genes (Wang et al., 2005). This process is further regulated post-transcriptionally through metalloregulatory riboswitches.

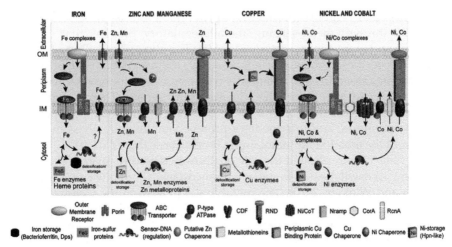

Figure 1. A representative model for homeostasis of biologically relevant metal ions such as Fe^{2+}, Zn^{2+}, Mn^{2+}, Cu^{2+}/Cu^+, Ni^{2+}, and Co^{2+} in bacteria. Some or all of the components of the homeostasis machinery are present in bacteria and not all bacteria have all the components shown here. Single headed arrows indicate unidirectional transport of metal ion, while the double headed arrows indicate that the metal ions can move in and out of target protein site in response to changes in intracellular metal ion concentrations. Reproduced with permissions from Ma et al. (2009). Copyright © 2009 American Chemical Society.

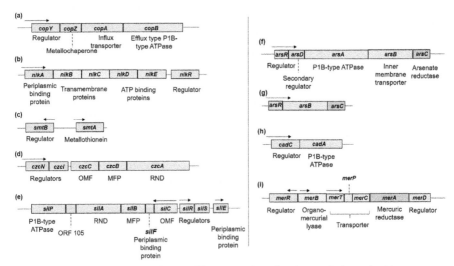

Figure 2. Schematic of operons for metal homeostasis and resistance in bacteria. (a) *cop* operon in *E. hirae*; (b) *nik* operon in *E. coli*; (c) *smt* operon in *Synechococcus* PCC 7942; (d) *czc* operon in *Cupriavidus metallidurans*; (e) *sil* operon on *Salmonella* 180-kb IncH1 silver resistance plasmid, pMG10; (f) *ars* operon on the R773 plasmid of *E. coli*; (g) *ars* operon on *E. coli* chromosome; (h) *cad* operon of pI258 plasmid in *S. aureus*; (i) *mer* operon of pPB plasmid in *P. stutzer*.

2.1 Role of metalloregulatory proteins in metal ion homeostasis

The expression of genes encoding metallochaperones, metal importers, and exporters is controlled by a panel of specialized transcriptional regulators known as metalloregulatory proteins or "metal sensor" proteins. In prokaryotes, the cellular response to perturbations in metal homeostasis is nearly exclusively transcriptional. The transcriptional response elicited by these metalloregulatory proteins depends on the intracellular metal activity or free metal concentration that dictates the activation or the inhibition of the downstream genes (Outten and O'Halloran, 2001). The metal mediated repression or derepression of a metal uptake or efflux gene depends on the kind of allosteric regulation exhibited by the metal sensor protein. In most of the instances, the transcriptional repressor binds to a specific DNA operator blocking the binding of RNA polymerase subsequently preventing the transcription of the gene (Cavet et al., 2002). An exception to this rule is exhibited by the MerR-family of regulators, where both the RNA polymerase and the metal bound MerR family member are bound simultaneously to the promoter, thereby activating the transcription of the efflux transporters (Outten et al., 1999). Thus, MerR family are transcriptional activators; ArsR/SmtB, CopY, and CsoR/RcnR families are transcriptional de-repressors; and the Fur, DtxR, and NikR families are the transcriptional co-repressors (Table 1).

MerR family

The MerR family of regulators are exclusive transcriptional activators that have similar N-terminal helix-turn-helix DNA binding regions and C-terminal effector binding regions that are specific to the effector. The first 100 amino acids of the N-terminal region are conserved and is the signature of this family of regulators (Brown et al., 2003). Within the family, the Hg^{2+} resistance regulator, MerR was the first metalloregulatory protein identified in transposons Tn501 from *P. aeruginosa* 278 and Tn21 from *Shigella flexneri* R100 plasmid (Barkay et al., 2003). Subsequently, various regulators belonging to this family were identified that recognised other metal ions such as ZntR for Zn^{2+}, CueR for Cu^+, GolS for Au^+, CadR for Cd^{2+}, and PbrR for Pb^+ (Ma et al., 2009).

ArsR/SmtB family

The most functionally diverse group of metalloregulatory proteins is the ArsR/SmtB family of transcriptional repressors that regulate genes involved in effluxing, scavenging, or detoxifying excess metal ions found in the cytosol (Campbell et al., 2007; Busenlehner et al., 2003). The family is named on the ArsR sensor of *E. coli* that recognises AsO_3^{3-}/SbO_3^{3-} and the SmtB sensor of *Synechococcus* PCC 7942 that recognises Zn^{2+} (Wu and Rosen, 1993; Morby et al., 1993). ArsR/SmtB proteins are involved in sensing a variety of metals such as Ni^{2+}, Zn^{2+}, AsO_3^{3-}, Cd^{2+}, and Pb^{2+}. The proteins of this family

Table 1. Families of metalloregulatory proteins in bacteria, their mode of action, cognate ligands and representative structure.

Family	Mode of action	Cognate ligand	DNA-binding domain	Examples	Structure	
ArsR	Derepression	Co, Ni, Cu, Zn, Cd, As, Sb, Bi, Hg, Pb	Winged-helix	ArsR, CadC, SmtB, NmtR, BxmR, CmtR		a
MerR	Activation	Co, Cu, Zn, Ag, Cd, Au, Hg, Pb	Winged-helix	MerR, ZntR, CueR, PbrR, CoaT, MtaN		b
CsoR	Derepression	Co, Ni, Cu	α-helical bundle	RcnR, CsoR		c
CopY	Derepression	Cu	Winged-helix	CopY, BlaI		d
Fur	Co-repression	Mn, Fe, Ni,	Winged-helix	Mur, Fur, Nur, Zur, Irr		e
DtxR	Co-repression	Mn, Fe	Winged-helix	AntR, IdeR, MntR, DtxR		f
NikR	Co-repression	Ni	Ribbon-helix helix	NikR		g

Metals up-regulate metal efflux/sequestration systems (rows ArsR–CopY)
Metals down-regulate metal uptake systems (rows Fur–NikR)

(a) Cyanobacterial SmtB; 1R1T; Eicken et al. (2003); (b) *E. coli* CueR; 1Q09; Changela et al. (2003); (c) *Streptomyces lividans* CsoR; 4ADZ; Dwarakanath et al. (2012); (d) *S. aureus* BlaI as model for CopY; 1XSD; Safo et al. (2005); (e) *Vibrio cholerae* Fur; 2W57; Sheikh and Taylor (2009); (f) *C. diphtheriae* DtxR; 1DDN; White et al. (1998); (g) *E. coli* NikR; 2HZV; Schreiter et al. (2006)

are dimeric and possess a similar fold with a winged helix-turn-helix motif for DNA binding. An interesting aspect of these proteins is the diverse metal-binding sites that have evolved at structurally distinct places on the same protein fold. The apo ArsR/SmtB proteins bind to the DNA operator partially overlapping the promoter and repressing the transcription of the downstream genes. Metal binding in the presence of excess metal ions results in dissociation of the metalloregulatory protein from the DNA resulting in transcriptional derepression of the downstream genes encoding the metal exporters (Ma et al., 2009).

CsoR/RcnR family

A new structural family of transcriptional repressors, CsoR/RcnR, that are characterized by a four-helix bundle and adopt an all α-helical dimer of dimers structure (Liu et al., 2007; Iwig et al., 2006). RciR, DmeR, InrS, CstR, and FrmR are some of the other structurally characterized members of this group. Members of this family coordinate Cu^+ or Ni^{2+}/Co^{2+} homeostasis or perform cysteine sulfur chemistry to overcome metabolite toxicity (Higgins and Giedroc, 2014). These regulators have been identified in a wide variety of bacteria, however, CsoR of *Mycobacterium tuberculosis* for Cu^+ sensing and the RcnR of *E. coli* for Ni^{2+}/Co^{2+} are considered as the representatives of this family (Liu et al., 2007; Iwig et al., 2006). The members of this family are classified into sub-families based on the presence of 3 to 4 conserved residues at various key positions that bind to the cognate metal ion. In the case of CsoR, the signature fingerprint is composed of x-Cys-His-Cys (where x is any amino acid), whereas in RcnR it is His-Cys-His-His with residues in the exact analogous positions relative to CsoR (Higgins and Giedroc, 2014; Ma et al., 2009).

CopY family

A Cu-specific metalloregulatory family of proteins, CopY, restricted largely to firmicutes was first identified in *Enterococcus hirae* (Liu et al., 2007; Strausak and Solioz, 1997). It was found to regulate the transcription of the *copYZBA* operon (Fig. 2a), consisting of CopA and CopB, the P-type ATPases for Cu influx and efflux, respectively, and the CopZ, a metallochaperone. In the absence of Cu^+, CopY-Zn^{2+} acts as a transcriptional repressor by binding to the operator-promoter region of the operon. CopZ transfers Cu^+ when present in the cell to the CopY-Zn^{2+} complex, where it replaces the Zn^{2+} to form CopY-Cu^+ complex. This results in the dissociation of CopY from the operon bringing about transcriptional derepression. CopYs are characterized by the presence of a conserved sequence -CXCXXXXCXC- motif close to the C-terminus (Cobine et al., 2002). The mechanism of allosteric regulation on DNA binding by Cu^+ and not by Zn^{2+} is not yet understood.

Fur family

The ferric uptake regulator (Fur) family of metalloregulatory proteins control transport and homeostasis of Fe^{2+}, Zn^{2+}, Mn^{2+}, and Ni^{2+}. Fur was first discovered in *E. coli* where it functions as a global transcriptional regulator of over 90 genes encoding both proteins and noncoding RNAs and is involved in Fe^{2+} homeostasis as well as oxidative stress and acid tolerance (Andrews et al., 2003). This class is present in the genomes of virtually every Gram-negative bacterium, except for *Rhizobium* species and other closely related α-proteobacteria (Johnston et al., 2007). Various Fur orthologs include Zur for Zn^{2+} sensing, Mur for Mn^{2+}/Fe^{2+} sensing, and Nur for Ni^{2+} sensing (Ma et al., 2009). Although they were typically considered transcriptional repressors when the apo form was bound to its cognate metal, however, recent studies have exhibited the ability of these Fe bound Fur proteins to act as transcriptional activators (Rajagopalan et al., 2013; Delany et al., 2001). For example, in *Helicobacter pylori*, when Fe^{2+} is abundant then Fe^{2+}-Fur holo form brings about transcriptional repression of genes involved in Fe^{2+} uptake such as *frpB1* gene (Delany et al., 2001). However, in the event of Fe^{2+} starvation, the apo form of Fur is responsible for the transcriptional repression of the genes involved in iron storage such as *pfr* encoding ferritin (Bereswill et al., 2000). In the former case, Fe^{2+} acts as the corepressor, while in the latter case it acts as the inducer. Thus, Fur family of proteins regulate gene expression by modulating their DNA binding affinity in response to the metal ion concentration, much akin to inducible or repressible ON-OFF switches (Agriesti et al., 2014).

DtxR family

The DtxR family is comprised of two subfamilies of metal sensing proteins viz., DtxR such as Fe^{2+} sensors and the MntR such as Mn^{2+} sensors (Merchant and Spatafora, 2014; Ma et al., 2009). The DtxR (diphtheria toxin repressor) were first studied in *Corynebacterium diphtheriae* as Fe^{2+} sensor proteins and since then several other members of this subfamily such as IdeR and SirR have also been studied (Gold et al., 2001; Hill et al., 1998). These proteins regulate genes involved in Fe^{2+} uptake and storage that are constitutively expressed under Fe^{2+} limiting conditions (Andrew et al., 2003). DtxR/IdeR/SirR mediated repression of these genes is observed in the presence of elevated levels of Fe^{2+} in the cytosol. The founding member of the MntR like subfamily is the *B. subtilis* MntR, that represses the expression of genes involved in Mn^{2+} uptake and is highly specific for Mn^{2+} and Cd^{2+} (Que and Helmann, 2000). ScaR and TroR are the other regulators belonging to this subfamily (Jakubovics et al., 2000; Posey et al., 1999), whereas SloR and MtsR are regulators that sense both, Fe^{2+} and Mn^{2+} (Merchant and Spatafora, 2014). DtxR like and MntR like regulators contain an N-terminal winged helix DNA binding domain followed by a helical dimerization domain and a C-terminal SH3-like domain. The C-terminal SH3-like domain enhances the DNA binding affinity by stabilizing intra- and/or inter-subunit protein-protein interactions (Liu et al., 2008).

NikR family

The nickel responsive regulatory protein, NikR, is the transcriptional repressor of genes involved in Ni^{2+} uptake and other Ni^{2+} requiring enzymes (Caballero et al., 2011). They belong to the ribbon-helix-helix (RHH) family of transcriptional regulators that consists of an N-terminal DNA-binding domain homologous to the other members of RHH regulators and a C-terminal domain that is required for binding nickel and tetramerization (Chivers and Sauer, 2002). NikR was first identified in *E. coli*, where it exhibits a Ni dependent repressor function by binding to a palindromic sequence in the promoter region of the *nik* operon. The operon consists of 6 genes, *nikABCDE* (Fig. 2b) that encode components of a typical ATP-dependent transport system. During Ni^{2+} starvation or when the cytosolic concentration of Ni^{2+} is low, the *nik* genes are expressed. However, when the cytosolic concentration of Ni^{2+} is high, the apo form of NikR binds Ni^{2+} (co-repressor) and brings about transcriptional repression of the *nik* operon which essentially results in decreased influx of the Ni^{2+} (Dosanjh and Michel, 2006).

2.2 Role or metalloregulatory riboswitches in metal ion homeostasis

Metal-sensing regulatory riboswitches for Mg^{2+}, Mn^{2+}, Ni^{2+}, and Co^{2+} have been identified in various bacteria (Price et al., 2015; Furukawa et al., 2015; Ramesh and Winkler, 2010; Wakeman et al., 2009; Dan III et al., 2007). Riboswitches are cis-acting RNAs present within the untranslated regions (UTRs) of the mRNAs that regulate the downstream gene expression by sensing specific cellular small molecule metabolites (Fig. 3). These RNAs contain aptamer regions which on binding to their cognate ligands undergo conformational changes that accomplish genetic regulation. They respond to a variety of small organic metabolites including amino acids, nucleosides, nucleobases, amino sugars, enzymatic cofactors, and metal ions (Furukuwa et al., 2015; Price et al., 2015; Dambach et al., 2015).

Magnesium transporters of the P-type ATPase family such as Mgt A/B and Mgt E, involved in Mg^{2+} import in various bacteria (Smith and Maguire, 1998), are regulated by a Mg-responsive riboswitch, M-box. In *Salmonella*, expression of *mgtA* and *mgtB* is under the control of metalloregulatory proteins PhoQ (membrane bound sensor kinase) and PhoP (transcriptional regulator) (Garcia et al., 1996), as well as post-transcriptional regulation through M-box (Fig. 4a). This M-box resides within the 5'UTR region of *mgtA* in *Salmonella enterica* (Cromie et al., 2006) and upstream of *mgtE* gene in *Bacillus subtilis* (Dann III et al., 2007). Under low Mg conditions, the magnesium responsive riboswitch of *S. enterica* undergoes conformational changes that favor transcription elongation, while in presence of excess Mg,

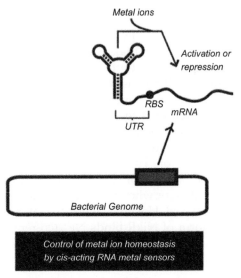

Figure 3. Schematic representation of regulation of gene expression by cis-acting metalloregulatory riboswitches residing in 5′ UTRs achieved by controlling transcription elongation or termination. Reproduced with permission from Ramesh and Winkler (2010). Copyright © 2010 Landes Bioscience.

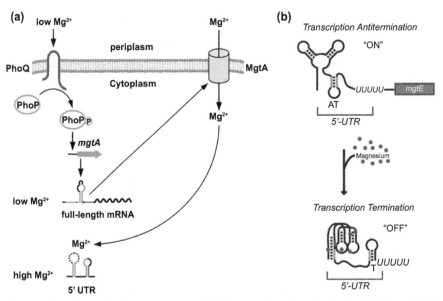

Figure 4. Regulation of downstream genes Mg^{2+} responsive riboswitches in (a) *Salmonella* (b) *B. subtilis*, where the M-box RNA act as an 'off' switch for transcription in the presence of excess Mg^{2+}. Reproduced with permissions from Cromie et al. (2006) copyright © 2006 Elsevier Inc. and Ramesh and Winkler (2010) copyright © 2010 Landes Bioscience.

the alternative conformation prevents transcription of downstream genes (Cromie et al., 2006). Similarly, the *B. subtilis* M-box RNA acts as an 'off' switch in the presence of excess Mg decreasing the expression of downstream genes (Fig. 4b). The Mg^{2+} binds to the most highly conserved region present within the Mg^{2+} sensing aptamer domain. Binding of Mg^{2+} to this aptamer domain stabilizes the conformation to form a compacted RNA by a suite of long-range base interactions that render a stretch of nucleotides inaccessible for the formation of the anti-terminator helix. This allows the formation of downstream intrinsic transcription terminator, thus preventing expression of downstream genes (Dan III et al., 2007).

The widespread *yybP-ykoY* riboswitch family has been implicated in maintaining Mn^{2+} homeostasis in bacteria such as *E. coli*, *B. subtilis*, and *Lactococcus lactis* (Price et al., 2015; Dambach et al., 2015). The Mn^{2+}-dependent transcription-ON riboswitch in *L. lactis* regulates the expression of YoaB, a P-type ATPase Mn^{2+} exporter. In the presence of elevated levels of Mn^{2+}, this riboswitch selectively binds to the metal ion at its aptamer region based on its charge, intracellular free ionic concentration, ligand hardness, preferred coordination geometry, and ionic radius. Two phosphate rich pockets are created within the aptamer region that binds the Mn^{2+} only after the complete dehydration of the metal ion. This brings about the conformational changes that result in the expression of *yoaB* gene (Fig. 5), which is not expressed in the

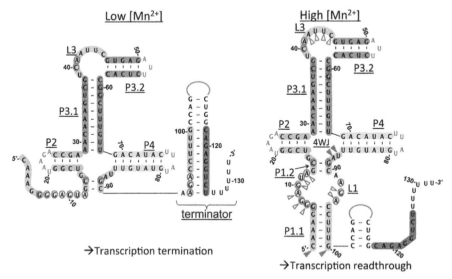

Figure 5. Secondary structure of the *L. lactis yybP-ykoY* riboswitch exhibiting the Mn^{2+}-dependent alternative structures. Underlined text indicates the conserved secondary structures, while the conserved bases are circled in red. The 3′ halves of L1 and P1.1 form a terminator helix with complementary downstream sequence in low Mn^{2+} conditions resulting in transcription termination, while in the presence of Mn^{2+} the P1.1 gets stabilized allowing transcription readthrough. Reproduced with permission from Price et al. (2015). Copyright © 2015 Elsevier Inc.

absence of Mn^{2+} (Price et al., 2015). Mn^{2+} sensing riboswitch present in *E. coli* and *B. subtilis* exhibit similar properties, except the gene regulated by these riboswitches is *mntP* gene that encodes the MntP manganese transporter (Dambach et al., 2015).

The concentrations of other heavy metals such as Ni^{2+} and Co^{2+} are even more tightly regulated as they are required in trace levels and hence the RNA regulators need to sense the low abundance of these heavy metals. This is achieved by a special class of riboswitches, 'NiCo', that bind their target ligands despite the presence of other metal ions at much higher concentrations (Furukawa et al., 2015). The NiCO aptamer responsible for metal sensing is specific for Ni/Co owing to the specific metal site geometry created by the positioning of nucleophilic groups from evolutionarily conserved guanine nucleotides. Cooperative Ni^{2+}/Co^{2+} binding stabilizes the antiterminator structure in the NiCo aptamer, resulting in the production of full length transcripts. A putative cation efflux gene, COG0053 of *Clostridium scindens*, present upstream of NiCo riboswitch on exposure to increasing concentrations of Ni^{2+}, exhibited a concomitant increase in the corresponding transcript (Furukawa et al., 2015). It is thus evident that bacteria employ highly selective metalloregulatory riboswitches that act in concert with metalloregulatory proteins for maintaining metal ion homeostasis.

Bacteria have evolved various metalloregulatory mechanisms for maintaining metal ion homeostasis which also participate in conferring metal resistance to overcome toxicity exerted by an increased metal ion concentration.

3. Metal resistance

Bacterial metal resistance systems exist for practically all toxic metals because of the selection pressure exerted by the metal-containing environment inhabited by them. These resistance mechanisms developed shortly after prokaryotic life began on earth. Bacteria carry most of the non-essential metal resistance determinants as operons (Fig. 6) on their plasmids which are often associated with antibiotic resistance. Essential metal resistance determinants, although, are usually chromosome based and more complex than plasmid systems. The resistance mechanisms that involve ion-efflux transporters are needed occasionally and are more likely to be plasmid borne for quick mobilization (Nies, 2003). The type of mechanisms for metal uptake, the role each metal plays in normal metabolism, and the presence of genes located on plasmids, chromosomes, or transposons that control metal resistance are some of the factors that determine the extent of metal resistance exhibited by a microorganism (Bruins et al., 2000). Five mechanisms are postulated to be involved in resistance to metals as described in the subsequent sections. The location of the metal sensitive component determines the mechanism that

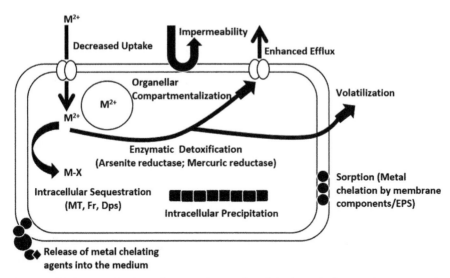

Figure 6. Various mechanisms that may be employed by bacteria for metal resistance. In response to metal toxicity, bacteria may exhibit sorption of metals, volatilization, release of metal chelating compounds in the medium, enhanced efflux, impermeability, decreased uptake, enzymatic detoxification, or intracellular chelation as mechanisms for metal resistance. With the exception of magnetosomes in magnetotactic bacteria, organellar compartmentalization is observed only in eukaryotes. Reproduced with permission from Srivastava and Kowshik (2013). Copyright © 2013 under creative commons attribution license.

may be utilized by the bacterium and a combination of one or more of these are needed by microbes to resist noxious metal (Fig. 6).

3.1 Permeability barrier for metal exclusion and extracellular sequestration

Bacteria prevent the entry of the toxic metal into the cell and protect the intracellular metal sensitive machinery by creating a permeability barrier. This may be achieved by altering the cell-wall, membrane, or envelope as these are the primary sites of interaction. In Gram-negative bacteria, it is the outer membrane that provides this kind of protection. This is exemplified by *E. coli* B where porins present in the outer membrane, that control the transport of various hydrophilic substances including metal ions, are altered by genetic mutation in the presence of Cu^{2+} thereby preventing its influx (Bruins et al., 2000). Besides, the outer envelope may also confer natural resistance to a barrage of metals through non-specific binding, however, this kind of resistance is transient and holds true only till the saturation limit is reached (Rouch et al., 1995).

Biosorption of metal ions by exopolysaccharides (EPS) elucidated by microbes is the primary line of defence exhibited by numerous bacteria.

EPS secretion occurs either as a response to environmental conditions such as nutrient starvation and presence of toxic conditions or to form biofilms (Sutherland, 2001). Organisms in biofilms exhibit increased resistance towards various metals as well as antibiotics. EPS are comprised of high molecular weight macromolecules such as polysaccharides, proteins, phospholipids, nucleic acids, and low molecular weight nonpolymeric constituents (Wingender et al., 1999). The EPS binds positively charged metal ions through electrostatic interaction with negatively charged functional groups such as uronic acids, phosphoryl groups associated with membrane components, or carboxylic groups of amino acids. In addition, there may also be cationic binding by positively charged polymers or coordination with hydroxyl groups. Several bacteria such as *Pseudomonas aeruginosa, Micrococcus* sp., *Ochrobactum* sp., *P. putida, Paenibacillus polymyxa,* and *Klebsiella aerogenes* exhibit a metal dependent secretion of EPS which increases with the extracellular concentration of the metal (Pal and Paul, 2008). In fact, extracellular sequestration by EPS creating a permeability barrier is the most often used mechanism to overcome Pb^{2+} toxicity by bacteria such as *Rhizobium etli* M4, *Paenibacillus jamilae* CECT 5266 and *Alteromonsa macleodii* subsp. *fijiensis.* The EPS of these bacteria preferentially bind Pb^{2+} over other toxic metals like Cd^{2+}, Cu^{2+}, Zn^{2+}, Ni^{2+}, or Co^{2+} (Jaroslawiecka and Piotrowska-Seget, 2014).

Bacteria may also secrete certain metal chelating agents such as polymers and inorganic molecules resulting in extracellular sequestration of metals, thereby preventing its entry into the cell. In presence of Pb^{2+}, *P. marginalis* elucidates an exopolymer that binds the metal ion through electrostatic interactions, exhibiting an extracellular mechanism of exclusion (Roanne, 1999). A strain of *K. aerogenes* precipitates Cd^{2+} as CdS extracellularly by secreting sulfur to limit metal uptake (Scott and Palmer, 1990).

3.2 Intracellular sequestration

An increase in the intracellular metal ion concentration triggers its chelation within the cytoplasm to protect the essential cellular components. It entails sequestration of metals by proteins such as metallothioneins (MTs), ferritins (Fr), and Dps (DNA binding protein from starved cells) or precipitation as phosphates or polyphosphates. MTs are small molecular weight genetically coded polypeptides that are classified based upon the number of cysteine residues (Cobbett and Goldbrough, 2002). They typically have two cysteine rich domains that bind heavy metals through mercaptide bonds giving these proteins a dumbbell shaped conformation comprising of a N-terminal β domain that usually binds 3 metal ions and a C-terminal α domain that binds 4 metal ions (Wang et al., 2006). Although commonly used by eukaryotes to sequester heavy metals, MTs are relatively rare in bacteria. Bacterial MTs were first discovered in marine cyanobacterium *Syenochococcus* sp. strain RRIMP N1, freshwater cyanobacterium *Syenochococcus* Tx-20, and the γ-proteobacterium *P. putida*. These MTs were induced when the cells were grown in presence of Cd^{2+}. Bacterial MTs differ considerably from the previously characterized eukaryotic MTs, and contain aromatic amino acid residues such as His. The His present in the pseudothioneins of *P. putida* is involved in binding Cd^{2+} (Blindauer, 2011). The SmtA, a MT from *Syenchococcus* PCC 7942, encoded by the *smt* operon is under the control of the metalloregulatory protein SmtB (Fig. 2c), and chelates Zn^{2+} and Cd^{2+}. The Cu^+ binding MT of *Mycobacterium smegmatis* binds 6 Cu^+ ions to confer resistance (Blindauer, 2011).

Ferritins (Fr) are a superfamily of iron storage proteins that carry out the basic function of supplying cells with the requisite amount of iron. They control the reversible phase transition between hydrated Fe^{2+} in solution and the solid ferrihydrite mineral core inside its cavity (Carrondo, 2003). Bacterial ferritins that possess a haem are known as bacterioferritins (BFr). Besides the primary function of iron storage, these proteins also undertake iron detoxification functions under extreme conditions in several bacteria including *E. coli, Rhodobacter capsulatus, Listeria innocua* and *Desulfovibirio desulfuricans* (Chiancone et al., 2004). Both Fr and BFr have the same structure that assembles into a 24-mer cluster to form a hollow roughly spherical cage (Fig. 7a,b) with a 432-point symmetry. The iron storage cavity has a diameter of ~80 Å and binds up to 4500 Fe^{3+} as ferrihydrite complex (Carrondo, 2003). Prior to iron storage, the Fe^{2+} is converted to Fe^{3+} at the ferroxidase centre of the Fr, a binuclear di-iron centre following which the Fe^{3+} enters the core to be stored as the mineral (Carrondo, 2003). The BFrs, however, are heteropolymers with one subunit exhibiting ferroxidase activity and the other haem binding ability (Carrondo, 2003). Dps have also been identified

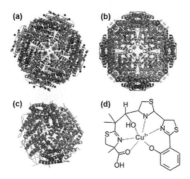

Figure 7. (a) Structure of 24-meric ferritin from *E. coli,* where the helices are represented as ribbons and bound Fe^{2+} are shown as spheres (4REU; Thiruselvam et al., 2014); (b) Structure of 24-meric bacterioferritin from *E. coli* (3E1O; Crow et al., 2009); (c) Structure of 12-meric Dps from *L. innocua*. View along the 3-fold axis (1QGH; Ilari et al., 2000); (d) Yersiniabactin, an extracellular copper-binding molecule, has three heterocyclic nitrogens (two thiazoline and one thiazolidine) forming a square planar complex with Cu^{2+} involving the phenolate oxygen. Reproduced with permissions from Koh and Henderson (2015). Copyright © 2015 American Society for Biochemistry and Molecular Biology.

as ferritin-like proteins due to their ability to oxidize Fe^{2+} to Fe^{3+} (Chiancone and Ceci, 2010). Li Dps of *Listeria innocua* exhibits a 12-mer cluster with a 23-point symmetry unlike the 24-subunit ferritin architecture (Fig. 7c). It displays the functionally relevant structural features of the ferritin 24-mer, namely the negatively charged channels along the three-fold symmetry axes that serve for iron entry into the cavity and a negatively charged internal cavity for iron deposition (Ilari et al., 2000). The other members of Dps family exhibit a similar di-iron ferroxidase centre with 12 subunit architecture that assembles into a spherical hollow cage with a diameter of 40–50 Å for iron storage (Andrews et al., 2003). These proteins are also involved in Fe^{2+} and Ni^{2+} detoxification in several bacteria including *E. coli, Streptococcus suis,* and *Anabaena* sp. PCC 7120 (Haikarainen et al., 2011; Calhoun and Kwon, 2011; Wei et al., 2007). More recently, yersiniabactin, a siderophore produced by *Yersinia* species that is conventionally involved in Fe^{2+} acquisition was found to bind Cu^{2+} (Fig. 7d), thus conferring copper tolerance (Chaturvedi et al., 2012).

Intracellular precipitation of heavy metals to phosphates and polyphosphates prevents the sensitive cellular components from the toxic effect of metal. Several bacterial species including *S. aureus, Citrobacter freundii, Vibrio harveyi,* and *B. megaterium* lower the free ionic concentration of Pb^{2+} by precipitating it as phosphate salt (Jaroslawiecka and Piotrowska-Seget, 2014). Cd^{2+} resistance in some bacteria such as *Pseudomonas* sp. H1, *B. cereus, Luteibacter* sp. II-116-7, and *Massilia* sp. III–116-18 entail intracellular precipitation as polyphosphates (Hrynkiewicz et al., 2015).

3.3 Increased efflux of metals

Export systems represent the largest system of metal resistance mechanisms exhibited by bacteria. Microbes utilize active transport for extrusion of the toxic metal from the cytoplasm. The toxic metal ions do not have dedicated specific importers, and are usually imported within the cell through the nutrient uptake transporters. For instance, AsO_4^{3-} is taken up by Pit (phosphate inorganic transport), a low affinity transporter and/or by ABC-type transporter, Pst (phosphate affinity transport), a high affinity transporter. However, once taken up, the toxic metal is quickly exported out of the cell through RND (resistance-nodulation-cell division), CDF (cation diffusion facilitator), or P-type ATPase type superfamily.

RND family

The RND family of transporters found in all three domains of life represent the first level of heavy metal resistance and are involved in export of superfluous cations (Nies, 2003). They were first described as a related group of bacterial transport proteins involved in heavy metal resistance (*Cupriavidus metallidurans*), nodulation (*Mesorhizobium loti*) and cell division (*E. coli*). RND superfamily is involved in export of heavy metals, hydrophobic compounds, amphiphiles, nodulation factors, and proteins (Nies, 2003). The CzcA (for $Co^{2+}/Zn^{2+}/Cd^{2+}$ efflux), SilA (Ag^+-specific exporter), and CusA (Cu^+/Ag^+ effluxer) are components of tripartite CBA-type efflux complexes responsible for metal resistance mostly in Gram-negative bacteria (Silver, 2003). The CzcA protein is encoded by the *czcCBA* operon (Fig. 2d), which also encodes for the CzcC, the outer membrane factor (OMF) and CzcB, the membrane fusion protein (MFP) (van der Lelie et al., 1997). RND proteins employ proton motive force to achieve cation efflux (Kim et al., 2011). The homotrimers or heterotrimers forming the RND proteins are composed of two different RND polypeptides in a 2:1 ratio (Kim et al., 2011). Each RND monomer has 12 transmembrane α-helices that span the inner membrane. The RND trimer contains a large hydrophilic portion that extends into the periplasmic space (Murakami et al., 2006), and connects to the second component of the CBA system, the trimeric OMF. The three subunits of OMF span the outer membrane as β-barrel (Koronakis et al., 2000). The MFP is periplasmic and forms a hexa/trimeric ring around the RND and the OMF to complete the CBA system (Fig. 8) (Akama et al., 2004). The active transport of heavy metals out of the cell results in decreased intracellular concentrations thereby conferring resistance.

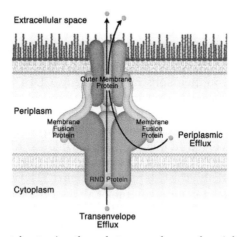

Figure 8. Model of metal extrusion through transenvelope and periplasmic efflux by RND transporters. The superfluous/toxic metal ions present either in cytoplasm or periplasm are transported to the extracellular space through the RND-driven complex protein. Reproduced with permissions from Kim et al. (2011). Copyright © 2011 American Society of Microbiology.

Interestingly, a three protein chemiosmotic RND Ag^+/H^+ exchange system, SilCBA in concert with SilP, a P1B-type ATPase confers resistance to Ag^+ as well as Hg^{2+} and TeO_3^{2-}. The genes for these transporters are part of the *sil* operon, discovered in *Salmonella* 180-kb IncH1 silver resistance plasmid, pMG101 (Gupta et al., 1999). The *sil* operon has a total of nine genes (Fig. 2e), of which seven have been annotated as structural genes (SilE, SilC, SilF, SilB, SilA, ORF105 and SilP) and two (SilR and SilS) encode a putative two-component regulatory circuit (Silver et al., 2006; Gupta et al., 1999). Except for SilE, the other genes have been annotated based on their amino acid sequence similarity to other metal resistance operons like *czc*. SilE has been determined as the periplasmic Ag^+-binding protein. SilC, SilB and SilA have been annotated as OMF, MFP, and RND respectively, while SilF has been designated as another periplasmic Ag^+-binding protein. SilRS is the sensor/responder pair, forming a two-component signal transduction machinery (Silver, 2003; Gupta et al., 1999). Thus, the proteins encoded by the *sil* operon may mediate silver resistance by preventing toxic build-up of Ag^+ in the cell through a combination of silver sequestration in the periplasm (via SilE and SilF binding) and active efflux (via SilCBA, and SilP) (Randall et al., 2015; Silver, 2003).

CDF family

The CDF family of transporters serve as the second level of resistance against toxic metals due to their role as secondary cation filters in bacteria. These proteins have been reported in all three domains of life (Paulsen and Saier, 1997). Although CDF transporters primarily recognize Zn^{2+}, other cations such as Hg^{2+}, Pb^{2+}, Co^{2+}, Ni^{2+}, Cd^{2+}, and Fe^{2+} are also detoxified by these proteins (Nies, 2003). They are classified as Zn^{2+}-CDF, Fe^{2+}/Zn^{2+}-CDF, and Mn^{2+}-CDF, based upon their substrate specificity. CDFs are composed of six transmembrane domains (TMDs) and a cytoplasmic N- and C-terminal with a histidine loop of variable length located between TMD 4 and 5 (Fig. 9) (Kolaj-Robin et al., 2015). The most conserved amphipathic domains TMD 1, 3, 5, and 6 are involved in metal transfer, while the hydrophobic TMD 2 and 4 are crucial for Zn^{2+} specificity and mutations within these domains alter substrate specificity. The C-terminal domain (CTD) is involved in cation efflux and is the characteristic signature of all the proteins of this family. The cation is transferred from the metallochaperone to the CTD, from where it is translocated across the membrane to TMD 2 and 4. An exception to this is the MmCDF3, a Zn^{2+} and Cd^{2+} transporting CDF from the marine bacterium, *Maricaulis maris* MCS10 that lacks the CTD, instead it has slightly elongated N-terminus as compared to the classical members of this family. The elongated N-terminus carries an additional metal binding site and exhibits

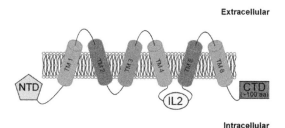

Figure 9. Schematic model of a prokaryotic CDF. Transmembrane (TM) domains are numbered 1–6, where metal specific residues are found in TMs 2 and 5, while residues predicted to form a hydrophobic gate are present in TM 3. NTD: N-terminal domain; IL2: histidine-rich interconnecting loop; and CTD: C-terminal domain. Reproduced with permissions from Kolaj-Robin et al. (2015). Copyright © 2015 Federation of European Biochemical Societies. Published by Elsevier B.V.

transport of more than one divalent cation making these CTD lacking CDFs more versatile efflux systems (Kolaj-Robin et al., 2015).

P1B-type ATPase family

The P1B-type ATPases are a superfamily of integral membrane proteins that couple ATP hydrolysis to metal cation transport and are significant to all three domains of life (Inesi et al., 2014). P-type ATPases are involved in trafficking of numerous heavy metal ions including Cu^+, Ag^+, Zn^{2+}, Pb^{2+}, and Cd^{2+} by generating and maintaining electrochemical gradient across membranes. They represent the third level of heavy metal resistance providing the basic defence against heavy metal cations (Nies, 2003). Transporters that import macromolecules such as Mg^{2+} are known as importing P1B-type ATPases, whilst those involved in efflux and detoxification of heavy metals are called exporting P1B-type ATPases (Nies, 2003). Based on their substrate specificity the exporting P1B-type ATPases may further be classified as Cu-P1B-type ATPases and Zn-P1B-type ATPases. The Cu-P1B-type ATPases are involved in the transport of Cu^+ and Ag^+ and have been characterized in several bacteria such as *Synechocystis*, *B. subtilis*, *Lactobacillus sakei*, *Rhizobium leguminosarum* and *P. putida*. Zn^{2+}, Cd^{2+}, and Pb^{2+} are recognized and transported by the Zn-P1B-type ATPases. These transporters have been identified in bacteria such as *S. aureus*, *B. subtilis*, *Listeria monocytogenes*, *P. putida*, *Stenotrophomonas maltophilia* and *E. coli*. The first P1B-type ATPases to be identified was CadA, a cadmium resistance-mediating protein encoded on a plasmid from *S. aureus* (Nies, 2003). The structure of P1B-type ATPases (Fig. 10) includes (i) a catalytic headpiece at the cytosol side with domains for ATP-binding (N domain), phosphoryl transfer (P domain) and, catalytic activation (A domain) and; (ii) transmembrane helices with cation-binding

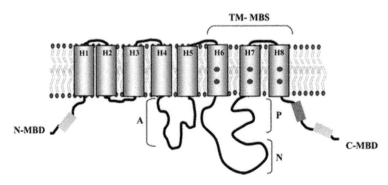

Figure 10. Schematic representation of the P1B-type ATPases. H1–H8 represent the transmembrane segments of the protein. Grey dots in H6, H7, and H8 represent the conserved amino acids that form the transmembrane metal binding sites (TM-MBS). The catalytic activation, phosphorylation, and the ATP-binding domains are represented as A, P, and N, respectively. The N- and C-terminal metal binding domains (MBD) are represented as light and dark grey rectangles. Reproduced with permissions from Argüello et al. (2007). Copyright © 2007 Springer Science+Business Media B.V.

sites (transmembrane metal binding sites-TM MBS) for catalytic activation and cation translocation (Argüello et al., 2007). ATP utilization results in the generation of a phosphoenzyme intermediate, following which the bound metal ion is displaced from the TM-MBS to the lumenal membrane surface without H⁺ exchange (Inesi et al., 2014).

ArsA, the AsO_3^{2-}/SbO_3^{2-} induced ATPase is encoded by the *ars* operon, that confers $AsO_4^{3-}/AsO_3^{2-}/SbO_3^{2-}$ resistance (Branco et al., 2008). The *ars* operon of the *E. coli* plasmids R773 and R46 (Fig. 2f) is transcribed as a single polycistronic mRNA in the order of *arsRDABC* (Wu and Rosen, 1991). ArsA is the P1B-type ATPase and ArsB is the protein involved in the transport of AsO_3^{2-} across the inner membrane. ArsC is arsenate reductase that reduces AsO_4^{3-} to AsO_3^{2-}. ArsR is the trans-acting regulator of the ArsR/SmtB family of metalloregulatory proteins, while the ArsD acts as the secondary regulator and is not essential for arsenicals resistance (Cervantes et al., 1994). Interestingly, the ArsB protein of the R773 *ars* operon is capable of active efflux of AsO_3^{2-} even in the absence of ArsA ATPase suggesting that besides ATP hydrolysis, membrane potential can also stimulate the function of the ArsB membrane pump (Dey and Rosen, 1995). Consistent to this, the genes *arsA* and *arsD* are absent from the *ars* operons in the Gram-positive staphylococcal plasmids pI258 and pSX267. Similarly, *E. coli* harbors another *arsRBC* operon in the chromosome (Fig. 2g) and lacks the *arsA* and *arsD* genes. Such operons are present in the chromosomes of numerous Gram-negative bacteria and are an integral component of the arsenic detoxification system (Cai et al., 1998).

CadA, a Cd^{2+} effluxing P1B-type ATPase is encoded by the *cadCA* operon (Fig. 2h), which is the best characterized cadmium resistance system in bacteria. It was discovered on the pI258 plasmid of *S. aureus* (Tsai and Linet, 1993). CadC is the transcriptional repressor of the operon and Cd^{2+} binding brings about transcriptional derepression (Endo and Silver, 1995). However, the CadC of *S. aureus* ATCC12600, and MRSA do not exhibit transcriptional repression of the *cad* operon (Hoogewerf et al., 2015). PbrA, a Zn-P1B-type ATPase, in concert with PbrB, an undecaprenyl pyrophosphate phosphatase confers lead resistance to *C. metallidurans* (Hynninen et al., 2009). These proteins are encoded by the *pbrTRABCD* operon on a plasmid, that also encode PbrR, a transcription factor of the MerR family; PbrT, a putative Pb^{2+} uptake protein; PbrB/PbrC, a predicted integral membrane protein and a putative signal peptidase; and PbrD, a putative intracellular Pb-binding protein (Borremans et al., 2001). Therefore, it follows that Pb^{2+} imported by PbrT is exported from the cytoplasm by PbrA following which in the periplasm it is sequestered with the inorganic phosphate generated by PbrB as phosphate salt (Hynninen et al., 2009).

3.4 Enzymatic detoxification

Microbes detoxify heavy metals by enzymatically converting the toxic form to non-toxic form. Mercury resistance accomplished by gene products of *mer* operon (Fig. 2i) exemplifies this model of resistance and is widely distributed in both Gram-positive and Gram-negative bacteria (Nascimento and Chartone-Souza, 2003). The genes involved are *merD/merR* for detection, *merP/merT/ merC* for transport, and *merB/merA* for enzymatic detoxification of inorganic and organic mercurials (Schelert et al., 2004). MerR is expressed in response to Hg^{2+} exposure that further regulates the expression of downstream genes of the operon, including MerA. Mercury reductase (MerA), is essentially a NAD(P)H-dependent flavin oxidoreductase that is responsible for the reduction of highly toxic Hg^{2+} to less toxic and volatile Hg^0 (Mathema et al., 2011). As this enzyme is located intracellularly, the Hg^{2+} that diffuses across the outer membrane, binds to a pair of cysteine residues on the MerP protein located in the periplasm, from where it is transferred to a pair of cysteine residues on MerT, a cytoplasmic membrane protein, and finally to a cysteine pair at the active site of MerA. The non-toxic volatile Hg^0 on reduction is released into the cytoplasm and volatilizes from the cell. MerB is an organomercurial lyase that cleaves the C-Hg bonds of organomercurials, to generate Hg^{2+} which is reduced to Hg^0 by MerA (Begley et al., 1986). Thus, the presence of both *merA/merB* confers a broad-spectrum resistance to organisms against a variety of mercurials. Another enzymatic detoxification system entails the plasmid mediated AsO_4^{3-} resistance in bacteria such as

B. subtilis and *E. coli*. ArsC encoded by the *ars* operon (Fig. 2f,g) is an arsenate reductase which needs reduced glutathione (GSH) and the small thiol transfer protein and glutaredoxin (Grx) for its activity (Mukhopadhyay and Rosen, 1998). Other details of the *ars* operon are given in the previous Section (3.3).

3.5 Decreasing the metal sensitivity of cellular targets

A degree of natural protection against toxic metals is achieved by various bacteria by altering the sensitivity of the essential cellular components. This may be a consequence of genetic mutations that either results in an increased synthesis of a cellular component to nullify the metal inactivation or a decreased sensitivity without any changes in the basic function. Nucleic acids, especially DNA are a common target for heavy metal toxicity, which may be protected by the inherent DNA repair mechanisms of the bacteria. For instance, on exposure to Cd^{2+}, *Caulobacter crescentus* exhibited an upregulation of genes for oxidative-stress management and those involved in replenishing the precursors of DNA (Hu et al., 2005). Similarly, chromate exposure has been known to induce upregulation of DNA repair enzymes such as recA in *E. coli* and *Caulobacter crescentus* (Hu et al., 2005; Aiyer et al., 1989). Continuous growth in the presence of toxic metals may also allow the organism to adapt to the heavy metal. For example, *E. coli* on first exposure to Cd^{2+} exhibits extensive DNA damage, however on adaptation (frequent sub-culture in the presence of sub-lethal concentrations of metal), the genes involved in DNA repair mechanisms are activated with a concomitant decrease in the sensitivity of the cellular targets to heavy metal toxicity, thus inducing resistance to Cd^{2+} (Bruins et al., 2000; Rouch et al., 1995). However, this alone does not offer complete resistance, which entails activation of the enzymatic detoxification and/or increased efflux of the toxic metal ion(s) from within the cell (Bruins et al., 2000).

4. Conclusion

Bacteria possess sophisticated mechanisms for optimal utilization of essential and biologically relevant metals, while efficiently detoxifying the non-essential, toxic metals. The presence of metals since the origin of life has resulted in the evolution of specific genes and genetic determinants that tightly control homeostasis of essential metals with the help of metalloregulatory riboswitches and proteins. The presence of toxic metals trigger mechanisms that include extracellular sequestration by biomolecules, increased efflux by transporters such as RND/CDF/P1B-type ATPases, intracellular sequestration, enzymatic detoxification, and lowering the sensitivity of the target cellular components. Two or more of these mechanisms may be functional at the same time to efficiently detoxify the metal. Thus, bacteria exhibit numerous molecular mechanisms to regulate the concentrations of both essential and non-essential metals within the cell.

References

Agriesti F., Roncarati D., Musiani F., Del Campo C., Iurlaro M., Sparla F., Ciurli S., Danielli A., Scarlato V. FeON-FeOFF: The *Helicobacter pylori* Fur regulator commutates iron-responsive transcription by discriminative readout of opposed DNA grooves. Nucleic Acids Res. 42, 3138–3151, 2014.

Aiyar J., Borges K.M., Floyd R.A., Wetterhahn K.E. Role of chromium(V), glutathione thiyl radical and hydroxyl radical intermediates in chromium(VI)-induced DNA damage. Toxicol. Environ. Chem. 22, 135–148, 1989.

Akama H., Kanemaki M., Yoshimura M., Tsukihara T., Kashiwagi T., Yoneyama H., Narita S., Nakagawa A., Nakae T. Crystal structure of the drug discharge outer membrane protein, OprM, of *Pseudomonas aeruginosa*: dual modes of membrane anchoring and occluded cavity end. J. Biol. Chem. 279, 52816–52819, 2004.

Andrews S.C., Robinson A.K., Rodríguez-Quiñones F. Bacterial iron homeostasis. FEMS Microbiol. Rev. 27, 215–237, 2003.

Argüello J.M., Eren E., González-Guerrero M. The structure and function of heavy metal transport P1B-ATPases. Biometals 20, 233–248, 2007.

Barkay T., Miller S.M., Summers A.O. Bacterial mercury resistance from atoms to ecosystems. FEMS Microbiol. Rev. 27, 355–384, 2003.

Begley T.P., Walts A.E., Walsh C.T. Mechanistic studies of a protonolytic organomercurial cleaving enzyme: bacterial organomercurial lyase. Biochemistry 25, 7192–7200, 1986.

Bereswill S., Greiner S., van Vliet A.H., Waidner B., Fassbinder F., Schiltz E., Kusters J.G., Kist M. Regulation of ferritin-mediated cytoplasmic iron storage by the ferric uptake regulator homolog (Fur) of *Helicobacter pylori*. J. Bacteriol. 82, 5948–5953, 2000.

Blindauer C.A. Bacterial metallothioneins: past, present, and questions for the future. J. Biol. Inorg. Chem. 16, 1011–1024, 2011.

Borremans B., Hobman J.L., Provoost A., Brown N.L., van Der Lelie D. Cloning and functional analysis of the pbr lead resistance determinant of *Ralstonia metallidurans* CH34. J. Bacteriol. 183, 5651–5658, 2001.

Branco R., Chung A.-P., Morais P.V. Sequencing and expression of two arsenic resistance operons with different functions in the highly arsenic-resistant strain *Ochrobactrum tritici* SCII24. BMC Microbiol. 8, 95–107, 2008.

Brown N.L., Stoyanov J.V., Kidd S.P., Hobman J.L. The MerR family of transcriptional regulators. FEMS Microbiol. Rev. 27, 145–163, 2003.

Bruins M.R., Kapil S., Oehme F.W. Microbial resistance to metals in the environment. Ecotoxicol. Environ. Saf. 45, 198–207, 2000.

Busenlehner L.S., Pennella M.A., Giedroc D.P. The SmtB/ArsR family of metalloregulatory transcriptional repressors: structural insights into prokaryotic metal resistance. FEMS Microbiol. Rev. 27, 131–143, 2003.

Caballero H.R., Campanelloa G.C., Giedroc D.P. Metalloregulatory proteins: metal selectivity and allosteric switching. Biophys. Chem. 156, 103–114, 2011.

Cai J., Salmon K., DuBow M.S. A chromosomal *ars* operon homologue of *Pseudomonas aeruginosa* confers increased resistance to arsenic and antimony in *Escherichia coli*. Microbiology 144, 2705–2713, 1998.

Calhoun L.N., Kwon Y.M. Structure, function and regulation of the DNA-binding protein Dps and its role in acid and oxidative stress resistance in *Escherichia coli*: a review. J. Appl. Microbiol. 110, 375–386, 2011.

Campbell D.R., Chapman K.E., Waldron K.J., Tottey S., Kendall S., Cavallaro G., Andreini C., Hinds J., Stoker N.G., Robinson N.J., Cavet J.S. Mycobacterial cells have dual nickel-cobalt sensors: sequence relationships and metal sites of metal-responsive repressors are not congruent. J. Biol. Chem. 282, 32298–32310, 2007.

Carrondo M.A. Ferritins, iron uptake and storage from the bacterioferritin viewpoint. EMBO J. 22, 1959–1968, 2003.

Cavet J.S., Meng W., Pennella M.A., Appelhoff R.J., Giedroc D.P., Robinson N.J. A nickel-cobalt-sensing ArsR-SmtB family repressor. Contributions of cytosol and effector binding sites to metal selectivity. J. Biol. Chem. 277, 38441–38448, 2002.

Cervantes C., Ji G., Ramirez J.L., Silver, S. Resistance to arsenic compounds in microorganisms. FEMS Microbiol. Rev. 15, 355–367, 1994.

Changela A., Chen K., Xue Y., Holschen J., Outten C.E., O'Halloran T.V., Mondragón A. Molecular basis of metal-ion selectivity and zeptomolar sensitivity by CueR. Science 301, 1383–1387, 2003.

Chaturvedi K.S., Hung C.S., Crowley J.R., Stapleton A.E., Henderson J.P. The siderophore yersiniabactin binds copper to protect pathogens during infection. Nat. Chem. Biol. 8, 731–736, 2012.

Chiancone E., Ceci P., Ilari A., Ribacchi F., Stefanini S. Iron and proteins for iron storage and detoxification. Biometals 17, 197–202, 2004.

Chiancone E., Ceci P. The multifaceted capacity of Dps proteins to combat bacterial stress conditions: Detoxification of iron and hydrogen peroxide and DNA binding. Biochim. Biophys. Acta. 1800, 798–805, 2010.

Chivers P.T., Sauer R.T. NikR repressor: High-affinity nickel binding to the C-terminal domain regulates binding to operator DNA. Chemistry & Biology 9, 1141–1148, 2002.

Cobbett C., Goldsbrough P. Phytochelatins and metallothioneins: roles in heavy metal detoxification and homeostasis. Annu. Rev. Plant Biol. 53, 159–182, 2002.

Cobine P.A., George G.N., Jones C.E., Wickramasinghe W.A., Solioz M., Dameron C.T. Copper transfer from the Cu(I) chaperone, CopZ, to the repressor, Zn(II)CopY: Metal coordination environments and protein interactions. Biochemistry 41, 5822–5829, 2002.

Cromie M.J., Shi Y., Latifi T., Groisman E.A. An RNA sensor for intracellular Mg(2+). Cell 125, 71–84, 2006.

Dambach M., Sandoval M., Updegrove T.B., Anantharaman V., Aravind L., Waters L.S., Storz G. The ubiquitous *yybP-ykoY* riboswitch is a manganese-responsive regulatory element. Mol. Cell. 57, 1099–1109, 2015.

Dann C.E. III, Wakeman C.A., Sieling C.L., Baker S.C., Irnov I., Winkler W.C. Structure and mechanism of a metal-sensing regulatory RNA. Cell 130, 878–892, 2007.

Delany I., Pacheco A.B., Spohn G., Rappuoli R., Scarlato V. Iron-dependent transcription of the frpB gene of *Helicobacter pylori* is controlled by the Fur repressor protein. J. Bacteriol. 183, 4932–4937, 2001.

Dey S., Rosen B.P. Dual mode of energy coupling by the oxyanion-translocating ArsB protein. J. Bacteriol. 177, 385–389, 1995.

Dosanjh N.S., Michel S.L.J. Microbial nickel metalloregulation: NikRs for nickel ions. Curr. Opin. Chem. Biol. 10, 123–130, 2006.

Endo G., Silver S. CadC, the transcriptional regulatory protein of the cadmium resistance system of *Staphylococcus aureus* plasmid pI258. J. Bacteriol. 177, 4437–4441, 1995.

Furukawa K., Ramesh A., Zhou Z., Weinberg Z., Vallery T., Winkler W.C., Breaker R.R. Bacterial riboswitches cooperatively bind Ni^{2+} or Co^{2+} ions and control expression of heavy metal transporters. Mol. Cell. 57, 1088–1098, 2015.

Garcia Vescovi E., Soncini F.C., Groisman E.A. Mg2+ as an extracellular signal: environmental regulation of *Salmonella* virulence. Cell 84, 165–174, 1996.

Gold B., Rodriguez G.M., Marras S.A., Pentecost M., Smith I. The *Mycobacterium tuberculosis* IdeR is a dual functional regulator that controls transcription of genes involved in iron acquisition, iron storage and survival in macrophages. Mol. Microbiol. 42, 851–865, 2001.

Gupta A., Matsui K., Lo J.F., Silver S. Molecular basis for resistance to silver cations in *Salmonella*. Nat. Med. 5, 183–188, 1999.

Haikarainen T., Thanassoulas A., Stavros P., Nounesis G., Haataja S., Papageorgiou A.C. Structural and thermodynamic characterization of metal ion binding in *Streptococcus suis* Dpr. J. Mol. Biol. 405, 448–460, 2011.

Higgins K.A., Giedroc D. Insights into protein allostery in the CsoR/RcnR family of transcriptional repressors. Chem. Lett. 43, 20–25, 2014.

Hill P.J., Cockayne A., Landers P., Morrissey J.A., Sims C.M., Williams P. SirR, a novel iron-dependent repressor in *Staphylococcus epidermidis*. Infect. Immun. 66, 4123–4129, 1998.

Hoogewerf A.J., Dyk L.A., Buit T.S., Roukema D., Resseguie E., Plaisier C., Le N., Heeringa L., Griend D.A. Functional characterization of a cadmium resistance operon in *Staphylococcus aureus* ATCC12600: CadC does not function as a repressor. J. Basic Microbiol. 55, 148–159, 2015.

Hrynkiewicz K., Złoch M., Kowalkowski T., Baum C., Niedojadło K., Buszewski B. Strain-specific bioaccumulation and intracellular distribution of Cd2+ in bacteria isolated from the rhizosphere, ectomycorrhizae, and fruitbodies of ectomycorrhizal fungi. Environ. Sci. Pollut. Res. Int. 22, 3055–3067, 2015.

Hu P., Brodie E.L., Suzuki Y., McAdams H.H., Andersen G.L. Whole-genome transcriptional analysis of heavy metal stresses in *Caulobacter crescentus*. J. Bacteriol. 187, 8437–8449, 2005.

Hynninen A., Touze T., Pitkanen L., Mengin-Lecreulx D., Virta, M. An efflux transporter PbrA and a phosphatase PbrB cooperate in a lead-resistance mechanism in bacteria. Mol. Microbiol. 74, 384–394, 2009.

Ilari A., Stefanini S., Chiancone E., Tsernoglou D. The dodecameric ferritin from *Listeria innocua* contains a novel intersubunit iron-binding site. Nat. Struct. Biol. 7, 38–43, 2000.

Image from the RCSB PDB (www.rcsb.org) of PDB ID 1DDN (White A., Ding X., vanderSpek J.C., Murphy J.R., Ringe D. Structure of the metal-ion-activated diphtheria toxin repressor/tox operator complex. Nature 394, 502–506, 1998).

Image from the RCSB PDB (www.rcsb.org) of PDB ID 1Q09 (Changela A., Chen K., Xue Y., Holschen J., Outten C.E., O'Halloran T.V., Mondragon A. Molecular basis of metal-ion selectivity and zeptomolar sensitivity by CueR. Science 301, 1383–1387, 2003).

Image from the RCSB PDB (www.rcsb.org) of PDB ID 1QGH (Ilari A., Stefanini S., Chiancone E., Tsernoglou D. The dodecameric ferritin from *Listeria innocua* contains a novel intersubunit iron-binding site. Nat. Struct. Biol. 7, 38–43, 2000).

Image from the RCSB PDB (www.rcsb.org) of PDB ID 1R1T (Eicken C., Pennella M.A., Chen X., Koshlap K.M., VanZile M.L., Sacchettini J.C., Giedroc D.P. A metal-ligand-mediated intersubunit allosteric switch in related SmtB/ArsR zinc sensor proteins. J. Mol. Biol. 333, 683–695, 2003).

Image from the RCSB PDB (www.rcsb.org) of PDB ID 1XSD (Safo M.K., Zhao Q., Ko T.-P., Musayev F.N., Robinson H., Scarsdale N., Wang A.H.-J., Archer G.L.J. Crystal structures of the BlaI repressor from *Staphylococcus aureus* and its complex with DNA: insights into transcriptional regulation of the *bla* and *mec* operons. Bacteriol. 187, 1833–1844, 2005).

Image from the RCSB PDB (www.rcsb.org) of PDB ID 2HZV (Schreiter E.R., Wang S.C., Zamble D.B., Drennan C.L. NikR-operator complex structure and the mechanism of repressor activation by metal ions. Proc. Natl. Acad. Sci. U. S. A. 103, 13676–13681, 2006).

Image from the RCSB PDB (www.rcsb.org) of PDB ID 2W57 (Sheikh M.A., Taylor G.L. Crystal structure of the *Vibrio cholerae* ferric uptake regulator (fur) reveals insights into metal co-ordination. Mol. Microbiol. 72, 1208–1220, 2009).

Image from the RCSB PDB (www.rcsb.org) of PDB ID 3E1O (Crow A., Lawson T.L., Lewin A., Moore G.R., Le Brun N.E. Structural basis for iron mineralization by bacterioferritin. J. Am. Chem. Soc. 131, 6808–6813, 2009).

Image from the RCSB PDB (www.rcsb.org) of PDB ID 4ADZ (Dwarakanath S., Chaplin A.K., Hough M.A., Rigali S., Vijgenboom E., Worrall J.A.R. The response to copper stress in *Streptomyces lividans* extends beyond genes under the direct control of a copper sensitive operon repressor protein (CsoR). J. Biol. Chem. 287, 17833–17847, 2012).

Image from the RCSB PDB (www.rcsb.org) of PDB ID 4REU (Thiruselvam V., Sivaraman P., Kumarevel T.S., Ponnuswamy M.N. Revelation of endogenously bound Fe(2+) ions in the crystal structure of ferritin from *Escherichia coli*. Biochem. Biophys. Res. Commun. 453, 636–641, 2014).

Inesi G., Pilankatta R., Tadini-Buoninsegni F. Biochemical characterization of P-type copper ATPases. Biochem. J. 463, 167–176, 2014.

Iwig J.S., Rowe J.L., Chivers P.T. Nickel homeostasis in *Escherichia coli*—the rcnR-rcnA efflux pathway and its linkage to NikR function. Mol. Microbiol. 62, 252–262, 2006.

Jarosławiecka A., Piotrowska-Seget Z. Lead resistance in micro-organisms. Microbiology 160, 12–25, 2014.

Jakubovics N.S., Smith A.W., Jenkinson H.F. Expression of the virulence-related Sca (Mn2+) permease in *Streptococcus gordonii* is regulated by a diphtheria toxin metallorepressor-like protein ScaR. Mol. Microbiol. 38, 140–153, 2000.

Johnston A.W., Todd J.D., Curson A.R., Lei S., Nikolaidou-Katsaridou N., Gelfand M.S., Rodionov D.A. Living without fur: the subtlety and complexity of iron-responsive gene regulation in the symbiotic bacterium *Rhizobium* and other alpha-proteobacteria. Biometals 20, 501–511, 2007.

Kim E.H., Nies D.H., McEvoy M.M., Rensing C. Switch or funnel: how RND-type transport systems control periplasmic metal homeostasis. J. Bacteriol. 193, 2381–2387, 2011.

Koh E.-I., Henderson J.P. Microbial copper-binding siderophores at the host-pathogen interface. J. Biol. Chem. 290, 18967–18974, 2015.

Kolaj-Robin O., Russell D., Hayes K.A., Pembroke J.T., Soulimane T. Cation diffusion facilitator family: structure and function. FEBS Lett. 589, 1283–1295, 2015.

Koronakis V., Sharff A., Koronakis E., Luisi B., Hughes C. Crystal structure of the bacterial membrane protein TolC central to multidrug efflux and protein export. Nature 405, 914–919, 2000.

Liu C., Mao K., Zhang M., Sun Z., Hong W., Li C., Peng B., Chang Z. The SH3-like domain switches its interaction partners to modulate the repression activity of mycobacterial iron-dependent transcription regulator in response to metal ion fluctuations. J. Biol. Chem. 283, 2439–2453, 2008.

Liu T., Reyes-Caballero H., Li C., Scott R.A., Giedroc D.P. Multiple metal binding domains enhance the Zn(II) selectivity of the divalent metal ion transporter AztA. Biochemistry 46, 11057–11068, 2007.

Lu M., Fu D. Structure of the zinc transporter YiiP. Science 317, 1746–1748, 2007.

Ma Z., Jacobsen F.E., Giedroc D.P. Coordination chemistry of bacterial metal transport and sensing. Chem. Rev. 109, 4644–4681, 2009.

Mathema V.B., Thakuri B.C., Sillanpaa M. Bacterial *mer* operon-mediated detoxification of mercurial compounds: a short review. Arch. Microbiol. 193, 837–844, 2011.

Merchant A.T., Spatafora G.A. A role for the DtxR family of metalloregulators in gram-positive pathogenesis. Mol. Oral Microbiol. 29, 1–10, 2014.

Morby A.P., Turner J.S., Huckle J.W., Robinson N.J. SmtB is a metal-dependent repressor of the cyanobacterial metallothionein gene *smtA*: identification of a Zn inhibited DNA-protein complex. Nucleic Acids Res. 21, 921–925, 1993.

Mukhopadhyay R., Rosen B.P. *Saccharomyces cerevisiae* ACR2 gene encodes an arsenate reductase. FEMS Microbiol. Lett. 168, 127–136, 1998.

Murakami S., Nakashima R., Yamashita E., Matsumoto T., Yamaguchi A. Crystal structures of a multidrug transporter reveal a functionally rotating mechanism. Nature 443, 173–179, 2006.

Nascimento A.M., Chartone-Souza E. Operon *mer*: bacterial resistance to mercury and potential for bioremediation of contaminated environments. Genet. Mol. Res. 2, 92–101, 2003.

Nies D.H. Microbial heavy-metal resistance. Appl. Microbiol. Technol. 51, 730–750, 1999.

Outten C.E., O'Halloran T.V. Femtomolar sensitivity of metalloregulatory proteins controlling zinc homeostasis. Science 292, 2488–2492, 2001.

Outten C.E., Outten F.W., O'Halloran T.V. DNA distortion mechanism for transcriptional activation by ZntR, a Zn(II)-responsive MerR homologue in *Escherichia coli*. J. Biol. Chem. 274, 37517–37524, 1999.

Pal A., Paul A.K. Microbial extracellular polymeric substances: central elements in heavy metal bioremediation. Indian J. Microbiol. 48, 49–64, 2008.

Paulsen I.T., Saier Jr. M.H. A novel family of ubiquitous heavy metal ion transport proteins. J. Membr. Biol. 156, 99–103, 1997.

Pennella M.A., Giedroc D.P. Structural determinants of metal selectivity in prokaryotic metal-responsive transcriptional regulators. Biometals 18, 413–428, 2005.

Posey J.E., Hardham J.M., Norris S.J., Gherardini F.C. Characterization of a manganese-dependent regulatory protein, TroR, from *Treponema pallidum*. Proc. Natl. Acad. Sci. U. S. A. 96, 10887–10892, 1999.

Price I.R., Gaballa A., Ding F., Helmann J.D., Ke A. Mn(2+)-sensing mechanisms of *yybP-ykoY* orphan riboswitches. Mol. Cell. 57, 1110–1123, 2015.

Que Q., Helmann J.D. Manganese homeostasis in *Bacillus subtilis* is regulated by MntR, a bifunctional regulator related to the diphtheria toxin repressor family of proteins. Mol. Microbiol. 35, 1454–1468, 2000.

Rajagopalan S., Teter S.J., Zwart P.H., Brennan R.G., Phillips K.J., Kiley P.J. Studies of IscR reveal a unique mechanism for metal-dependent regulation of DNA binding specificity. Nat. Struct. Mol. Biol. 20, 740–747, 2013.

Ramesh A., Winkler W.C. Magnesium-sensing riboswitches in bacteria. RNA Biol. 7, 77–83, 2016.

Randall C.P., Gupta A., Jackson N., Busse D., O'Neill A.J. Silver resistance in Gram-negative bacteria: a dissection of endogenous and exogenous mechanisms. J. Antimicrob. Chemother. 70, 1037–1046, 2015.

Roanne T.M. Lead resistance in two bacterial isolates from heavy metal-contaminated soils. Microb. Ecol. 37, 218–224, 1999.

Rouch D.A., Lee B.T., Morby A.P. Understanding cellular responses to toxic agents: a model for mechanism-choice in bacterial metal resistance. J. Ind. Microbiol. 14, 132–141, 1995.

Schelert J., Dixit V., Hoang V., Simbahan J., Drozda M., Blum P. Occurrence and characterization of mercury resistance in the hyperthermophilic archaeon *Sulfolobus solfataricus* by use of gene disruption. J. Bacteriol. 186, 427–437, 2004.

Scott J.A., Palmer S.J. Sites of cadmium uptake in bacteria used for biosorption. Appl. Microbiol. Biotechnol. 33, 221–225, 1990.

Silver S. Bacterial silver resistance: molecular biology and uses and misuses of silver compounds. FEMS Microbiol. Rev. 27, 341–353, 2003.

Silver S. Genes for all metals—a bacterial view of the periodic table. J. Ind. Microbiol. Biot. 20, 1–12, 1998.

Silver S., Phung L.T., Silver G. Silver as biocides in burn and wound dressings and bacterial resistance to silver compounds. J. Ind. Microbiol. Biotechnol. 33, 627–634, 2006.

Smith R.L., Maguire M.E. Microbial magnesium transport: unusual transporters searching for identity. Mol. Microbiol. 28, 217–226, 1998.

Smith S.R. A critical review of the bioavailability and impacts of heavy metals in municipal solid waste composts compared to sewage sludge. Environ. Int. 35, 142–156, 2009.

Solovieva I.M., Entian K.D. Investigation of the *yvgW Bacillus subtilis* chromosomal gene involved in Cd^{2+} ion resistance. FEMS Microbiol. Lett. 208, 105–109, 2002.

Srivastava P., Kowshik M. Mechanisms of metal resistance and homeostasis in haloarchaea. Archaea 2013, 16, 2013.

Strausak D., Solioz M. CopY is a copper-inducible repressor of the *Enterococcus hirae* copper ATPases. J. Biol. Chem. 272, 8932–8936, 1997.

Sutherland I.W. Biofilm exopolysaccharides: a strong and sticky framework. Microbiology 147, 3–9, 2001.

Tchounwou P.B., Yedjou C.G., Patlolla A.K., Sutton D.J. Heavy metal toxicity and the environment. EXS 101, 133–164, 2012.

Tottey S., Harvie D.R., Robinson N.J. Understanding how cells allocate metals using metal sensors and metallochaperones. Acc. Chem. Res. 38, 775–783, 2005.

Tsai K.-J., Linet A.L. Formation of a phosphorylated enzyme intermediate by the *cadA* Cd^{2+}-ATPase. Arch. Biochem. Biophys. 305, 267–270, 1993.

van der Lelie D., Schwuchow T., Schwidetzky U., Wuertz S., Baeyens W., Mergeay M., Nies, D.H. Two component regulatory system involved in transcriptional control of heavy metal homoeostasis in *Alcaligenes eutrophus*. Mol. Microbiol. 23, 493–503, 1997.

Wakeman C.A., Ramesh A., Winkler W.C. Multiple metal-binding cores are required for metalloregulation by M-box riboswitch RNAs. J. Mol. Biol. 392, 723–735, 2009.

Waldron K.J., Robinson N.J. How do bacterial cells ensure that metalloproteins get the correct metal? Nat. Rev. Microbiol. 7, 25–35, 2009.

Wang H., Zhang Q., Cai B., Li H., Sze K.H., Huang Z.X., Wu H.M., Sun H. Solution structure and dynamics of human metallothionein-3 (MT-3). FEBS Lett. 580, 795–800, 2006.

Wang Y., Hemmingsen L., Giedroc D.P. Structural and functional characterization of *Mycobacterium tuberculosis* CmtR, a PbII/CdII-sensing SmtB/ArsR metalloregulatory repressor. Biochemistry 28, 8976–8988, 2005.

Wei X., Mingjia H., Xiufeng L., Yang G., Qingyu W. Identification and biochemical properties of Dps (starvation-induced DNA binding protein) from cyanobacterium *Anabaena* sp. PCC 7120. IUBMB Life 59, 675–681, 2007.

Wingender J., Neu T.R., Flemming H.C. What are bacterial extracellular polymeric substances? *In*: Wingender J., Neu T.R., Flemming H.C. (eds.). Microbial Extracellular Polymeric Substances. Springer, Berlin, pp 1–20, 1999.

Wu J., Rosen B.P. Metalloregulated expression of the *ars* operon. J. Biol. Chem. 268, 52–58, 1993.

CHAPTER 3

Microbial Metalloproteins-Based Responses in the Development of Biosensors for the Monitoring of Metal Pollutants in the Environment

Elvis Fosso-Kankeu

1. Introduction

The accumulation of heavy metals and metalloids in the water systems and top soils to a toxic concentration level may lead to human poisoning and/or ecological damage. Exposure of humans to heavy metal may occur through ingestion or contact with contaminated water, crop, and soil (McLaughlin et al., 2000a,b). Most terrestrial and aquatic animal and plant species are sensitive to increased concentration of metals in the environment which can therefore result into the imbalance of the biodiversity and the alteration of food availability. To mitigate the impact of heavy metals on humans as well as the aquatic life, water authorities in various countries and international environmental agencies have established guidelines recommending the occurrence of limited amounts of heavy metals in drinking water and irrigation water. This serves for effective monitoring of water quality through regular detection of heavy metals; examples of techniques currently used for the detection of heavy metals in water include electrochemical, optical, piezoelectric and ion selective electrode (ISE) sensors (Aragay et al., 2011; Tekaya et al., 2013; Mehta et al., 2016).

The field of sensor development is continuously evolving to address issues such as poor accuracy, selectivity and reversibility; recent advances have considered the use of biological materials as receptors in detection

School of Chemical and Minerals Engineering, North–West University, South Africa.
Email: elvisfosso.ef@gmail.com

devices. Metalloproteins or metalloenzymes are particularly preferred in the fabrication of sensors because they are capable of distinguishing among metals during the binding process (Ramos et al., 1993). Metalloproteins have affinity for specific metals to a certain extent, allowing them to bind the metals and move them across the cell (Ma et al., 2009; Fosso-Kankeu et al., 2014). Typical example of metalloproteins often found in yeast, fungi, and algae are low molecular weight metallothioneins with high cysteine content (Ibrahim et al., 2001; Fosso-Kankeu et al., 2014); Metalloproteins with a higher affinity and metal binding capacity can be either designed *de novo* or selected by screening peptides libraries and then expressed in transformed strains (Mejare and Bulow, 2001). The functions of metalloprotein in microbial cells are distinguishable but all contribute in ensuring metal homeostasis in the cell by influencing uptake, efflux, intracellular trafficking within compartments, and storage (Tottey et al., 2008; Waldron and Robinson, 2009).

Metalloproteins as a part of the microbial whole cell or as an isolated biomolecule has been targeted for the sensing of heavy metals when incorporated in electrochemical and optical biosensor devices.

2. Sources of metal pollutants and impact on the environment

The occurrence of heavy metals in the environment may be through natural/ biogenic sources such as volcanic rocks, weathering and erosion of bed rocks and ore deposits, marine sedimentary rocks, and fossil fuels including coal and petroleum (Korte and Fernando, 1991; Smedley and Kinniburgh, 2002) or anthropogenic sources; the latter has been found as the major contributor to pollution since effluents from mining and hydrometallurgical plants, photography industries, electroplating, leaded gazoline and paints, coal combustion residues, spillage of petrochemicals, coke factories, power plants, garbage incinerators, land application of fertilizers and pesticides, and cement are dispersed into the receiving environment (Demirbas et al., 2004; Khan et al., 2008; Fosso-Kankeu et al., 2009, 2010, 2011, 2014; Zhang et al., 2010; Wuana and Okieimen, 2011; Mittal et al., 2013). Other major sources of heavy metals in developing countries include the discharge of agricultural, untreated domestic and industrial waste waters in the environment (Gupta, 2008). High concentrations of toxic heavy metals such as Fe, Mn, As, Pb, Zn and Co were found in the stream of Peddavagu around the Patancheru industrial development area (India). They were mostly derived from the untreated effluents of the Central Effluent Treatment Plant (CETP) which were released into the environment (Krishna et al., 2009). Gowd et al. (2010) have reported the tangible case of pollution from leather processing clusters of tannery industries contributing to high BOD, high pH, high total dissolved solid, and mostly high concentrations of toxic hexavalent chromium in soil and surface waters at Jajmau and Unnao industrial areas of the Ganga Plain,

Uttar Pradesh, India. They further uncovered the contribution of cotton and textile mills, large fertilizer factories, and several arms factories which arbitrarily dump hazardous waste and discharge effluents in the environment. In some instances, the metals' occurrence and distribution have been mostly dominated by geogenic influence; in an investigation conducted by Singh et al. (2003) on the distribution of heavy metals in sediments of the Ganges River, India was found out that variation in heavy metals in the Ganges River basin was mostly the result of extensive physical weathering of the Himalayas and monsoon-controlled fluvial process. According to Roman-Ross et al. (2002), change in the local environment such as acid deposition and climate change may enhance the weathering of metal-bearing minerals contributing to geogenic inputs of trace elements to the water system. High concentrations of heavy metals in river and canal sediments of the Jakarta region have been reported to be significantly controlled by the precursor volcanic rocks as well as weathering in the catchment area due to strong seasonal rainfall (Sindern et al., 2016). Chemical weathering and dissolution of the bedrock has also been implicated in the mobility of cations from plagioclase and biotite to the rivers in Kaelia region and Kola Peninsula, NW Russia (Zakharova et al., 2007). Filgueiras et al. (2004) used chemometric analysis to evaluate the distribution and mobility of heavy metals in surficial sediments of Louro River (Galicia, Spain). They found that the sources of Ni varied depending on the site; at certain sites, high Ni contents were ascribed to geogenic sources such as the occurrence of metamorphic rocks including calcite, while at other sites high Ni contents could be associated with anthropogenic sources such as the proximity to industrial areas along with sewage discharge. According to a study conducted by Wang and Mulligan (2006) on the sources of arsenic in the environment, biogenic sources of arsenic in water could be ascribed to reductive dissolution of arsenic-bearing iron, desorption of arsenic from (hydro)oxides, oxidation of arsenic-bearing sulfides, release of arsenic from geothermal water, and evaporative concentration (Kim et al., 2000; Bennett and Dudas, 2003). The same study revealed that anthropogenic activities such as pesticide production and application, nonferrous metal mining and smelting, wood preserving, fossil fuel processing and combustion, disposal, and incineration of municipal and industrial wastes contribute to the release of metals into the environment (Popovic et al., 2001; Prosun et al., 2002). Atmospheric pollution of heavy metals also plays an important role in the dispersion of metals in the environment; the action of wind on disturbed land, stockpiles of ore and waste materials from mines, and unpaved road soils result in the permanent introduction of dust into the atmosphere. Heavy metals originating from wear and tear of tires, alloys, wires, and brake parts that have been reported in vehicular emissions as lead, copper, zinc, iron, and cadmium, generally, accumulate in soils along the roadsides of unpaved roads (Ugwu et al., 2011; Raj and Ram, 2013). Studying the atmospheric pollution with heavy metals due to copper mining near Radovis in the Republic of Macedonia, Balabanova et al. (2011) found that the chemical elements present in the attic dust are derived from both geogenic (Ca, Li, Mg, Mn, Sr, Co, Cr and Ni) and anthropogenic (As, Cu and lead) sources. The occurrence of Pt close to a major highway in Oslo, Norway, was related to anthropogenic source traffic emissions (automotive exhaust catalytic converters) and geogenic source namely the Cambro-Silurian sediments (Reimann et al., 2007). After detailed investigation of the sources of toxic metals present in the urban soils and roadside dust in Shangai, China, Shi and colleagues (2008) observed that the traffic contaminants such as vehicle exhaust particles, lubricating oil residues, tire wear particles, weathered street surface particles, and brake lining wear particles contributed to the presence of Pb, Zn and Cu in the atmospheric dust, while Ni was mainly associated to natural geogenic sources.

3. Metal uptake by microorganisms

Based on the physiology of microorganisms, mechanisms involved in the uptake of metal pollutants have been identified as passive and active. The passive mechanism is also known as metabolism-independent mechanism because it does not require consumption of energy to take place (Ahluwalia and Goyal, 2007). In this case, metals are removed from solution through complexation with extracellular biological chelates or by binding to the cell surfaces. Extracellular polymeric substances excreted by microorganisms are responsible for extracellular complexation as they bind to metals coming into contact with the microorganisms. The EPS in microbial biofilms can strongly bind to metals, releasing colloids materials such protein; the ecotoxicity, bioavailability, mobility, and chemical forms of the metals in solution will therefore be affected by the EPS characteristics which influence significantly the adsorption onto the microorganisms (Choi et al., 2009; Wu et al., 2012). The binding of metals to cell surfaces involves adsorption process such as ionic, physical, and chemical adsorption. Ionic process occurs as peptidoglycan, techoic acid, alginates, or any other polysaccharides at the surface of microorganisms exchange their counter ions against divalent metal ions in solution. During physical adsorption, weak forces such as electrostatic interactions and van der Waals' forces assist in the binding of the metal ions to the microorganism surfaces (Ahalya et al., 2003; Gavrilescu, 2004). The active or metabolism-dependent mechanism is exclusively found in living microorganisms which spent energy in the form of ATP to transport metal across the cell membrane resulting into their intracellular accumulation. In principles, ligands or small chelating agents excreted by microorganisms into the external environment are responsible for the binding and transportation of metals into the cells. This mechanism is often associated with the active defense system of the microorganisms as they produce compounds which favor the precipitation process in the presence of toxic metals. According to previous researches, metal accumulation in microorganisms involves four major steps (Scheerer et al., 2010; Garcia-Garcia et al., 2012, 2016): (i) metal ion uptake by plasma membrane transporters that have essential heavy and non-heavy metal ions as physiological substrates; (ii) increased activities due to enhanced gene transcription of plasma membrane sulfate transporters (PMST), metal ion plasma membrane transporters, ATP sulfurylase, adenosine 5'-phosphosulfate reductase (APSR), and γ-glutamylcysteine synthetase (γ-ECS); (iii) activation of phytochelatin synthase (PCS) by heavy metals; (iv) compartmentalization into vacuoles or other organelles of metal-glutathione or metal-phytochelatin complexes; and (v) formation of Metal-High Molecular Weight Complexes (Me-HMWC) in cytosol and inside organelles. The major disadvantage in using living cell for adsorption of metal ions is the risk of inhibition in the presence of relatively high concentration of metals. The metal adsorption capacity and mechanism of microorganisms can be affected in the presence of relatively high concentration of metals in the solution. This can result into the inhibition of the microorganisms which develop a strategy to respond to the change in the surrounding and maintain low intracellular concentration of metals. The ability of microorganisms to respond to such external aggression in the milieu varies depending on the group and species of the microorganisms as well as the source of the microorganisms (Wood and Wang, 1983). *Micrococcus* sp. and *Aspergillus* sp. isolated from the soil samples of electroplating industry by Congeevaram et al. (2007) exhibited relative tolerance to chromium and were successfully used for the removal of chromium and nickel from industrial wastewater. These isolates were found to express higher amount of proteins and polypeptides in the presence of relatively toxic concentrations of metals. In the recent work of Fosso-Kankeu et al., 2016, the bacterial species *Bacillaceae bacterium* was transformed with the plasmid carrying the (pECD312)-based cnr operon that encodes nickel and cobalt resistance; the transformed strain was found to tolerate higher concentration of nickel as compared to the wild type and its adsorption capacity was improved by 37%. Some metal-binding molecules such as metallothioneins and phytochelatins have been found to play an important role in metal detoxification of some microorganisms and have been expressed in non-producing microbial sorbents to enhance their metal tolerance (Fosso-Kankeu et al., 2016).

4. Microbial metalloproteins

Although some metals are considered as trace elements required for the growth of microorganisms, the presence of relatively large concentrations of these metals and relatively low concentration of toxic non-nutritional metals in the milieu are likely to inhibit the growth of microorganisms which have the potential to sense changes in the environment through membrane receptors. Sensing of the danger may trigger physiological changes in the microorganisms as they prepare to develop mechanisms that will allow them to resist the threat caused by the metals against their survival. A balance in the requirement of metals and their toxicity is achieved by the microorganisms through two main genetic mechanisms namely the active efflux pumping of the toxic metals out of the cell by systems involved in transport of nutrient cations or oxyanions and detoxification strategy by conversion of toxic to less available metal-ion species.

4.1 *Metal transporters and efflux pump pathways per microorganisms*

The metal transport system in microorganisms is well organized and governed by either peptides or small proteins expressed by specific genes from both plasmid and chromosomal systems which are regulated to ensure that the metal ions are transported inside the cell when needed or outside the cell when toxicity level is reached. Functions such as metal ion uptake, storage, detoxification, and homeostasis systems are encoded by specific genes whose expression is controlled by metalloproteins (O'Halloran, 1993) (Table 1). Resistance mechanisms, the transport system, nature of protein or gene expressed, and functions may vary depending on the microbial species as well as the type of metal ion involved. Silver and Phung (2005) have considered seven efflux pump systems as significant: (1) P-type ATPases (dominantly single polypeptide determinants), (2) ABC ATPases which is a family with both uptake pumps and efflux pumps (not for metal ions); (3) the cation diffusion facilitator family (CDF) which was first described for CzcD Cd^{2+} and Zn^{2+} efflux system of *R. metallidurans* (Nies, 2003; Haney et al., 2005); (4) the single membrane polypeptides of the Major Facilitator Superfamily (MFS); (5) the CBA family of three polypeptide chemiosmotic antiporters such as the CzcCBA complex that function as an ion/proton exchanger to efflux Cd^{2+}, Co^{2+}, and Zn^{2+} which also belongs to the superfamily Resistance-Nodulation-Cell Division (RND) as some members are involved in bacterial nodulation formation by Rhizobium and others are involved in cell division as in the case of *E. coli*, few genes including *ncc* (Ni^{2+} Co^{2+} resistance), *czr* (Cd^{2+} Zn^{2+} resistance), and *cnr* (Co^{2+} Ni^{2+} resistance) (Hassan et al., 1999; Legatzki et al., 2003; Mergeay et al., 2003; Nies, 2003); (6) ChrA (CHR) chromate efflux system; (7) ArsB, the arsenite [As(III)] and antimonite [Sb(III)] efflux system that can function alone as a chemiosmostic efflux system or with a second ArsA subunit functions with energy from ATP.

The transport mechanisms of metal ions such as zinc, copper, and cadmium have been well studied in various microorganisms and discussions around the expression of specific proteins in response to their requirement or tolerance could allow covering to a certain extent the implications of metalloproteins in the binding of metal ions. Bacterial metal transporters belonging to the cluster 9 family of ABC transporters have been identified in pathogenic streptococci and in cyanobacteria, and are members of

Table 1. Microbial systems involving genes and proteins for metal tolerance mechanisms.

System	Metal	Microorganism	Gene operon	Peptide/protein expressed	References
Efflux pump	Cu	*Synechococcus* PCC 7942	*pacS*	P1-type ATPase	Kanamaru et al., 1994
	Zn	*Synechococcus* PCC 6803	*zia*A	P1-type ATPase	Thelwell et al., 1998
	Zn	*Escherichia coli*	*znu*	ZnuACB	Rensing et al., 1999
	Cd	*S. aureus*	*cad*A, *cad*C	P1-type ATPase	Odermatt et al., 1993
	Zn	*Escherichia coli*	*znt*A	P1-type ATPase	Sofia et al., 1994
	Cr	*R. metallidurans*	*chr*	chrA	Juhnke et al., 2002
	Co and Ni	*R. metallidurans*	*cnr*	cnrYXH	Nies, 2003
	Co, Zn and Cd	*R. metallidurans*	*czc*	czcCBA	Nies, 2003
	Cu	*Enterococcus hirae*	*cop*	CopA and CopB	Cobine et al., 1999
	Ag	*Salmonella*	*sil*E	SilCBA	Gupta et al., 1999
Metallo protein	Zn	*Synechococcus* PCC 7942	*smt*A, *smt*B	Metallothionein	Huckle et al. 1993; Busenlehner et al., 2003
	Cd, Cu, Zn, Hg	*Tetrahymena*	TpyrMT-1, TpyrMT-2	Metallothionein (CdMT)	Fu and Miao, 2006
	Cd, Pb, As, Cu, Zn, Ni	*Tetrahymena*	TtheMTT1	Metallothionein (CdMT)	Diaz et al., 2007
	Cd, Zn, As, Ni, Cu, Pb	*Tetrahymena*	TtheMTT3	Metallothionein (CdMT)	Guttierez et al., 2011
	Pb, As, Cd, Cu, Zn, Ni	*Tetrahymena*	TtheMTT5	Metallothionein (CdMT)	Guttierez et al., 2011
	Pb, Cd, As, Zn, Cu, Ni	*Tetrahymena*	TrosMTT1	Metallothionein (CdMT)	Amaro et al., 2008
	Cu, Zn, Cd, Pb, Ni	*Tetrahymena*	TrosMTT2	Metallothionein (CuMT)	Guttierez et al., 2011
	Hg, Pb, Cd, Cu, Zn	*Tetrahymena*	TpigMT-1	Metallothionein (CdMT)	Guo et al., 2008

manganese- and zinc-binding-protein-dependent transport systems (Claverys, 2001). The znuABC genes which consists of the periplasmic binding protein ZnuA, the membrane permease ZnuB, and the ATPase ZnuC encode the aforementioned high-affinity zinc-uptake system. Xiong and coworkers (2011) isolated a novel and multiple metal(oid)-resistant strain *Comamonas testosteroni* S44 from which 9 putative Zn^{2+} transporters (4 znt operons encoding putative Zn^{2+} translocating P type ATPases and 5 czc operons encoding putative RND-driven (resistance, nodulation, cell division protein family)] tripartite protein complexes). Plant pathogens including *E. coli* and *Pseudomonas syringae* which were routinely exposed to copper-containing antimicrobial agents developed a resistance to copper and the operons involved were sequenced. The resistance mechanisms require two proteins for copper-responsive transcriptional activation of the resistance genes. The corresponding regulatory genes in *E. coli* are *pco*R and *pco*S which have been sequenced by Lee and colleagues (Bryson et al., 1993) while *cop*R and *cop*S are the regulatory genes of the *P. syringae* systems and have been sequenced by Cooksey and colleagues (Mills et al., 1993). The copper transport and resistance system in *Enterococcus hirae* and cyanobacterium *Synechococcus* PCC7942 has been well studied (Solioz et al., 1994; Phung et al., 1994). In the first microorganism, the uptake and efflux P-type ATPases are found in a single operon and are determined by two genes namely *cop*A and *cop*B; while in cyanobacteria, the copper uptake and efflux ATPases are dissimilar allowing for separate gene regulation. A number of bacteria including *Stenotrophomonas maltophilia* (Alonso et al., 2000), *Listeria monocytogenes* (Lebrun et al., 1994), *Bacillus subtilis* (Tsai et al., 1992), *Pseudomonas putida* (Lee et al., 2001), *Staphylococcus aureus* (Nucifora et al., 1989), and *Helicobacter pylori* (Herrmann et al., 1999) are among those with the ability to express a cadmium resistance mechanism provided by the *cadA* operon of the efflux system in which a P-type ATPase is involved in metal ion transport across the cell membrane.

The CADA ATPase which is an example of a system now widely found in Gram-positive bacteria represents the 727 amino acid Cd^{2+} efflux ATPase. It was first identified on *Staphylococcal* plasmid pI258 (Silver, 1996). Naz et al. (2005) investigated cadmium resistance among a mixture of sulphate reducing bacteria (SRB) and identified two isolates (*Desulfuvibrio desulfuricans* DSM 1926 and *Desulfococcus multivorans* DSM 2059) with highest resistance to cadmium; using DNA hybridization technique, they found the genes *smt*AB, *cad*AC, and *cad*D in the SRB tested. Studying the cadmium resistance properties of a newly isolated *Lactococcus lactis* subsp. *lactis*, Sheng and coworkers (2016) found that cadmium stress led to the identification of 12 over-expressed proteins and activation of the antioxidant capacity of the strain. Furthermore, they found that the up-regulated *cad*A was associated with the activated P-type ATPases which constitutes part of the efflux pump system necessary for metal resistance.

Detoxification of metal by microorganisms could also occur through the use of intracellular metal resistance mechanisms that involve metal binding or sequestration by metalloproteins such as metallothioneins and phytochelatins. Metallothioneins are cystein-rich proteins of small size whose biosynthesis is regulated at the transcriptional level and stimulated under the stress conditions of heavy metals (Zn, Cd, Hg, Ag, Cu, Au, Bi, Co, and Ni) (Kagi and Schaffer, 1988). Multiple copies of gene producing metallothionein have been found in selected cells grown on increasingly high levels of Cd^{2+} (Gupta et al., 1992). Bacterial metallothionein polypeptides are not homologous in sequence; in the cyanobacteria strain *Synechococcus*, they are 56 amino-acid-long encoded by the *smt*A gene and contain nine cysteine residues clustered in two groups of four and five (Silver and Phung, 1996; Mejare and Bulow, 2001). Apart from the efflux pump mechanism, Zn can be maintained to non-toxic level by lowering its free concentration within the cytoplasm; few bacteria achieve this by encoding metallothionein protein which is a metal cation-binding protein of small size (Blindauer et al., 2002; Cavet et al., 2003). In *Synechococcus* spp., the resistance to cadmium and zinc is mediated by the *smt* locus which consists of two divergently transcribed genes, *smt*A and *smt*B (Turner et al., 1995). Zinc resistance in *Synechococcus* PCC 7942 was significantly decreased after deletion of the *smt*A gene (Turner et al., 1993). To tightly bind and sequester surplus atoms of Zn, some cyanobacteria and *Pseudomonas* express metallothioneins (MT) under the control of Zn sensors, such as *smt*B (Huckle et al., 1993; Waldron and Robinson, 2009). The SmtB protein mediates Zn-responsive regulation of a bacterial MT gene involved in Zn-Cd tolerance. A structure involving nine cysteine thiols found in a 56 amino acid polypeptide of the cyanobacterial metallothionein was found to bind 4 Zn^{2+} cations (Blindauer et al., 2002; Cavet et al., 2003). Cyanobacterial metallothionein also exhibit affinity for metals and preferably bind cations in the order $Zn^{2+} > Cd^{2+} > Cu^{2+}$ (Silver, 1996).

Phytochelatins (PC) are low molecular weight cysteine-rich peptides enzymatically synthesized and have been shown to have high binding affinity for metals; they are commonly found in plants and other eukaryotes including fungi and consist of the repeating γ-Glu-Cys dipeptide unit terminated by Gly residue. PC was first discovered by Hayashi and his group in the Cd-binding complexes produced in fission yeast, *Schizosaccharomyces pombe* were exposed to Cd^{2+} ions and named "Cadystins", the name phytochelatin was later given by Grill et al. (1985) who found the same peptides in various cells of plants (Murasugi et al., 1981; Inouhe, 2005). PC biosynthesis is induced under heavy metals' (Cd, Hg, Pb, and Cu) stress conditions; under such conditions the enzyme γ-glutamylcysteinyl dipeptidyl transpeptidase (PC synthase) is activated and catalyzes the transfer of γ-Glu-Cys from glutathione (GSH) to another GSH or other PCs (Mejare and Bulow, 2001; Kang et al., 2007). The first genetical identification of the PC synthase gene *cad*1 was done in *Arabidopsis*, while the expression of cDNA libraries from *Arabidopsis* and wheat in *S. cerevisiae* allowed to identify two genes (AtPCS1 and TaPCS1) conferring increase in Cd resistance (Clemens et al., 1999; Vatamaniuk et al., 1999). Through their thiolate complexes, PCs have been found to have high binding-affinity for metals such as Pb, Cd, Hg, and As. PC deficiency and hypersensitivity to Cd were observed in GSH-deficient mutants of *S. pombe* and *Arabidopsis* during genetic studies (Cobbett and Goldsbrough, 2002). Compared to MTs, PCs have been found to offer many advantages due to their unique structural characteristics (repeated γ-Glu-Cys) allowing them to exhibit higher metal-binding capacity than MTs on a per cysteine basis (Bae et al., 2000).

The metal-binding capacity, tolerance, or accumulation of bacteria has been increased through the introduction and/or overexpression of metal-binding proteins. Studying the effect of cadmium and lead ions on *Escherichia coli* expressing human metallothionein gene (MT-3), Adam et al. (2014) found that the cloned species tolerated higher concentration of cadmium (IC50 of 95.5 µM) but not lead (IC50 of 207 µM) compared to the wild type (IC50 of 352.5 µM). In a separate study, Ma et al. (2011) investigated Cd and As bioaccumulation capacity of the recombinant *Escherichia coli* expressing the human MT (hMT-1A) gene as well as glutathione S-transferase gene fused with the hMT-1A gene, and found that the later exhibited the highest bioaccumulation ability (6.36 mg Cd/g fry cells and 7.59 mg As/g dry cells). Many studies have covered the expression of phytochelatins in engineered microorganisms. Aiming to develop microbial sorbents for Cd removal, Kang et al. (2007) overexpressed PC synthase from *Schizosaccharomyces pombe* in *Escherichia coli*, and achieved a Cd accumulation level in the recombinant strain that was 25-fold higher than the wild strain. *Arabidopsis thaliana phytochelatin* synthase expressed in the yeast *Saccharomyces cerevisiae* for enhanced arsenic accumulation and removal resulted in significant increase (six times) of As accumulation by the engineered strain compared to the wild strain (Singh et al., 2008). Synthetic genes encoding for several metal-chelating phytochelatin analog were expressed in *Escherichia coli* by Bae et al. (2000) who found that the engineered cells transformed with the longer Glu-Cys unit (EC20) accumulated more cadmium (60 nmoles/mg dry cells) than the cells displaying EC8 (18 nmoles/mg dry cells).

5. Concept of metal detection by sensors

The recurrence of incidents related to heavy metal pollution of environmental water has prompted researchers to develop the most effective approach to address the pressing issue. Limitations associated with the conventional analyses based on standard spectroscopic techniques such as inductively coupled plasma optical mass spectrometry, atomic absorption spectrometry, and ion chromatography (D'Ilio et al., 2008; Parham et al., 2009) can be listed as: (1) The samples collected need to be treated to solubilize the metal ions; (2) There is a possibility to compromise the integrity of the sample during transportation and storage; (3) These techniques require expensive instruments and specialized personnel to carry out the operational procedures; (4) Significant labor is required during analysis carried out in centralized laboratories making the process time consuming; (5) The conventional analytical methods provide results that represent the amount of total metals in the sample which is not always comparable to the actual toxicity of the sample (Kim et al., 2008; Priyadarshini and Pradhan, 2017). These limitations have been overcome by the new approach using the sensor technique which allows a real time monitoring of the water quality in the field, is cost effective, is simple, and gives an estimation of the exact toxicity of the sample as it quantifies the bioavailable metal species. The major sensoring systems can be classified as electrochemical and optical which have been used for several decades in the monitoring of the environmental water quality.

5.1 Electrochemical sensors

Electrochemical systems provide leading edge compared to optical systems in terms of simplicity and cost, and generally have high sensitivity and selectivity; their ability to detect chemicals is based on the measurement of electrical quantities such as charge or potential and their relationship to chemical parameters (Brett, 2001; Hanrahan et al., 2004). Electrochemical techniques are divided into three, however, for environmental monitoring of heavy metals, most electrochemical devices reported have been mainly either in the category of amperometry and voltammetry or potentiometry (Hanharan et al., 2004; Aragay and Merkoci, 2012). In the voltammetry techniques the electrode surface is used for quantification of an analyte, the movement of electrons crossing the electrode-solution interface after oxidation/reduction of an electroactive species generates an amount of current which is measured and is proportional to the target analyte's concentration. Because of their high sensitivity and selectivity, the anodic stripping voltammetry techniques have been the most common voltammetry techniques used for the detection of trace heavy metals (Wang, 2007; Merkoci and Alegret, 2007). In this type of analytical method, reduced metals that have progressively accumulated at the electrode are reoxidized; the accumulation of metal at the electrode over the time ensures the remarkable sensitivity of the technique and detection limits that are well below the levels permitted by environmental legislation (Paneli and Voulgaropoulos, 1993; Brainina and Neyman, 1994). Effective sensors based on stripping voltammetry analysis have been developed by improving the quality of electrodes and shape or size of the device. These changes include, among other, the replacement of the traditional laboratory-based mercury electrodes due to the toxicity of mercury and difficulties associated with its handling, storage and disposal. According to Wang et al. (2000) the common mercury-film electrodes can be surpassed by Bismuth-coated carbon electrodes which display an attractive stripping voltammetric performance. A novel miniaturized electrochemical carbon modified sensor for on-stripping analysis of trace heavy metals was developed by Palchetti and coworker (2005). In this study, the traditional electrode was replaced by a graphite working electrode that was modified with a cellulose-derivative mercury coating. The developed sensor was used to test samples from polluted soils and sediments in the mining site of Aznalcollar (Spain); a linear correlation of results obtained with the newly developed sensor and those obtained with ICP-MS revealed that the new sensor was sufficiently robust to yield useful results for the determination of Cu, Cd and Pb. In a separate study, Pan and coworkers (2009) developed a nanomaterial/ionophore-based electrode for anodic stripping voltammetric determination of lead. To assemble such electrode, the ionophore was used because of its excellent selectivity toward lead while nanosized hydroxyapatite (NHAP) was used to improve the sensitivity for the detection of lead. Implementation of the proposed method allowed determining trace levels of lead in real water samples exhibiting higher sensitivity and relatively lower detection limit. A graphene-based nanocomposite was used as enhanced sensing platform in the development of a sensor for ultrasensitive stripping voltammetric detection of Hg (II) (Gong et al., 2010); to construct the platform, the two-dimensional (2D) graphene nanosheet matrix was homogenously covered with Au nanoparticles (AuNPs).

5.2 Optical sensors

According to Jeronimo et al. (2007), optical sensors represent a group of chemical sensors in which electromagnetic radiation is used to generate the analytical signal in a transduction element; they can be categorized on the basis of optical principles allowing to identify colorimetric, fluorescent, surface plasmon resonance (SPR), and surface enhanced Raman scattering (SERS) sensors (Li et al., 2013). Optical sensors have been effectively used in the field of biology and chemistry including the detection of heavy metals from the environment. For these sensors, the effective attachment of the targeted reagent and the size of the sensing area are critical for the intensity of the signal and therefore the response time. With this type of sensors, to ensure the effective transmission of electromagnetic radiation to and from a sensing area that is in direct contact with the sample, optical fibres are commonly employed as preferred platform (Wolfbeis, 1991; Tan et al., 1992). To increase the surface size of the sensing area, researchers have now been focusing on using nanomaterials and nanoarchitecture in sensors. A host of design methods have been considered in order to ensure the applicability of sensing devices to real-world samples. For optical sensors, it is important that the probes used have reduced or have no spectral interferences related to other molecules with similar properties to the molecular probe; investigation by Casay and coworkers (1996) to minimise such interferences consisted synthesizing a tetra-substituted aluminum 2, 3-naphthalocyanine dyes by means of homogeneous phase reaction to improve product yields, higher purity, and effective use for the determination of metal ions. The synthesized dye was entrapped in a permeable polymer and, the polymer attached to the probe support, which was successfully used to detect metal ions via steady-state fluorescence using both fibre optic system and a commercially available instrument. To detect heavy metal ions Vaughan and Narayanaswamy (1998) developed optical fibre based reflectance sensors using the immobilized reagent Br-PADAP. They employed physical adsorption and membrane entrapment as immobilization methods. It was found that the Br-PADAP immobilized onto XAD-4 could detect relatively low level (31 ppb) of Zn and exhibited a reproducible and reversible response to heavy metal ions, while irreversible but quick and reproducible responses were achieved by Br-PADAP/PVC membranes. Yusof and Ahmad (2003) developed a method for the detection of lead by immobilizing the spectrometric reagent 7-dimethylamino-4-hydroxy-3-oxo-phenoxazine-1-carboxylic acid (gallocynine) on copolymer XAD 7 as the sensing phase in an optical fibre chemical sensor for lead. Although the response was reproducible and reversible, metals such Hg(II) and Ag(I) were found to significantly interfere during the determination of lead. A paired emitter-detector light emitting diodes (LED) was developed by Lau et al. (2006) for the measurement of lead and cadmium; the malachite green-potassium iodide method which is sensitive to lead and selective to cadmium was chosen as the chemistry for the detection of lead and cadmium. The developed LED-based device was found to have a performance comparable to the most expensive bench top UV-vis instruments.

6. Metalloproteins-based biosensors for metal detection

According to the definition from the International Union of Pure and Applied Chemistry (divisions of Analytical Chemistry and Physical Chemistry), an electrochemical biosensor is a self-contained integrated device, which is capable of providing specific quantitative and semi-quantitative analytical information using a biological recognition element which is retained in direct spatial contact with an electrochemical transduction element (Thevenot et al., 2001). However, some researchers (Malhotra, 2005; Turner, 2013; Kurbanoglu et al., 2016) define biosensors as compact devices and analytical tools incorporating biological material in an intimate contact with a suitable transducer device that converts the biochemical signal into quantitative electric signal. From these definitions, it is clear that the important parts of the biosensor include the receptor and the transducer; the receptor is basically the part of the biosensor responsible for the recognition of the analyte, while the transducer ensures the transfer of the signal from the output of the receptor system to the electrical section of the biosensor. Based on the concept of

Cammann (1977) and Turner (1987), a biosensor should have a biological receptor system that utilizes a biochemical mechanism in the recognition of analytes. Differentiation of biosensors can depend on the type of receptor or transducer element fitted in the device. Bio-receptors that have been used in the assemblage of biosensors include:—Nucleic acids, mostly DNA (DNA microchip has emerged as useful technology for sensing devices) (Sharma et al., 2003);—Enzymes, which are currently the most solicited recognition system (Fennouh et al., 1998; Thompson et al., 1998);—proteins such as metal binding proteins and metalloproteins are also often used in biosensors (Bontidean et al., 2003); whole cells (microorganisms such as bacteria, fungi, and other eukaryotic cells) (Corbisier et al., 1999; Wittman et al., 1997); the sensing potential of microorganisms is a spontaneous response resulting from the influence by external factors which induce physiological mechanisms including recognition by protein binding molecules. It is therefore important in the scope of this study to not only consider metalloproteins, but also whole-cell biosensors as well as other metal-binding protein molecules (Table 2). The difference between metalloproteins and metal-binding proteins is reflected at the structural level as the first contains appropriate type and number of ligands which promote higher-affinity interactions when they bind to metals compared to the metal-binding proteins (Herald et al., 2003; Garcia et al., 2005). An important step in the assemblage of biosensors is to ensure proper incorporation of the biological molecule in an intimate contact with a suitable transducer device through the use of effective immobilization technique. A successful immobilization of the receptor contributes to an increase in its stability, enhanced recognition ability, and ensures the reusability of the biosensor. Several techniques are available for the immobilization of biomolecules at the surface of the transducer during the construction of a specific biosensor. These immobilization techniques are mainly classified as chemical and physical immobilizations (Turner et al., 1987; Lei et al., 2006; Datta et al., 2013; Kurbanoglu et al., 2017). Chemical immobilization techniques include covalent bonding which involves functional groups at the surfaces of the receptors and the support material and cross-linking by means of bifunctional reagents. The physical immobilization techniques mostly used include entrapment and adsorption. Physical entrapment in polymeric gel or within microcapsules takes place when biomolecules or whole cells are mixed with a solution of polymers such as k-carrageenan, colagen, polyacrylamide, starch, polyurethane, chitosan, and Ca-alginate or when semipermeable membrane are used for entrapment of the receptor. However, entrapment methods may suffer disadvantages such as leakages leading to the loss of receptors, diffusion limitation to the transport of analytes leading to reaction retardation, lower sensitivity, and detection limit; unfavorable microenvironmental conditions can also affect the performance of the receptor. Physical adsorption at solid surface involving weak forces such as Van der Waal's forces, hydrogen bonds, ionic bonds, hydrophobic

Table 2. Characteristics of biosensors using protein-like sensing part and their performance.

Bio-receptor type	Bioreceptor	Analyte	Signal gene/proteins Sensing	Signal gene/proteins Reporter	Transducer type	Immobilization support	Detection limit	Sensing system	References
Whole cell	*A. eutrophus* AE1239	Cu	GST-SmtA	*lux*	Optic fibres	Alginate and agarose	20 uM	Optical	Corbisier et al., 1999
	S. cerevisiae	Cu	*CUP1*	*lacZ*	Amp.	Polyvinyl alcohol	0.5 mM	Electrochem	Lehmann et al., 2000
	E. coli RBE23-17	Cd	*zntA*	*lacZ*	Amp.	Polystyrene	25 nM	Electrochem	Biran et al., 2000
	A. eutrophus AE104	Cr(VI)	*chr*	*lux*	NA	Polystyrene	10 uM	Optical	Peitzsch et al., 1998
	E. coli JM109	Sb and As	*ars*	*lacZ*	Volt.	Gold foil	0.1 uM	Electrochem	Scott et al., 1997
	E. coli HMS174	Hg	*mer*	*lux*	NA	NA	0.5 nM	Optical	Selifonova et al., 1993
	Staphylococcus RN4220, *B. subtilis* BR151	Cd and Pb	*cadA*	*lucFF*	NA	NA	10 nM (Cd) and 33 nM (Pb)	Optical	Tauriainen et al., 1998
	E. coli MC1061	Hg	*mer*	*lucFF*	NA	NA	0.1 fM	Optical	Virta et al., 1995
	E. coli JM109	Hg	*mer*	*lucFF*	Optic fibres	NA	0.01 uM	Optical	Tescione and Belfort, 1993
	Staphylococcus RN4220	As, Sb, Cd	*ars*	*lucFF*	NA	NA	100 nM (As), 33 nM (Sb), 330 nM (Cd)	Optical	Tauriainen et al., 1997

Table 2 contd. ...

...Table 2 contd.

| Bio-receptor type | Bioreceptor | Analyte | Signal gene/proteins | | Transducer type | Immobilization support | Detection limit | Sensoring system | References |
			Sensing	Reporter					
Whole cell	*B. subtilis* BR151	As, Sb, Cd	*ars*	*lucFF*	NA	NA	3.3 nM (As), 330 nM (Sb), 330 nM (Cd)	Optical	Tauriainen et al., 1997
	E. coli MC1061	As, Sb, Cd	*ars*	*lucFF*	NA	NA	3.3 uM (As), 3.3 uM (Sb), 33 uM (Cd)	Optical	Tauriainen et al., 1997
	E. coli MC1061	Hg	*mer*	*lux*	NA	NA	0.167 pM	Optical	Roda et al., 2001
	E. coli pCHRGFP1	Cr	*chr*	*gfp*	NA	NA	0.5 uM	Optical	Coehlo et al., 2015
	O. tritici pCHRGFP2	Cr	*chr*	*gfp*	NA	NA	2 uM	Optical	Coehlo et al., 2015
	E. coli DH5a	As	*ars*	*lacZ*	NA	NA	10 ug/L	Optical	Huang et al., 2015
Metal-binding peptide	CP1 (dansyl-Cys-Pro-Gly-His-Glu)	Zn	NA	*dansyl*	NA	NA	8 uM	Optical	Joshi et al., 2007
	Fluorescent peptidyl chemisensor	Cu	NA	*dansyl*	NA	NA	32 ug/L	Optical	White and Holcombe, 2007

	Protein	Metal	Gene		Method	Electrode	Detection limit	Type	Reference
	Fluorescent peptide sensor	Hg, Cd, Hg, Cu, Zn and Ag		dansyl	NA	NA	2.8 ug/L (Zn), 4.3 (Ag) ug/L, 5 ug/L, 15 ug/L (Pd), 5.2 ug/L (Hg)	Optical	Joshi et al., 2009
	Fluorescent peptide probedansyl-Cys-X-Gly-His-X-Gly-Glu-NH2	Zn and Cu	NA	dansyl	NA	NA	0.25 uM	Optical	Joshi and Lee, 2008
Metallo Protein	Tn501MerR	Cu, Zn, Hg and Cd	mer	NA	Volt.	Gold rods	1 fM	Electrochem	Bontidean et al., 2000
	MerR	Cu, Cd and Hg	mer	NA	Volt.	Gold	1 fM	Electrochem	Corbisier et al., 1999
	GST-SmtA	Cu	smt	NA	Volt.	Gold	1 fM	Electrochem	Corbisier et al., 1999
	GST-SmtA	Cu, Zn, Hg and Cd	smt	NA	Volt.	Gold rods	1 fM	Electrochem	Bontidean et al., 2000
	Phytochelatin	Cd, Zn and Pt	NA	NA	Volt.	Graphite	1 pM (Cd), 13.3 pM (Zn), 1.9 pM (Pt)	Electrochem	Adam et al., 2005
	Phytochelatin (EC20)	Cu, Zn, Hg, Pb and Cd	pMC20	NA	Volt.	Gold	100 fM	Electrochem	Bontidean et al., 2003
	Metallothionein	Cd, Zn and Ni	NA	NA	Surface plasmon resonance (refract)	Carboxymethylated chips	< 200 ppb	Optical biosensor	Wu and Lin, 2004

interaction, and multiple salt linkages is a simple technique which minimizes diffusion limitation but also faces the challenge of receptor leakage.

6.1 Whole cell biosensors

In response to environmental changes, microorganisms have developed physiological mechanisms allowing them to regulate the transport of foreign substances within the cell. These mechanisms are governed by genetic disposition as well as cytosolic and membrane protein receptors capable of sensing the presence of toxins or useful substances, therefore prompting adequate responses such as oxygen consumption, surface chemical potential, or genetic activity by the microorganism. This natural advantage is being exploited to develop biosensors using adequate microorganisms as receptors; however, the required metabolic pathways and uptake systems can be induced in unusual microbial hosts through genetic modification to obtain effective receptors. A number of advantages could arise from the use of whole microbial cells as genetic-based biosensors namely their wide repertoire of genetic responses to their environment given the multitude of microbial groups (e.g., prokaryotes and eukaryotes), their physical robustness as they have adapted to a wider range of environmental conditions, and the fact that they are easy to handle. Furthermore, such biosensors have the potential to provide more sophisticated sensing systems containing multiple integrated molecular components at competitive costs. Microbial biosensors have been extensively applied to detect environmental toxins, with the living cells providing the opportunity to directly reveal toxic conditions. In this chapter, the focus will mainly be on the detection of metal pollutants by microbial biosensors using either electrochemical or optical sensing technique. Most microbial biosensors constructed have been bioluminescence-based biosensors because they provide the advantage of a real-time process monitoring during wastewater biotreatment, avoiding the cost and time involved in reactivating or reinitiating the wastewater biotreatment plant after shut down as early detection of upsets resulting from sudden inflow of toxic heavy metals can be achieved (Dolatabadi and Manjulakumari, 2012).

Bioluminescent-based biosensors are constructed by incorporating genetically engineered microorganisms as receptors which are modified using a specific gene sequence (fused with luminescent reporter gene) that is induced by the presence of target analyte (Roda et al., 2001). Previously used bioluminescent reporters include marine bacteria, *Renilla reniformis* luciferases, lux, luc, and ruc genes encoding firefly (Garrison et al., 1996; Lorenz et al., 1996).

Microbial biosensors have been developed for the detection of a number of metals including cadmium, mercury, copper, cobalt, and chromium from wastewater; however, particular attention has been paid to mercury which is one of the most toxic metals because of its tendency to react and strongly bind to organic molecules. Through genetic modification of *Vibrio fischeri* bacteria, consisting of the fuse of the lux genes to the mercury resistance operon of Tn21 mercury ion, biosensors have been constructed (Geiselhart et al., 1991; Selifinova et al., 1993). Using *E. coli* as a host organism and firefly luciferase luc gene as a reporter under the control of the mercury-inducible promoter from the mer operon from transposon Tn21, Roda et al. (2001) constructed an easy, sensitive, and rapid luminescent microbial biosensor for the determination of urinary mercury (II). A conductometric biosensor was constructed by Chouteau et al. (2005), by immobilizing *C. vulgaris* on a bovine serum albumin membrane deposited on a platinum interdigitated electrode as bioreceptor; they achieved a detection limit of 10 ppb for both Cd^{2+} and Zn^{2+} after 30 minutes of exposure to the biosensor. In a separate study, Lehmann et al. (2001) developed an amperometric microbial biosensor for the detection of Cu^{2+}; the amperometric detection of Cu^{2+} was achieved using recombinant *Saccharomyces cerevisiae* strains as biocomponent in the microbial biosensor. The authors engineered the wild strains by using plasmids with the Cu^{2+}-inducible promoter of the CUP1-gene from *S. cerevisiae* fused to the *lacZ*-gene from *E. coli*. The developed biosensor measured Cu^{2+} in a concentration range between 0.5 and 2 mM $CuSO_4$. A voltammetric microbial biosensor based on *Circinella* sp. modified carbon paste electrode was developed by Alpat et al. (2008) for the detection of preconcentrated Cu^{2+} and achieved a detection limit of 54 nM Cu^{2+}. Fusion of lux genes to a number of metal ion-responsive promoters from *Alcaligenes eutrophus* had allowed the development of biosensors for the detection of nickel, zinc, copper, chromate, and thallium (Collard et al., 1994).

6.2 Metal-binding proteins' biosensors

Microbial-based biosensors often suffer issues related to selectivity and low sensitivity which are very important parameters to consider when designing a sensor device; for these reasons selective macromolecules such as DNA, enzyme, binding proteins, and/or metalloproteins have been mostly considered as bioreceptors. Metal binding proteins which could be easily synthesized have been demonstrated as viable alternative. According to Hellinga and Marvin (1998), the design of the modular protein-engineering systems for incorporation in biosensors follows two strategies namely:

1. After identification of a protein with the appropriate specificity, a signal-transduction function must be inserted.

2. A protein with a well-behaved intrinsic signal-transduction function must be identified for the construction of appropriate binding sites.

Earlier investigation by Gershon and Khilko (1995) allowed immobilizing four recombinant proteins (tagged at either terminus with oligohistidine) in the flow cell of the "BIAcore" surface plasmon resonance (SPR) biosensor.

A fluorescent peptide probe was designed and synthesized by Joshi et al. (2007) for the detection of Zn^{2+} in aqueous solution; the synthesized peptide had a unique amino acid sequence and exhibited good selectivity for Zn^{2+} in multi-metals aqueous solution. Using Fmoc solid-phase peptide approach, White and Holcombe (2007) synthesized a fluorescent peptidyl chemosensor for Cu^{2+} ions with fluorescence resonance energy transfer (FRET) capability. The synthesized metal chelating unit consisted the amino acids glycine and aspartic acid (Gly-Gly-Asp-Gly-Gly-Asp-Gly-Gly-Asp-Gly-Gly-Asp-Gly-Gly) flanked by the fluorophores tryptophan (donor) and dansyl chloride (acceptor). The Cu^{2+}-induced quenching of the acceptor dye could be used to monitor the concentrations of Cu^{2+} ions with a detection limit of 32 µg/L. In a transition metal matrix at pH 7.0, the newly developed fluorescent peptidyl chemosensor was found to be sensitive and selective towards Cu^{2+}.

6.3 Metalloproteins-based biosensors

Metalloproteins have a particular structure with sequential arrangement of amino acids promoting higher selectivity to specific metals compared to other metal-binding proteins. This attribute makes them very attractive as bioreceptors for biosensors (Fig. 1). Bontidean et al. (1998) developed sensors based on proteins (GST-SmtA and MerR) with distinct binding sites for heavy metal ions by overexpressing the above proteins in *E. coli* and immobilizing the pure proteins at surface of gold electrode. Following exposure to zinc, cadmium, copper, and mercury ions, the selectivity and sensitivity of the two protein-based biosensors were measured; the MerR-based electrodes showed an accentuated selectivity for mercury only, while the GST-SmtA electrodes were able to sense all the four heavy metals. Corbisier et al. (1999)

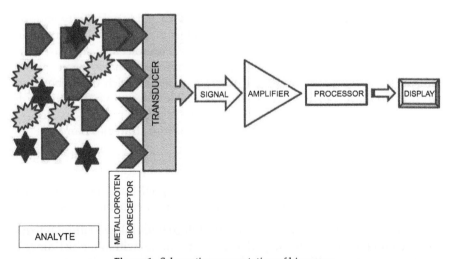

Figure 1. Schematic representation of biosensor.

overexpressed the fusion protein GST-SmtA, containing glutathione-S-transferase linked to the *Synechococcal metallothionein* protein in *E. coli* from an expression vector pGEX3X containing SmtA, the fusion protein was then immobilized on gold surface to prepare the biosensor.

The detection capacity of the developed GST-SmtA biosensor was tested for cadmium, copper, zinc, and mercury and it was found that the biosensor had broad selectivity towards the heavy metals and could be regenerated upon treatment with 1 mM EDTA. In a separate study, the same prokaryotic metallothionein SmtA was used by Bontidean et al. (2000) to construct the GST-SmtA electrode and the relative response of the developed biosensor to Cu(II), Zn(II), Cd(II), and Hg(II) was tested across a 10^5-fold range from 10^{-15} M. Furthermore, the sensitivity of the GST-SmtA biosensor at various pH and in different buffers (HEPES and borate buffers) was investigated; the investigators found that both Cu(II) and Hg(II) produced better response at pH 8.75 than at pH 8.0 in borate buffer, while Hg(II) gave a better response in borate buffer than in HEPES at pH 8.0. The developed GST-SmtA biosensor was also found to be relatively robust as its capacitance in the presence of HEPES was relatively constant across three cycles of exposure to 10^{-4} M Hg(II).

Due to their unique structural characteristics and particularly the continuously repeating γGlu-Cys units, phytochelatins have higher metal-binding capacity than metallothionein. A capacitance biosensor based on synthetic phytochelatin $(Glu-Cys)_{20}Gly$ (EC20) was developed by Bontidean et al. (2003) for quantitative determination of heavy metals; EC20 expressed in *E. coli* JM105 carrying pMC20 plasmid which allows the cytoplasmic expression of EC20 as a fusion of the maltose binding protein (MBP) was used for the construction of the biosensor. In solutions containing 100 fM – 10 mM concentrations of Cu^{2+}, Pb^{2+}, Hg^{2+}, Zn^{2+}, and Cd^{2+} ions the newly developed biosensor achieved detection sensitivity in the order $S_{Zn} > S_{Cu} > S_{Hg} > S_{Cd} = S_{Pb}$. The biocomponent of the sensor was regenerated in the EDTA and the biosensor could be stored for 15 days. Phytochelatin was immobilized on mercury drop electrode to construct a heavy metal voltammetric biosensor (Adam et al., 2005) and the sensitivity for Cd^{2+} and Zn^{2+} in human urine and Pt in pharmaceutical drug was tested. The newly constructed biosensor achieved a detection limit of 1.0, 13.3 and 1.9 pmole in 5 µL for Cd^{2+}, Zn^{2+}, and Pt respectively.

7. Conclusion

Industrialization and the risk of dispersion of metal pollutants in water sources have increased tremendously over the years. It is therefore imperative for water treatment plant to complement the conventional spectroscopic methods with real time detection methods such as sensing for the real-time estimation of the exact toxicity of samples and effective monitoring of water.

The sensoring systems (whether electronic or optical) discovered few years ago were predominantly using chemical receptors in the early days; however, in the recent years there has been a considerable shift toward the use of biological component because of their fast response, specificity, and low cost. Most of the developed heavy metal biosensors have been using whole microbial cells with natural potential to synthesize macromolecules required for resistance mechanisms against toxic heavy metals. Further investigations have allowed improving immobilization techniques to enhance the activity of the microbial receptors by providing suitable microenvironment and ideal interaction with the transducer for effective transfer of the signal. Optimization strategies also consisted, among others, to modify unusual microbial host through molecular engineering therefore creating effective bioreceptor with desired metabolic pathways and uptake systems; genetic transformations were also applied for the insertion of luminescent reporter gene in microbial receptor such as to provide the advantage of real-time process monitoring during wastewater treatment. The limitations of microbial-based biosensor related to low sensitivity, slow response, and poor selectivity have motivated development of protein-based biosensors using metalloproteins such as metallothionein and phytochelatins as well as other metal-binding peptides for the detection of toxic heavy metals in solution. Having high metal binding capacities, metalloproteins have contributed to the development of extremely sensitive biosensors capable of the detection of heavy metals present at concentrations as low as femtomolar. Through the use of DNA recombinant technology, the capacity of the metalloprotein-based biosensors were further improved by constructing fusion protein bioreceptors providing better sensitivity, selectivity, and robustness (opportunity of biosensor reuse). The development of metalloprotein-based biosensors have contributed to significant advancement in the field of sensing system and has increased the feasibility of implementing biosensors for the monitoring of toxic heavy metals' occurrence in water sources.

References

Adam V., Chudobova D., Tmejova K., Cihalova K., Krizkova S., Guran R., Kominkova M., Zurek M., Kremplova M., Jimenez A.M.J., Konecna M., Hynek D., Pekarik V., Kizek R. An effect of cadmium and lead ions on *Escherichia coli* with the cloned gene for metallothionein (MT-3) revealed by electrochemistry. Electrochimica Acta 140, 11–19, 2014.

Adam V., Zehnalek J., Petrlova J., Potesil D., Sures B., Trnkova L., Jelen F., Vitecek J., Kizek R. Phytochelatin modified electrode surface as a sensitive heavy-metal ion biosensor. Sensors 5, 70–84, 2005.

Ahalya N., Ramachandra T.V., Kanamadi R.D. Biosorption of heavy metals. Res. J. Chem. Environ. 7, 71–79, 2003.

Ahluwalia S.S., Goyal D. Microbial and plant derived biomass for removal of heavy metals from wastewater. Bioresource Technol. 98, 2243–2257, 2007.

Alonso A., Martinez J.L. Cloning and characterization of SmeDEF, a novel multidrug efflux pump from *Stenotrophomonas maltophilia*. Antimicrob. Agents Chemother. 44, 3079–3086, 2000.

Alpat S., Alpat S.K., Cadirci B.H., Yasa I., Telefoncu A. Sens. Actuators B. 134, 175–181, 2008.

Amaro F., Pilar de Lucas M., Martin-González A., Gutierrez J.C. Two new members of the Tetrahymena multi-stress inducible metallothionein family: Characterization and expression analysis of *T. rostrata* Cd/Cu metallothionein genes. Gene 423, 85–91, 2008.

Aragay G., Merkoçi A. Nanomaterials application in electrochemical detection of heavy metals. Electrochimica Acta 84, 49–61, 2012.

Aragay G., Pons J., Merkoci A. Recent trends in macro-, micro-, and nanomaterial based tools and strategies for heavy-metal detection. Chem. Rev. 111, 3433–3458, 2011.

Bae W., Chen W., Mulchandani A., Mehra R.K. Enhanced bioaccumulation of heavy metals by bacterial cells displaying synthetic phytochelatins. Biotechnol. Bioeng. 70, 518–524, 2000.

Balabanova B., Stafilov T., Sajn R., Baceva K. Distribution of chemical elements in attic dust as reflection of their geogenic and anthropogenic sources in the vicinity of the copper mine and flotation plant. Arch. Environ. Contam. Toxicol. 61, 173–184, 2011.

Bennett B., Dudas M.J. Release of arsenic and molybdenum by reductive dissolution of iron oxides in a soil with enriched levels of native arsenic. J. Environ. Eng. Sci. 2, 265–272, 2003.

Biran I., Babai R., Levcov K., Rishpon J., Ron E.Z. Online and *in situ* monitoring of environmental pollutants: electrochemical biosensing of cadmium. Environ. Microbiol. 2, 285–290, 2000.

Blindauer C.A., Harrison M.D., Robinson A.K., Parkinson J.A., Bowness P., Sadler P., Robinson N.J. Multiple bacteria encode metallothioneins and SmtA-like zinc fingers. Mol. Microbiol. 45, 1421–1432, 2002.

Boeris P.S., Agustin M.D.R., Acevedo D.F., Lucchesi G.I. Biosorption of aluminum through the use of non-viable biomass of *Pseudomonas putida*. J. Biotechnol. 236, 57–63, 2016.

Bontidean I., Ahlqvist J., Mulchandani A., Chen W., Bae W., Mehra R.K., Mortari A., Csoregi E. Novel synthetic phytochelatin-based capacitive biosensor for heavy metal ion detection. Biosens. Bioelect. 18, 547–553, 2003.

Bontidean I., Berggren C., Johansson G., Csoregi E., Mattiasson B., Lloyd J.R., Jakeman K.J., Brown N.L. Detection of heavy metal ions at femtomolar levels using protein-based biosensors. Anal. Chem. 70, 4162–4169, 1998.

Bontidean I., Lloyd J.R., Hobman J.L., Wilson J.R., Csoregi E., Mattiasson B., Brown N.L. Bacterial metal-resistance proteins and their use in biosensors for the detection of bioavailable heavy metals. J. Inorg. Biochem. 79, 225–229, 2000.

Brainina Kh.Z., Neyman E. Electroanalysis Stripping Methods. Wiley, New York, 1994.

Brett C.M.A. Electrochemical sensors for environmental monitoring. Strategy and examples. Pure Appl. Chem. 73, 1969–1977, 2001.

Bryson J.W., O'Halloran T.V., Rouch D.A., Brown N.L., Camakaris J., Lee B.T.O. Chemical and genetic studies of copper resistance in *E. Coli. In*: Karlin K.D., Tyklar Z. (eds.). Bioinorganic Chemistry of Copper. Chapman & Hall, New York, pp. 101–109, 1993.

Busenlehner L.S., Pennella M.A., Giedroc D.P. The SmtB/ArsR family of metalloregulatory transcriptional repressors: Structural insights into prokaryotic metal resistance. FEMS Microbiol. Rev. 27, 131–143, 2003.

Cammann K. Biosensors based on ion-selective electrodes. Fresenius Z. Anal. Chem. 287, 1, 1977.

Casay G.A., Narayanan N., Evans III L., Czuppon T., Patonay G. Near-infrared tetra-substituted aluminum 2, 3-naphthalocyanine dyes for optical fiber applications. Talanta 43, 1997–2005, 1996.

Cavet J.S., Borrelly G.P., Robinson N.J. Zn, Cu and Co in cyanobacteria: selective control of metal availability. FEMS Microbiol. Rev. 27, 165–181, 2003.

Choi A., Wang S., Lee M. Biosorption of cadmium, copper, and lead ions from aqueous solutions by *Ralstonis* sp. and *Bacillus* sp. isolated from diesel and heavy metal contaminated soil. Geosci. J. 13, 331–341, 2009.

Chouteau C., Dzyadevych S., Durrieu C., Chovelon J.M. A bi-enzymatic whole cell conductometric biosensor for heavy metal ions and pesticides detection in water samples. Biosens. Bioelectron. 21, 273–281, 2005.

Claverys J.P. A new family of high-affinity ABC manganese and zinc permeases. Res. Microbiol. 152, 231–243, 2001.

Clemens S., Kim E.J., Neumann D., Schroeder J.I. Tolerance to toxic metals by a gene family of phytochelatin synthases from plants and yeast. EMBO J. 18, 3325–3333, 1999.

Cobbett C. Phytochelatins and their roles in heavy metal detoxification. Am. Soc. Plant Biol. 123, 825–832, 2000.

Cobine P., Wickramasinghe W.A., Harrison M.D., Weber T., Solioz M., Dameron C.T. The *Enterococcus* hirae copper chaperone CopZ delivers copper(I) to the CopY repressor. FEBS Lett. 445, 27–30, 1999.

Coehlo C., Branco R., Natal-da-Luz T., Sousa J.P., Morais P.V. Evaluation of bacterial biosensors to determine chromate bioavailability and to assess ecotoxicity of soils. Chemosphere 128, 62–69, 2015.

Collard J.M., Corbisier P., Diels L., Dong Q., Jeanthon C., Mergeay M., Taghavi S., Vander L.D., Wilmotte A., Wuertz S. Plasmids for heavy metal resistance in *Alcaligenes eutrophus* CH34: mechanisms and applications. FEMS Microbiol. Rev. 14, 405–414, 1994.

Congeevaram S., Dhanarani S., Park J., Dexilin M., Thamaraiselvi K. Biosorption of chromium and nickel by heavy metal resistant fungal and bacterial isolates. J. Hazard. Mater. 146, 270–277, 2007.

Cooksey D.A. Copper uptake and resistance in bacteria. Mol. Microbiol. 7, 1–5, 1993.

Corbisier P., van der Lelie D., Borremans B., Provoost A., de Lorenzo V., Brown N.L., Lloyd J.R., Hobman J.L., Csöregi E., Johansson G., Mattiasson B. Whole cell- and protein-based biosensors for the detection of bioavailable heavy metals in environmental samples. Anal. Chim. Acta 387, 235–244, 1999.

D'Ilio S., Petrucci F., D'Amato M., Di Gregorio M., Senofonte O., Violante N. Method validation for determination of arsenic cadmium, chromium and lead in milk by means of dynamic reaction cell inductively coupled plasma mass spectrometry. Anal. Chim. Acta. 624, 59–67, 2008.

Datta S., Christena L.R., Rajaram Y.R.S. Enzyme immobilization: an overview on techniques and support materials. Biotechnology 3, 1–9, 2013.

Demirbas E., Kobya M., Senturk E., Ozkan T. Adsorption kinetics for the removal of chromium (VI) from aqueous solutions on the activated carbons prepared from agricultural wastes. Water SA 30, 533–540, 2004.

Diaz S., Amaro P., Rico D., Campos V., Benitez L., Martin-Gonzalez A., Hamilton E., Orias E., Gutierrez J.C. Tetrahymena metallothioneins fall into two discrete subfamilies. PLoS One 2, e291, 2007.

Dolatabadi S., Manjulakumari D. Microbial biosensors and bioelectronics. Res. J. Biotechnol. 7, 102–108, 2012.

Fennouh S., Casimiri V., Geloso-Meyer A., Burstein C. Kinetic study of heavy metal salt effects on the activity of L-lactate dehydrogenase in solution or immobilized on an oxygen electrode. Biosens. Bioelectron. 13, 903–909, 1998.

Filgueiras A.V., Lavilla I., Bendicho C. Evaluation of distribution, mobility and binding behaviour of heavy metals in surficial sediments of Louro River (Galicia, Spain) using chemometric analysis a case study. Sci. Total Environ. 330, 115–129, 2004.

Fosso-Kankeu E., Mulaba-Bafubiandi A.F., Mamba B.B., Barnard T.G. Mitigation of Ca, Fe, and Mg loads in surface waters around mining areas using indigenous microorganism strains. J. Phys. Chem. Earth. 34, 825–829, 2009.

Fosso-Kankeu E., Mulaba-Bafubiandi A., Mamba B.B., Marjanovic L., Barnard T.G. A comprehensive study of physical and physiological parameters that affect biosorption of metal pollutants from aqueous solutions. J. Phys. Chem. Earth. 35, 672–678, 2010.

Fosso-Kankeu E., Mulaba-Bafubiandi A.F. Implication of plants and microbial metalloproteins in the bioremediation of polluted waters. J. Phys. Chem. Earth. 67–69, 242–252, 2014.

Fosso-Kankeu E., Mulaba-Bafubiandi A.F., Mamba B.B., Barnard T.G. Prediction of metal-adsorption behaviour in the remediation of water contamination using indigenous microorganisms. J. Environ. Manage. 92, 2786–2793, 2011.

Fosso-Kankeu E., Mulaba-Bafubiandi A.F., Piater L.A., Tlou M.G. Cloning of the cnr operon into a strain of *Bacillaceae* bacterium for the development of a suitable biosorbent. World J. Microbiol. Biotechnol. 32, 114–123, 2016.

Fu C., Miao W. Cloning and characterization of a new multi-stress inducible metallothionein gene in *Tetrahymena pyriformis*. Protist 157, 193–203, 2006.

Garcia J.S., de Magalhaes C.S., Arruda M.A.Z. Trends in metal-binding and metalloprotein analysis. Talanta 69, 1–15, 2005.

Garcia-Garcia J.D., Olin-Sandoval V., Saavedra E., Girard L., Hernandez G., Moreno-Sanchez R. Sulfate uptake in photosynthetic *Euglena gracilis*. Mechanisms of regulation and contribution to cysteine homeostasis. Biochim. Biophys. Acta 1820, 1567–1575, 2012.

Garcia-Garcia J.D., Sanchez-Thomas R., Moreno-Sanchez R. Bio-recovery of non-essential heavy metals by intra- and extracellular mechanisms in free-living microorganisms. Biotechnol. Adv. 34, 859–873, 2016.

Garrison P.M., Tullis K., Aarts J.M.M.J.G., Brouwer A., Giesy J.P., Denison M.S. Species-specific recombinant cell lines as bioassay systems for the detection of 2, 3, 7, 8-Tetrachlorodibenzo-p-dioxin-like chemicals. Fundam. Appl. Toxicol. 30, 194–203, 1996.

Gavrilescu M. Removal of heavy metals from the environment by biosorption. Eng. Life Sci. 4, 219–232, 2004.

Geiselhart L., Osgood M., Holmes D.S. Construction and evaluation of a self-luminescent biosensor. Ann. NY Acad. Sci. 646, 53–60, 1991.

Gershon P.D. and Khilko S. Stable chelating linkage for reversible immobilization of oligohistidine tagged proteins in the BIAcore surface plasmon resonance detector. J. Immun. Met. 183, 65–76, 1995.

Gong J., Zhou T., Song D., Zhang L. Monodispersed Au nanoparticles decorated graphene as an enhanced sensing platform for ultrasensitive stripping voltammetric detection of mercury (II). Sens. Act. B: Chem. 150, 491–497, 2010.

Gowd S.S., Reddy M.R., Govil P.K. Assessment of heavy metal contamination in soils at Jajmau (Kanpur) and Unnao industrial areas of the Ganda Plain, Uttar Pradesh, India. J. Hazard. Mater. 174, 113–121, 2010.

Grill E., Winnacker E.L., Zenk M.H. Phytochelatins: the principal heavy metal complexing peptides of higher plants. Science 230, 674–676, 1985.

Guo L., Fu C., Miao W. Cloning, characterization, and gene expression analysis of a novel cadmium metallothionein gene in *Tetrahymena pigmentosa*. Gene 423, 29–35, 2008.

Gupta A., Matsui K., Lo J.-F., Silver S. Molecular basis for resistance to silver cations in *Salmonella*. Nature America Inc. 5, 183–188, 1999.

Gupta A., Whitton B.A., Morby A.P., Huckle J.W., Robinson N.J. Amplification and rearrangement of a prokaryotic metallothionein locus *smt* in *Synechococcus* PCC 6301 selected for tolerance to cadmium. Proc. R. Soc. Lond. Ser. B. 248, 273–281, 1992.

Gupta D. Implication of environmental flows in river basin management. Phys. Chem. Earth. 33, 298–303, 2008.

Gutierrez J.C., Amaro F., Diaz S., de Francisco P., Cubas L.L., Martin-Gonzalez A. Ciliate metallothioneins: Unique microbial eukaryotic heavy-metal-binder molecules. J. Biol. Inorg. Chem. 16, 1025–1034, 2011.

Haney C.J., Grass G., Franke S., Rensing C. New developments in the understanding of the cation diffusion facilitator family. J. Ind. Microbiol. Biotechnol. 32, 215–226, 2005.

Hanrahan G., Patil D.G., Wang J. Electrochemical sensors for environmental monitoring: design, development and applications. J. Environ. Monitor. 6, 657–664, 2004.

Hassan M.T., van der Lelie D., Springael D., Romling U., Ahmed N., Mergeay M. Identification of a gene cluster, czr, involved in cadmium and zinc resistance in *Pseudomonas aeruginosa*. Gene 238, 417–425, 1999.

Hellinga H.W. and Marvin J.S. Protein engineering and the development of generic biosensors. TIBTECH. 16, 183–189, 1998.

Herald V.L., Heazlewood J.L., Day D.A., Millar A.H. Proteomic identification of divalent metal cation binding proteins in plant mitocondria. FEBS Lett. 537, 96–100, 2003.

Hermann L., Schwan D., Garner R., Mobley H.L.T., Haas R., Schafer K.P., Melchers K. *Helicobacter pylori* cadA encodes an essential Cd(II)-Zn(II)-Co(II) resistance factor influencing urease activity. Mol. Microbiol. 33, 524–536, 1999.

Huang C.-W., Wei C.-C., Liao H.-C. A low cost color-based bacterial biosensor for measuring arsenic in groundwater. Chemosphere 141, 44–49, 2015.

Huckle J.W., Morby A.P., Turner J.S., Robinson N.J. Isolation of a prokaryotic metallothionein locus and analysis of transcriptional control by trace metals. Mol. Microbiol. 7, 177–187, 1993.

Ibrahim Z., Azlina W., Baba A.B. Bioaccumulation of silver and the isolation of metal-binding protein from P. dimunita. Brazilian Archives of Biology and Technology. 44(3), 223–225, 2001.

Inouhe M. Phytochelatins. Braz. J. Plant Physiol. 17, 65–78, 2005.

Jeronimo P.C.A., Araujo A.N., Montenegro M.C.B.S.M. Optical sensors and biosensors based on sol-gel films. Talanta 72, 13–27, 2007.

Joshi B.P., Lee K.-H. Synthesis of highly selective fluorescent peptide probes for metal ions: Tuning selective metal monitoring with secondary structure. Bioinorg. Med. Chem. 16, 8501–8509, 2008.

Joshi B.P., Cho W.-M., Kim J., Yoon, Lee K.-H. Design, synthesis, and evaluation of peptidyl fluorescent probe for Zn2+ in aqueous solution.Bioorg. Med. Chem. 17, 6425–6429, 2007.

Joshi B.P., Wan In Lee, Lee K.-H. Ratiometric and turn-on monitoring for heavy and transition metal ions in aqueous solution with a fluorescent peptide sensor. Talanta 78, 903–909, 2009.

Juhnke S., Peitzsch N., Hübener N., Große C., Nies D.H. New genes involved in chromate resistance in *Ralstonia metallidurans* strain CH34. Arch. Microbial. 179, 15–25, 2002.

Kägi J.H.R., Schäffer A. Biochemistry of metallothionein.Biochemistry 27, 8509–8515, 1988.

Kanamaru K., Kashiwagi S., Mizuno T. A copper-transporting P-type ATPase found in the thylakoid membrane of the cyanobacterium *Synechococcus* species PCC 7942. Mol. Microbiol. 13, 369–377, 1994.

Kang S.H., Singh S., Kim J.-Y., Lee W., Mulchandani A., Chen W. Bacteria metabolically engineered for enhanced phytochelatin production and cadmium accumulation. Appl. Environ. Microbiol. 73, 6317–6320, 2007.

Khan S., Cao Q., Zheng Y.M., Huang Y.Z., Zhu Y.G. Health risks of heavy metals in contaminated soils and food crops irrigated with wastewater in Beijing, China. Environ. Poll. 152, 686–692, 2008.

Kim M.J., Nriagu J.O., Haack S.K. Carbonate ions and arsenic dissolution by groundwater. Environ. Sci. Technol. 34, 3094–3100, 2000.

Kim T.H., Kim S.H., Tan L.V., Dong Y., Kim H., Kim J.S. Diazo-coupled calix[4]arenes for qualitative analytical screening of metal ions. Talanta 74, 1654–1658, 2008.

Korte N.E., Fernando Q. A review of arsenic (III) in groundwater. Crit. Rev. Environ. Control. 21, 1–39, 1991.

Krishna A.K., Satyanarayanan M., Govil P.K. Assessment of heavy metal pollution in water using multivariate statistical techniques in an industrial area: A case study from Patancheru, Medak District. J. Hazard. Mater. 167, 366–373, 2009.

Kurbanoglu S., Ozkan S.A., Merkoci O.A. Nanomaterials-based enzyme electrochemical biosensors operating through inhibition for biosensing applications. Biosens. Bioelectron. 89, 886–898, 2017.

Lau K.-T., McHugh E., Baldwin S., Diamond D. Paired emitter-detector light emitting diodes for the measurement of lead (II) and cadmium (II). Anal. Chem. Acta 569, 221–226, 2006.

Lebrun M., Audurier A., Cossart P. Plasmid-borne cadmium resistance genes in *Listeria monocytogenes* are similar to *cad*A and *cad*C of *Staphylococcus aureus* and are induced by cadmium. J. Bacteriol. 176, 3040–3048, 1994.

Lee S.W., Glickmann E., Cooksey D.A. Chromosomal locus for cadmium resistance in *Pseudomonas putida* consisting of a cadmium-transporting ATPase and a MerR family response regulator. Appl. Environ. Microbiol. 67, 1437–1444, 2001.

Legatzki A., Grass G., Anton A., Rensing C., Nies D.H. Interplay of the Czc system and two P-type ATPases in confering metal resistance to *Ralstonia metallidurans*. J. Bacteriol. 185, 4354–4361, 2003.

Lehmann M., Riedel K., Adler K., Kunze G. Amperometric measurement of copper ions with a deputy substrate using a novel *Saccharomyces cerevisiae* sensor. Biosens. Bioelectron. 15, 211–219, 2000.

Lei Y., Chen W., Mulchandani A. Microbial biosensors. Anal. Chim. Acta 568, 200–210, 2006.

Li M., Gou H., Al-Ogaidi I., Wu N. Nanostructured sensors for detection of heavy metals: A review. ACS Sust. Chem. Eng. 1, 713–723, 2013.

Lorenz W.W., Cormier M.J., O'Kane D.J., Hua D., Escher A.A., Szalay A.A. Expression of the *Renilla reniformis Luciferase* Gene in mammalian cells. J. Biolumin. Chemilumin. 11, 31–37, 1996.

Ma Y., Lin J., Zhang C., Ren Y., Lin J. Cd(II) and As(III) bioaccumulation by recombinant *Escherichia coli* expressing oligomeric human metallothioneins. J. Hazard. Mater. 185, 1605–1608, 2011.

Malhotra B.D., Singhal R., Chaubey A., Sharma S.K., Kumar A. Recent trends in biosensors. Curr. Appl. Phys. 5, 92–97, 2005.

McLaughlin M.J., Hamon R.E., McLaren R.G., Speir T.W., Rogers S.L. Review: a bioavailability-based rationale for controlling metal and metalloid contamination of agricultural land in Australia and New Zealand. Aust. J. Soil Res. 38, 1037–1086, 2000.

McLaughlin M.J., Zarcinas B.A., Stevens D.P., Cook N. Soil testing for heavy metals. Comm. Soil Sci. Plant Anal. 31, 1661–1700, 2000.

Mehta J., Bhardwaj S.K., Bhardwaj N., Paul A.K., Kumar P., Kim K.-H., Deep A. Progress in the biosensing techniques for trace-level heavy metals. Biotechnol. Adv. 34, 47–60, 2016.

Mejare M., Bulow L. Metal-binding proteins and peptides in bioremediation and phytoremediation of heavy metals. Trends Biotechnol. 19(2), 67–73, 2001.

Mergeay M., Monchy S., Vallaeys T., Auquier V., Benotmane A., Bertin P., Taghavi S., Dunn J., van der Lelie D., Wattiez R. *Ralstonia metallidurans*, a bacterium specifically adapted to toxic metals: towards a catalogue of metal-responsive genes. FEMS Microbiol. Rev. 27, 385–410, 2003.

Merkoci A., Alegret S. (eds.). Comprehensive Analytical Chemistry, Volumen 49, Elsevier, 2007.

Mills S.D., Jasalavich A., Cooksey D.A. A two-component regulatory system required for copper resistance operon of Pseudomonas syringae. J. Bacteriol. 175, 1656, 1993.

Mittal H., Fosso-Kankeu E., Mishra S.B., Mishra A.K. Biosorption potential of Gum ghatti-g-poly (acrylic acid) and susceptibility to biodegradation by *B. subtilis*. Int. J. Biol. Macromol. 62, 370–378, 2013.

Murasugi A., Wada C., Hayashi Y. Cadmium-binding peptide induced in fission yeast, *Schizosaccharomyces pombe*. J. Biochem. 90, 1561–1564, 1981.

Naz N., Young H.K., Ahmed N., Gadd G.M. Cadmium accumulation and DNA homology with metal resistance genes in sulfate-reducing bacteria. Appl. Environ. Microbiol. 71, 4610–4618, 2005.

Nies D.H. Efflux-mediated heavy metal resistance in prokaryotes. FEMS Microbiol. Rev. 27, 313–339, 2003.

Nucifora G., Chu G.L., Misra T.K., Silver S. *Stenotrophomonas maltophilia* D457R contains a cluster of genes from gram-positive bacteria involved in antibiotic and heavy metal resistance. Antimicrob. Ag. Chemother. 44, 1778–1782, 2000.

Nucifora G., Chu L., Misra T.K., Martinez S. Cadmium resistance from *Staphylococcus aureus* plasmid pI258 *cad*A gene results from a cadmium-efflux ATPase. Proc. Natl. Acad. Sci. USA. 86, 3544–3548, 1989.

Odermatt A., Suter H., Krapf R., Solioz M. Primary structure of two P-type ATPases involved in copper homeostasis in *Enterococcus hirae*. J. Biol. Chem. 268, 12775–12779, 1993.

O'Halloran T.V. Transition metals in control of gene expression. Science 261(5122), 715–725, 1993.

Palchetti I., Laschi S., Mascini M. Miniaturised stripping-based carbon modified sensor for in field analysis of heavy metals. Anal. Chim. Acta 530, 61–67, 2005.

Pan D., Wang Y., Chen Z., Lou T., Qin W. Nanomaterial/Ionophore-based electrode for anodic stripping voltammetric determination of lead: An electrochemical sensing platform toward heavy metals. Anal. Chem. 81, 5088–5094, 2009.

Paneli M., Voulgaropoulos A. Applications of adsorptive stripping voltammetry in the determination of trace and ultratrace metals. Electroanalysis 5, 355–373, 1993.

Parham H., Pourreza N., Rahbar N. Solid phase extraction of lead and cadmium using solid sulfur as a new metal extractor prior to determination by flame atomic adsorption spectrometry. J. Hazard. Mater. 163, 588–592, 2009.

Peitzsch N., Eberz G., Nies D. *Alcaligenes eutrophus* as a bacterial chromate sensor. Appl. Environ. Microbiol. 64, 453–458, 1998.

Phung L.T., Ajlani G., Haselkorn R. P-type ATPase from the cyanobacterium *Synechococcus* 7942 related to the human Menkes and Wilson disease products. Proc. Natl. Acad. Sci. USA. 91, 9651–9654, 1994.

Popovic A., Djordjevic D., Polic P. Trace and major element pollution originating from coal ash suspension and transport processes. Environ. Int. 26, 251–255, 2001.

Priyadarshini E., Pradhan N. Gold nanoparticles as efficient sensors in colorimetric detection of toxic metal ions: A review. Sens. Act. B238, 888–902, 2017.

Prosun B., Mukherjee A.B., Gunnar J., Nordqvist S. Metal contamination at a wood preservation site: characterization and experimental studies on remediation. Sci. Total Environ. 290, 165–180, 2002.

Raj S.P., Ram P.A. Determination and contamination assessment of Pb, Cd and Hg in roadside dust along Kathmandu-Bhaktapur road section of Amiko Highway, Nepal. Res. J. Chem. Sci. 3, 18–25, 2013.

Ramos J.A., Bermejo E., Zapardiel A., Perez J.A., Hernandez L. Direct determination of lead by bioaccumulation at a moss-modified carbon paste electrode. Anal. Chim. Acta 273, 219–227, 1993.

Reimann C., Arnoldussen A., Englmaier P., Filzmoser P., Finne T.E., Garrett R.G., Koller F., Nordgulen O. Element concentrations and variations along a 120-km transect in southern Norway—Anthropogenic vs. geogenic vs. biogenic elements sources and cycles. Appl. Geochem. 22, 851–871, 2007.

Rensing C., Ghosh M., Rosen B.P. Families of softmetal-ion-transporting ATPases. J. Bacteriol. 181, 5891–5897, 1999.

Roda A., Pasini P., Mirasoli M., Guardigli M., Russo C., Musiani M., Baraldini M. Sensitive determination of urinary mercury (II) by a bioluminescent transgenic bacteria-based biosensor. Anal. Lett. 34, 24–41, 2001.

Roman-Ross G., Depetris P.J., Arribeere M.A., Ribeiro Guevara S., Cuello G.J. Geochemical variability since the Late Pleistocene in Lake Mascardi sediments, northern Patagonia, Argentina. J. South Am Earth Sci. 15, 657–667, 2002.

Scheerer U., Haensch R., Mendel R., Kopriva S., Rennenberg H., Herschbach C. Sulphur flux through the sulphate assimilation pathways is differently controlled by adenosine 5'-phosphosulphate reductase under stress and in transgenic poplar plants overexpressing gamma-ECS, SO, or APR. J. Exp. Bot. 61, 609–622, 2010.

Scott D.L., Ramanathan S., Shi W., Rosen B.P., Daunert S. Genetically engineered bacteria: Electrochemical sensing systems for antimonite and arsenite. Anal. Chem. 69, 16–20, 1997.

Selifonova O., Burlage R., Barkay T. Bioluminescent sensors for detection of bioavailable Hg(II) in the environment. Appl. Environ. Microbiol. 59, 3083–3090, 1993.

Sharma S.K., Sehgal N., Kumar A. Biomolecules for development of biosensors and their applications. Curr. Appl. Phys. 3, 307–316, 2003.

Sheng P.X., Ting Y.-P., Chen J.P., Hong L. Sorption of lead, copper, cadmium, zinc and nickel by marine algal biomass: characterization of biosorptive capacity and investigation of mechanisms. J. Coll. Interf. Sci. 275, 131–141, 2004.

Sheng Y., Yang X., Lian Y., Zhang B., He X., Xu W., Huang K. Characterization of a cadmium resistance *Lactococcus lactis* subsp. *lactis* strain by antioxidant assays and proteome profiles methods. Environ. Toxicol. Pharmacol. 46, 286–291, 2016.

Shi G., Xu S., Zhang J., Wang L., Bi C., Teng J. Potentially toxic metal contamination of urban soils and roadside dust in Shangai, China. Environ. Poll. 156, 251–260, 2008.

Silver S., Phung Le T. A bacterial view of the periodic table: genes and proteins for toxic inorganic ions. J. Ind. Microbiol. Biotechnol. 32, 587–605, 2005.

Silver S. Bacterial heavy metal resistance: New surprises. Annu. Rev. Microbiol. 50, 753–789, 1996.

Sindern S., Tremohlen M., Dsikowitzky L., Gronen L., Schwarzbauer J., Siregar T.H., Ariyani F., Irianto H.E. Heavy metals in river and coast sediments of the Jakarta Bay region (Indonesia)—Geogenic versus anthropogenic sources. Mar. Poll. Bull. 110, 624–633, 2016.

Singh M., Muller G., Singh I.B. Geogenic distribution and baseline concentration of heavy metals in sediments of the Ganges River, India. J. Geochem. Explor. 80, 1–17, 2003.

Singh S., Lee W., Da Silva N.A., Mulchandani A., Chen W. Enhanced arsenic accumulation by engineered yeast cells expressing *Arabidopsis thaliana* phytochelatin synthase. Biotechnol. Bioeng. 99, 333–340, 2008.

Smedley P.L., Kinniburgh D.G. A review of the source, behavior and distribution of arsenic in natural waters. Appl. Geochem. 17, 517–568, 2002.

Sofia H.J., Burland V., Daniels D.L., Plunkett III G., Blattner F.R. Analysis of the *Escherichia coli* genome. V. DNA sequence of the region from 76.0–81.5 min. Nucleic Acids Res. 22, 2576–2586, 1994.

Solioz M., Odermatt A., Krapf R. Copper pumping ATPases: common concepts in bacteria and man. FEBS Lett. 346, 44–47, 1994.

Tan W., Shi Z.-Y., Kopelman R. Development of submicron chemical fiber optic sensors. Anal. Chem. 64, 2985–2990, 1992.

Tauriainen S, Karp M., Chang W., Virta M. Recombinant luminescent bacteria for measuring bioavailable arsenite and antimonite. Appl. Environ. Microbiol. 63, 4456–4461, 1997.

Tauriainen S., Karp M., Chang W., Virta M. Luminescent bacterial sensor for cadmium and lead. Biosens. Bioelectron. 13, 931–938, 1998.

Tekaya N., Saiapina O., Ben Ouada H., Lagarde F., Ben Ouada H., Jaffrezic-Renault N. Ultrasensitive conductometric detection of heavy metals based on inhibition of alkaline phosphatase activity from *Arthrospira platensis*. Bioelectrochemistry 90, 24–29, 2013.

Tescione L., Belfort G. Construction and evaluation of a metal ion biosensor. Biotechnol. Bioeng. 42, 945–952, 1993.

Thelwell C., Robinson N.J., Turner-Cavet J.S. An SmtB-like repressor from Synechocystis PCC 6803 regulates a zinc exporter. Proc. Natl. Acad. Sci. USA. 95, 10728–10733, 1998.

Thevenot D.R., Toth K., Durst R.A., Wilson G.S. Electrochemical biosensors: recommended definitions and classification. Biosens. Bioelectron. 16, 121–131, 2001.

Thompson R.B., Laliwal B.P., Felicia V.L., Fierke C.A., McCall K. Determination of picomolar concentrations of metal ions using fluorescence anisotropy: biosensing with a reagentless enzyme transducer. Anal. Chem. 70, 4717–4723, 1998.

Tottey S., Waldron K.J., Firbank S.J., Reale B., Bessant C., Sato K., Cheek T.R., Gray J., Banfield M.J., Dennison C., Robinson N.J. Protein-folding location can regulate manganese-binding versus copper- or zinc-binding. Nature 455, 1138–1142, 2008.

Tsai K.J., Yoon K.P., Lynn A.R. ATP-dependent cadmium transport by the cadA cadmium resistance determinant in everted membrane vesicles of *Bacillus subtilis*. J. Bacteriol. 174, 116–121, 1992.

Turner A.P. Biosensors: sense and sensibility. Chem. Soc. Rev. 42, 3184–3196, 2013.

Turner A.P.F., Karube I., Wilson G.S. (eds.). Biosensors, Fundamentals and Applications. Oxford University Press, Oxford, 1987.

Turner J.S., Morby A.P., Whitton B.A., Gupta A., Robinson N.J. Construction of Zn^{2+}/Cd^{2+} hypersensitive cyanobacterial mutants lacking a functional metallothionein locus. J. Biol. Chem. 268, 4491–4498, 1993.

Turner J.S., Robinson N.J., Gupta A. Construction of Zn^{2+}/Cd^{2+} tolerant cyanobacteria with a modified metallothionein divergon: further analysis of the function and regulation of smt. J. Ind. Microbiol. 14, 259–264, 1995.

Ugwu J.N., Okoye C.O.B., Ibeto C.N. Impacts of vehicle emissions and ambient atmospheric deposition in Nigeria on the Pb, Cd, and Ni content of fermented cassava flour processed by sun-drying. Human Ecol. Risk Ass. 17, 478–488, 2011.

Vatamaniuk O.K., Mari S., Lu Y.-P., Rea P.A. AtPCS1, a phytochelatin synthase from *Arabidopsis*: isolation and *in vitro* reconstitution. Proc. Natl. Acad. Sci. USA. 96, 7110–7115, 1999.

Vaughan A.A., Narayanaswamy R. Optical fibre reflectance sensors for the detection of heavy metal ions based on immobilised Br-PADAP. Sens. Act. B. 51, 368–376, 1998.

Virta M., Lampinen J., Karp M. A luminescent-based mercury biosensor. Anal. Chem. 67, 667–669, 1995.

Waldron K.J., Robinson N.J. How do bacterial cells ensure that metalloproteins get the correct metal? Nature Reviews 6, 25–35, 2009.

Wang J. Stripping Analysis.Encyclopedia of Electrochemistry, Willey VCH, 2007.

Wang J., Lu J., Hocevar S.B., Farias P.A.M. Bismuth-coated carbon electrodes for anodic stripping voltammetry. Anal. Chem. 72, 3218–3222, 2000.

White B.R., Holcombe J.A. Fluorescent peptide sensor for the selective detection of Cu^{2+}. Talanta 71, 2015–2020, 2007.

Wittman C., Riedel K., Schmid R.D. Microbial and enzyme sensors for environmental monitoring. *In*: Kress-Rogers E. (ed.). Handbook of Biosensors and Electronic Noses. CRC, Boca Raton, FL. pp. 299–332, 1997.

Wolfbeis O.S. Fiber Optic Chemical Sensors and Biosensors. CRC Press, Boca Raton, 1991.

Wood J.M., Wang H.-K. Microbial resistance to heavy metals. Environ. Sci. Technol. 17, 582–590, 1983.

Wu C.-M., Lin L.-Y. Immobilization of metallothionein as a sensitive biosensor chip for the detection of metal ions by surface plasmon resonance. Biosens. Bioelectron. 20, 864–871, 2004.

Wu Y., Li T., Yang L. Mechanisms of removing pollutants from aqueous solutions by microorganisms. Bioresource Technol. 107, 10–18, 2012.

Wuana R.A., Okieimen F.E. Heavy metals in contaminated soils: A review of sources, chemistry, risks and best available strategies for remediation. Int. Schol. Res. Net. 1–20, 2011.

Xiong J., Li D., Li H., He M., Miller S.J., Yu L., Rensing C., Wang G. Genome analysis and characterization of zinc efflux systems of a highly zinc-resistant bacterium, *Comamonas testosteroni* S44. Res. Microbiol. 162, 671–679, 2011.

Yusof N.A., Ahmad M. A flow-through optical fibre reflectance sensor for the detection of lead ion based on immobilised gallocynine. Sens. Act. B 94, 201–209, 2003.

Zakharova E.A., Pokrovsly O.S., Dupre B., Gaillardet J., Efimova L.E. Chemical weathering of silicate rocks in karelia region and kola peninsula, NW Russia: Assessing the effect of rock composition, wetlands and vegetation. Chem. Geol. 242, 255–277, 2007.

Zhang M.K., Liu Z.Y., Wang H. Use of single extraction methods to predict bioavailability of heavy metals in polluted soils to rice. Comm. Soil Sci. Plant Anal. 41, 820–831, 2010.

CHAPTER 4

Exploration and Intervention of Geologically Ancient Microbial Adaptation in the Contemporary Environmental Arsenic Bioremediation

Tanmoy Paul and *Samir Kumar Mukherjee**

1. Introduction

Arsenic (As) is one of the most hazardous environmental toxic metalloid, ranking 20th considering its abundance in the earth's crust (Mandal and Suzuki, 2002). The metalloid is well distributed in soil, sediments, water, air, and living organisms since its geologically ancient presence. Arsenic is abundant in > 200 different mineral forms.

The presence of As in soil is greater than in rocks and also depends on the geographic region to which the soil belongs. Global distribution of As reveals a very large range between 0.5 and 5000 ppb prevalent as natural As contamination in > 70 countries (Ravenscroft et al., 2009). In the context of global As contamination, it is evident that although the provisional guideline values for groundwater are commonly set at 10 ppb but it can reach upto 50 ppb. Comparison of As distribution portrays that As contamination reaches statistically non-permissible limits in many parts of the world. People living in contaminated areas often suffer from serious health problems throughout the world. Recent studies report that about 6 million people of 2600 villages in 74 As-affected blocks of West Bengal, India are prone to risk with 8500 people out of 86,000 people examined are suffering from arsenicosis. This means that about 9.8% of the total population suffers from As in these areas.

Department of Microbiology, University of Kalyani, Kalyani 741235, India.
* Corresponding author: dr.samirmukherjee@gmail.com

The source of As can be oxidation of As rich pyrite or anoxic reduction of ferric iron hydroxides (Acharyya et al., 1999; Nickson et al., 1998) to ferrous iron in the sediments and thereby releasing the adsorbed As into groundwater (Mandal and Suzuki, 2002). Arsenic in these alluvial sediments is probably derived from sulphide deposits in the lower Gangetic plane of India and Bangladesh (Nickson et al., 1998). The source could be the copper belt of Bihar, India and the basins of the Damodar valley, India. These areas bear moderate level of As concentrations, which were drained by rivers flowing to the Ganges tributary system. The As content of drinking water in Bangladesh, for example, derives from eroded Himalayan sediments (Stolz et al., 2010). Arsenic contamination from both geogenic and anthropogenic sources literally affects millions of people around the globe but is especially marked in the deltaic regions of south Asia (*viz.* Bangladesh; West Bengal, India) and the surrounding regions (*viz.* Taiwan, Cambodia) (Oremland et al., 2009). Environmental As contamination over the years has dramatically increased resulting in its entry into the ecological food chain.

The adverse health impact of As contamination in terrestrial and aquatic ecosystems became a sensitive environmental as well as medical issue, extensively studied and periodically monitored by World Health Organization. Arsenic being a potent carcinogen in both its acute and chronic intake can cause skin, bladder, liver, and lung cancers (Yoshida et al., 2004; Tapio and Grosche, 2006). Epidemiological toxicity due to the formation of reactive oxygen species is also reported which affects the biological system (Wang et al., 2001; Shi et al., 2004). Health outcomes from As exposure depends on the dose and duration of exposure of As. Chemical species of As, more precisely the oxidation state of As is also an important determinant (Caussy et al., 2003). Acute exposure to high levels of inorganic As by humans can be fatal. Chronic effects observed with the ingestion of inorganic As include skin lesions, disturbances of the peripheral nervous system, anemia and leukopenia, hepatic disorder, circulatory disease, and carcinoma. At cellular level, As has been shown to induce chromosomal aberrations, interfere with DNA repair system, inhibition of p53, and telomerase activities (Chou et al., 2001). Many of these effects have been observed in populations that consumed contaminated water, including populations in Taiwan, Argentina, Bangladesh and West Bengal, India (IPCS, 2001; Chakraborti et al., 2009).

2. Arsenic as environmental pollutant

2.1 Prevalent environmental arsenic species

Arsenic in environment exists in different inorganic forms depending on their valence states. It is the type of As species that determines the bioavailability and toxicity of As. Arsenic toxicity is dependent on the chemical species of As as well as whether it is inorganic or organic. Arsenic readily changes its oxidation state and bonding configuration producing four oxidation states—

arsenite [As(III)], arsenate [As(V)], elemental As [As(0)], and arsine, with As(V) and As(III) being most abundant among others in nature (Cullen and Reimer, 1989). As(III) is 25–60 times more toxic than As(V) (Silver and Phung, 2005) because it is reactive and forms strong metal-like bonds with biological molecules. As(V) is less toxic and chemically similar to phosphate which can form esters similar to the phosphate esters (Zhu et al., 2014). The valence state of arsenic in arsine (AsH_3) is not clear as the electronegativity of both arsenic and hydrogen are almost identical. Predominance of As species in solution is determined by redox potential and pH.

Arsenic occurs naturally in soils as a result of the weathering of igneous rocks (1.5 to 3.0 $mg.kg^{-1}$ As) and sedimentary rocks (1.7 to 400 $mg.kg^{-1}$ As) (Smith et al., 1998). More than 300 As containing minerals are found in nature with approximately 60% being arsenates, 20% being sulphides and sulphosalts and 10% being oxides (Bowell and Parshley, 2001). The most abundant and widespread As containing mineral is arsenopyrite (FeAsS). Arsenic creates environmental hazard because of its relative mobility over a wide range of redox conditions (Smedley and Kinniburgh, 2002) affecting its toxicity in biological systems.

2.2 Biogeochemistry of As

Arsenic is widely distributed in the earth's continental crust (2.1 $mg.kg^{-1}$), soil (5 $mg.kg^{-1}$), and ocean water (0.0023 $mg.kg^{-1}$). Much higher concentration can be found where the environment has been impacted by geothermal sources, mining, smelting, and other anthropogenic activities which bring As into our environment (Fig. 1). Anthropogenic sources of As include both inorganic and organic forms. Arsenic serves as an active ingredient of various herbicides, insecticides, rodenticides, wood preservatives, animal feeds, dyes, and semiconductors. There are more than 200 arsenic-containing minerals formed in the high-temperature environment of earth's crust (Smedley and Kinniburgh, 2002). However, microbe-mediated formation of orpiment, an As sulfide mineral has also been reported (Newman et al., 1998).

In living organisms As is found only in the pentavalent and trivalent oxidation states. Inside living cells, As is predominantly in the As(III) state which enters via the aquaglyceroporin channels, while As(V) is less prevalent and enters into the cell via the phosphate transporters (Bhattacharjee et al., 2008; Yang et al., 2012; Zhao et al., 2010). As(V) is the predominant form in the oxygen rich environment such as in the surface water and aerobic soils. Reduction of As(V) to As(III) occurs under suboxic conditions making As(III) the predominant As species under reducing conditions, e.g., lowland flooded paddy fields or anaerobic sediments. Reduction of As(V) to As(III) leads to marked increased mobility of As due to the fact that As(III) is less strongly adsorbed in the solid phase (Fendorf et al., 2008). This process has important consequences in As mobilization in groundwater and increased As transfer from paddy soil to rice therefore increased As toxicity.

Arsenic present in the lithosphere is approximately five times larger than that present in the ocean or the terrestrial soil (Wenzel, 2013). Weathering of rocks, geothermal and volcanic activities, mining, and smelting release of As from the lithosphere to the terrestrial and oceanic environments. Arsenic also enters the biosphere and can be transferred through the food chain. The total amount of As accumulated in terrestrial plants is four times less than that in soil (Wenzel, 2013) indicating limited accumulation of As in terrestrial plants, likely due to low bioavailability of As in soil. There are number of plants named as As hyperaccumulators that can accumulate high levels of As without suffering from phytotoxicity (Ma et al., 2001; Zhao et al., 2002). The ratios of plant arsenic to soil concentrations are > 10 in the hyperaccumulators, whereas < 0.1 ratios in non-hyperaccumulators.

Microbes assist in cycling of various As oxidation states (Mukhopadhyay et al., 2002). As(V) entering the microbial cytosol through the phosphate transport system is reduced to As(III), which is then extruded out of the cell either through channels or secondary transporters (Rosen, 2002) leading to increased As(III) load in environment (Oremland and Stolz, 2005). As(III) is also generated by certain microbes that use As(V) as the terminal electron acceptor in anaerobic respiration (Oremland and Stolz, 2003). These As(V) respiring microbes can release As(III) from As(V) rich sediments.

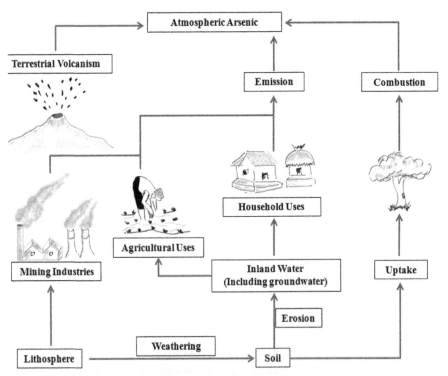

Figure 1. The biogeochemical cycles of environmental As.

Microbes can also convert inorganic As into gaseous methylated arsine (Bentley and Chasteen, 2002; Qin et al., 2006). Microbial activities are either directly involved or enhance these processes, e.g., in Mono and Searles Lakes in California where As concentrations are high enough supporting the biogeochemical cycle. Marine microorganisms can convert inorganic arsenicals to various water- or lipid-soluble organic As species which can be further degraded back to As by microbial metabolism completing the global As cycle (Dembitsky and Levitsky, 2004).

2.3 Biochemical circuitry of arsenic redox reactions

2.3.1 Arsenite oxidation

In addition to efflux, rapid As(III) oxidation has been observed to occur widely by microbial activities (Mukhopadhyay et al., 2002; Oremland and Stolz, 2003) (Fig. 2). The As(III) oxidation is carried out by periplasmic arsenite oxidase. The first arsenite oxidase was purified from the periplasm of the β-proteobacterium *Alcaligenes faecalis* (Anderson et al., 1992). All aerobic As(III) oxidation involves arsenite oxidases containing two heterologous subunits: AioA (AoxB) and AioB (AoxA) (Stolz et al., 2006; Zargar et al., 2010). Recently, the arsenite oxidase subunit AioA was shown to be involved in anaerobic oxidation of As(III) (Sun et al., 2010a, 2010b), however, the direct evidences are still lacking. Phylogenetic analysis of the two subunits of this heterodimeric enzyme suggests an early origin of arsenite oxidases probably before the divergence of archaea and bacteria (Lebrun et al., 2003). It is evolutionarily plausible that such an ancient oxidase function evolved before the atmosphere became oxidizing allowing As(III) oxidation to couple with energy production.

A photosynthetic purple sulfur bacterium (*Ectothiorhodospira* sp. strain PHS-1) capable of anoxic As(III) oxidation was isolated from a hot spring biofilm (Kulp et al., 2008). Another *arxA* gene identified in this organism which revealed a wide distribution of *arxA* like genes in As rich Mono Lake and Hot Creek sediments in California and alkaline microbial mats from Yellowstone National Park (Zargar et al., 2012). This new clade of arsenite oxidase genes has been proposed to have evolved in Archaean period and might be lineal to arsenite oxidases and respiratory arsenate reductases

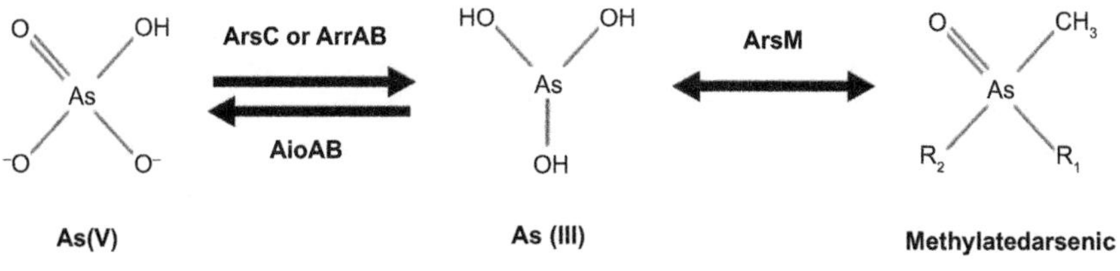

Figure 2. Arsenic species and their biotransformation. Responsible enzymes are indicated.

(Oremland et al., 2009; Zargar et al., 2012). In contrast, a recent phylogenetic study indicates that both Arx and respiratory As(V) reductases originated after the divergence of bacteria and archaea (van Lis et al., 2013).

2.3.2 Arsenate reduction

Origin of photosynthesis transforms the primordial atmosphere of earth which became oxidizing and in the oxidizing environment As(III) was oxidized to As(V). As(V) is less mobile than As(III) and is usually adsorbed strongly by iron and aluminum oxides (Fendorf et al., 2008). Thus respiratory reductase has evolved to use As(V) as a terminal electron acceptor in energy-generating respiratory chains. It has been postulated that thermodynamically dissimilatory As(V) reduction can provide enough reduction potential to sustain microbial life. Microorganisms isolated from As-contaminated sediments of the Aberjona watershed, Massachusetts was demonstrated to grow by reducing (respiring) As(V) (Ahmann et al., 1994). Such anaerobic dissimilatory As(V) reduction is catalyzed by the As(V) respiratory reductase (Arr) complex which consists of a large catalytic subunit (ArrA) and a small subunit (ArrB) (Krafft and Macy, 1998; Afkar et al., 2003; Saltikov and Newman, 2003; van Lis et al., 2013). The *arrA* gene encodes the large ArrA subunit of the reductase and this gene can be used as a reliable marker for As(V) respiration (Malasarn et al., 2004). For example, in arsenic-rich soda lakes in California, *arrA* was detected in sediment samples and its abundance corresponded with *in situ* rates of As(V) reduction (Kulp et al., 2006). The Arr enzymes have also been demonstrated to be bidirectional; it shows both As(V) reductase and As(III) oxidase activity *in vitro* (Richey et al., 2009). It has been difficult to monitor the existence and activity of As(V)-respiring bacteria in diverse environments. Multiple alignments of sequences related to ArrA reveals the phylogeny of Arr and closely associated enzymes (Duval et al., 2008). This analysis confirms the previously proposed proximity of Arr to the cluster of polysulfide/thiosulfate reductases and unravels a hitherto unrecognized clade even more closely related to Arr. The resulting phylogeny strongly suggests that Arr originated in bacteria subsequent to the generation of an oxidizing atmosphere after the bacteria/archaea divergence. Horizontal gene transfer might play important roles in laterally distributing the *arr* genes within the domain Bacteria. It is also suggested that an enzyme related to polysulfide reductase rather than As(III) oxidase may be the precursor of Arr.

Reduction of As(V) is therefore considered to be a key mechanism of its mobilization, such as in aquifer sediments causing As contamination in groundwater in South and Southeast Asia (Fendorf et al., 2010). Analysis of the microbial community of metal-reducing bacteria utilizes ArsC like As(V) reductase for dissimilatory As(V) reduction (Oremland and Stolz, 2005). To cope with the rise of As(V) produced by an oxidizing atmosphere, multiple microbial species independently evolved strategies to detoxify inorganic

As(V) by reduction (Mukhopadhyay and Rosen, 2002). Small As(V) reductase enzymes evolved a number of times from the ancestors of proteins such as redoxins and phosphatases.

One family of bacterial arsenate reductases, typified by the ArsC enzyme of plasmid R773 (Gladysheva et al., 1994; Oden et al., 1994), is closely related to the glutaredoxin family. This enzyme uses a cysteine residue in the protein and a glutathione molecule to reduce As(V) to As(III), producing oxidized glutathione that is re-reduced by the enzyme glutathione reductase using NADPH as the source of electrons. A second family of bacterial arsenate reductases, typified by the ArsC of *Staphylococcus aureus* (Ji et al., 1994), uses internal cysteine residues during the reduction of As(V) (Messens and Silver, 2006). Both the respiration and detoxification machinery facilitate reduction of As(V) by microbes but it is not known whether one is more prevalent in the environment than the other, nor is it clear which environmental factors modulate the predominance of the two mechanisms. The facts that arsenate reductases arose at least three times and that some are both reductases and phosphatases or can easily be changed from one activity to the other suggest that evolution of arsenate reductases is relatively rapid and uncomplicated.

3. Microbial arsenic resistance and arsenic resistance operon

In bacteria, As resistance genes are organized as *ars* operons. The majority of the *ars* operons have three genes constituting the *arsRBC*. Few operons have two more genes, *arsD* and *arsA* constructing the *arsRDABC* operon observed in *E. coli* plasmid R773. Cells expressing the five genes *arsRDABC* are more resistant to As(V) and As(III) than those expressing only the *arsRBC* genes (Ajees et al., 2011). Although the resistance genes were originally discovered on plasmids, they have also been found on the chromosomes of a diverse group of organisms including archaea, bacteria, yeasts, etc.

The information about microbial oxidation of As(III) was first documented in a bacillus in 1918 (Green, 1918; Stolz, 2012). Previously thought to be a mechanism for detoxification, this mechanism has been recently linked to energy generation. The phylogenetically widespread mechanism of As(III) oxidation occurs in several strains belonging to Crenarcheaota, Aquificales, and Thermus, as well as α-, β-, and γ-proteobacteria. In most cases, the organisms are aerobic heterotrophs or chemolithautotrophs and utilize oxygen as the electron acceptor for As(III) oxidation (Stolz et al., 2010).

Microorganisms adapted to function in As rich environments can achieve the As resistance by oxidizing uncharged As(III) ions to As(V) on the cell surface to avoid the passive uptake of As(III) by aquaglycerolporins. Whereas in cytoplasmic of As(V) accumulators, the As(V) resistance can be achieved by reduction of As(V) to As(III) through the ArsC system to facilitate As export out of the cell in a reaction that consumes ATP (Oremland et al., 2009) (Fig. 3).

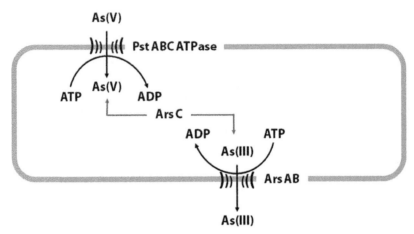

Figure 3. Pathways of As detoxification in prokaryotes depending on As redox reactions. Cellular transporter proteins are indicated, as well as the enzymes responsible for As(V) reduction.

3.1 Organization of arsenic resistance operon

3.1.1 arsR is the ars operon transcriptional regulator gene

The transcriptional regulator of the *ars* operon is the ArsR/SmtB family of metal(loid)-responsive element which is an As(III)-responsive transcriptional repressor (Xu and Rosen, 1999). The 117-residue As(III)-responsive ArsR repressor is encoded by both the *arsRDABC* operon of gram-negative bacteria (Wu and Rosen, 1991, 1993; Shi et al., 1994; Rosen, 1999) and the *arsRBC* operon of *E. coli* (Rosen, 1999). The number of identified members of this family has been reported to be 198, with 192 homologues in Gram-positive and negative bacteria, and six homologues in archaea. ArsR are homodimers in structure and repress transcription by binding to DNA in the absence of As(III) which dissociate from the DNA during As(III) abundance ultimately conferring As resistance.

3.1.2 arsB gene encodes protein acting as the arsenite efflux pump

Removal of As(III) from the cell interior is the most straightforward way toward As resistance. ArsB is an $As(OH)_3/H+$ antiporter that extrudes As(III) conferring As(III) resistance (Meng et al., 2004). Majority of prokaryote has membrane transport proteins (ArsB) catalyzing the As(III) efflux from the cytosol (Yang et al., 2012).

3.1.3 arsC gene encodes for arsenate reductase

ArsC is an arsenate reductase that converts As(V) to As(III). Organisms resistant to As(III) now can extrude As(III) from the cell utilizing the ArsB pump. This physiology extends the range of resistance to As(V) utilizing the same extrusion physiology targeted to remove As(III) (Mukhopadhyay and Rosen, 2002). Three families of reductase are reported till now. The *Staphylococcus aureus* plasmid pI258 reported to encode *arsC* gene product was found to be homologous to low molecular weight protein phosphotyrosine phosphatase. The other two are *Escherichia coli* plasmid R773 *arsC* arsenate reductase and *Saccharomyces cerevisae* Acr2p arsenate reductase, the only known eukaryotic arsenate reductase. Acr2p is homologous to cdc25a cell cycle protein tyrosine phosphatase and to Rhodanase-a thiosulfate sulfurtransferase.

3.1.4 arsA gene encodes for the function of arsenite active ATPase

ArsA is a 583-amino acid ATPase architecturally containing two pseudo-symmetric halves named A1 and A2 having consensus nucleotide-binding domains (NBDs) at their interface (Zhou et al., 2000) and connected together by a short linker. The metalloid binding domain (MBD) binds As(III) (Zhou et al., 2000, 2001) using conserved cystine residues (Ruan et al., 2006). Nucleotide binding at the NBDs stimulates metalloid binding which further stimulates ATP hydrolysis. ArsA and ArsB forms the ArsAB complex, a pump utilizing the energy of ATP hydrolysis for As(III) extrusion (Dey et al., 1994). The ArsAB complex serves the more efficient As(III) extrusion system in organisms possessing *arsRDABC* operon compared with those utilizing *arsRBC* operon.

3.1.5 arsD gene product plays the role of ars operon trans-acting repressor

Both prokaryotes and eukaryotes have metallochaperones that sequester metals in the cytoplasm, buffering their concentration, ArsD is such a metallochaperone that transfers As(III) to ArsAB As(III)-translocating ATPase (Lin et al., 2006). The *arsD* gene belongs to the five gene *ars* operon which are relatively rare, only 14 such operon have been found till now. It has been observed that ArsD increases ArsA's affinity for As(III). ArsD, encoded by *arsD* gene, is a homodimer of two 120-residue subunits and is a weak As(III)-responsive transcriptional repressor (Chen and Rosen, 1997). ArsD has three conserved cysteine residues—Cys12, Cys13 and Cys18 (Lin et al., 2007; Yang et al., 2010), forming a high affinity As(III) binding site required for the delivery of As(III). Thus, the affinity of ArsA for As(III) is increased producing increased efflux and resistance.

3.1.6 arsM arsenite S-adenosylmethyltransferase belongs to the arsM family

Methylation of environmental As by conversion to soluble and gaseous methylated species is a detoxifying process that contributes to the global cycling of As. In the organization of As resistance operon, there are subsets of these genes called *arsM* and their protein product ArsM, abbreviated for As(III) S-adenosyl methyl transferase. The uniqueness of these *arsM* genes among other homologues is that they are each downstream of the regulatory *arsR* gene, encoding the As responsive transcriptional repressor that controls expression of *ars* operon (Bhattacharjee et al., 1995; Bhattacharjee and Rosen, 2007) suggesting that these ArsMs evolved to confer As resistance.

4. Arsenic resistance as geologically ancient adaptation

4.1 Arsenic in the ancient life

The speculation of ancient life with unconventional biochemistry unrelated to ours is no doubt intriguing (Davies et al., 2009). Early life utilized hydrogen, carbon, nitrogen, oxygen, phosphorus, and sulfur. Microbes that use As(V) in place of phosphate constitute a shadow life, i.e., organisms relying on unconventional biochemistry based on As (Wolfe-Simon et al., 2010). Though this idea has been argued later (Rosen et al., 2011) and is still being considered as speculation, as there is little support found for the existence of organisms utilizing arsenic biochemistry.

Early life likely evolved in an environment rich in trivalent inorganic As(III) given that the atmosphere did not contain free oxygen for at least another billion years. As a consequence, early life form, mostly Archean ancestors, may have developed enzymes that oxidized As(III) to capture energy (Lebrun et al., 2003). Related enzymes that capture the reducing potential of As(V) by dissimilation may have evolved subsequently, recycling the arsenate to arsenite (van Lis et al., 2013). Extant microorganisms used to catalyze both types of reactions creating an arsenic redox cycle discussed in following sections.

4.2 Evolution of the arsenic resistance (ars) operon

The earliest ecosystems existed in an anoxic world ruled by anaerobic metabolism (Canfield et al., 2006). The first organisms on the planet Earth have been speculated to evolve approximately 3.5–3.8 billion years ago. It has been proposed that life evolved in an environment rich in As(III), since the ancient atmosphere did not contain oxygen for at least one billion years required for oxidation reactions. Life if originated in anoxic, metal-confluent waters of hot springs, then the resistance to As in the prevailing anoxic reducing environment would have been essential to the evolution of early life forms (Rensing et al., 1999; Rosen, 1999). Thus, the evolution of As detoxification strategies were of biological certainty. Furthermore, the archean ancestors must have developed mechanisms to avert As toxicity quite early. The chromosomal metal resistance genes are likely the precursors of genes of plasmid (Carlin et al., 1995). Study of the arsenic resistance (*ars*) operon located on the chromosomes of archaea and bacteria may therefore provide an insight to the evolutionary origin of the As-resistance genes. High concentration of As in primordial earth was believed to be linked with early means of energy generation on the ancient earth (Oremland et al., 2009). Depletion of volatile elements from the juvenile earth crust shaped the contemporary lower concentration of As (Kargel and Lewis, 1993). During the cooling and differentiation phase of earth, As was significantly concentrated within its core and mantle due to the sinking of these denser metal(loid)-sulfides. Arsenic is again brought back to the earth's surface by volcanism, during Archean (~3.8 Ga) times (Oremland et al., 2009), thus creating a biochemical opportunity for the early emergent life to explore As.

Many organisms, living in As rich ecological niche, has one or more As detoxification pathways (Bhattacharjee and Rosen, 2007) as their evolutionary legacy. In bacteria and archaea, being closer to the direct line with the last unified common ancestor (LUCA), As resistance genes are organized in operons conferring As tolerance. The most universal and well-characterized As resistance mechanism is conferred by the *ars* operon (Oremland and Stolz, 2003). Nearly all bacteria and archaea have As-resistance (*ars*) operons conferring resistance to different species of As (Rosen, 1999). The content and organization of *ars* operons exhibit great diversity and complexity, although can be unified by the ability of As resistance conferred by them. To date more than 400 *ars* operons have been identified (Ajees et al., 2011) which is also indicative of widespread occurrence of As in the environment.

The presence of arsenic resistance (*ars*) operons in the genome indicates primarily that *ars* genes are ancient in origin. Environmental ubiquity of As in present days provides the selective pressure that maintains these genes in present-day organisms. The content and organization of the *ars* system vary greatly between strains. More than one operon can be found in single organism as part of the duplication process. Most core genes in *ars* operon contain *arsR*, *arsB*, and *arsC* genes constituting the most common organization of the *ars* cluster, the *arsRBC* operon. Other genes such as *arsA*, *arsD*, *arsT*, *arsX*, *arsH*, and *arsN* have also been reported (Wang et al., 2016) with various functions related to the efficacy of the resistance phenotype. Arsenic resistance operons have the basic architecture of genes encoding for ArsR (As(III)-sensitive regulator) and ArsB (passive As(III)-tranporter protein). ArsB homologues across various taxonomic lineages indicated a clear relationship supporting the hypothesis that As resistance evolved early in the history of life (Gihring et al., 2003). The complexity of the *ars* system in diverse bacteria raises the question of its origin and evolution.

(a) *arsR*: The repressor protein of the *ars* operon, ArsR encoding *arsR* genes is thought to be the most ancient to have evolved. It is the common candidate in both the *arsRBC* and *arsRDABC* operons probably due to the functional role it plays in the *ars* operon today. Structural analysis often indicate the phenomenon of evolutionary convergence as the explanation of the evolution of different ArsR repressors probably as the result of high evolutionary selection pressure existing in As rich environment (Shi et al., 1996; Qin et al., 2007).

(b) *arsB*: ArsB is a secondary efflux protein coupled to the proton-motive force conferring As(III) resistance. ArsB associates with the ArsA ATPase to form a pump that confers high-level resistance. As ArsB is sufficient to provide resistance to As(III) to primitive cell, it has been speculated that the ancestral resistant life form was composed of only an *arsB* gene. Ancestral As resistance probably evolved with the formation of two gene operon comprising the essential As(III) pump, ArsB needed to detoxify cell anoxic atmosphere where As(III) prevailed and ArsR acting as the regulator of operon. A two-gene *arsRB* operon still exists today in *Bacillus subtilis* which can be considered as reminiscent of the ancestral function.

(c) *arsC*: It is thought that arsenate reductase arose independently three times by evolutionary convergence which shows the essential role of this protein in As rich environment. It seems that the horizontal gene transfer played a major role in the origin and evolution of *arsC* genes in the early times (Wang et al., 2016) along with the phenomenon of evolutionary convergence. Arsenate reductase is unconventional among well studied enzyme classes in that there is not a single family of evolutionarily related sequences which eventually rules out the duplication and divergence as the underlying evolutionary mechanism. There are several independent families and from the evolutionary standpoint may be considered as clades rather than trees. The gene *arsC* encoding the enzyme arsenate reductase have originated several times indicating the strong selection pressure on microbes for these functions.

(d) *arsA*: Phylogenetic analyses of ArsA-related proteins suggest unique evolutionary lineages for these proteins offering insights into the formation of the *arsA* gene (Gihring et al., 2003). It is probably associated with the later arsenic resistant life forms as the *arsA* gene act as an ATP driven pump that highly elevates efficiency of ArsB as As(III) transporter.

(e) *arsD*: The three conserved sites of ArsD is similar in nature but not sequence to the ArsA As(III) binding site (Bhattacharjee et al., 1995). Arsenic-rich environment provides bacterial and archaeal *ars* operons containing *arsD* gene a survival advantage. It has been noted that structurally the *arsA* and *arsD* are nearly always in tandem with each other. This tandem genetic organization of *arsA* and *arsD* genes might be a result of co-evolution of these genes leading to subsequent development of a genetic unit, the *ars* operon, or might be speculated to exhibit linked biochemical functions in As detoxification (Ajees et al., 2011). A number of lines of evidence support this proposition. The sole cause of the evolution in the first place is yet to be known.

Different evolutionary theories put forward that the efflux pump protein (ArsB) and As(V) reductase (ArsC) may have evolved through convergent evolution as evidenced by sequence analysis (Mukhopadhyay et al., 2002; Wang et al., 2016). Parsimonious evolutionary principle indicates the probable existence of the common ancestor with oxyanion binding site as all the three kinds of As metabolizing enzymes—arsenate reductases, phosphatases, and sulfurtransferases utilize anions as their substrate for the possible exploitation of As modification and therefore optimum As utilization. It warrants further study to resolve whether the common ancestor was contemporary to the LUCA or not.

Geologically, As(V) reduction could arise only after sufficient As(V) became available, presumably, after aerobic oxidation of As(III). The utilization of As(III) as the electron donor in anoxygenic phototrophy, however, provides a mechanism for generating As(V) in the absence of oxygen prevailing atmosphere. Once the primordial atmosphere became oxidizing, the majority of aqueous As(III) started to get oxidized to As(V). This was an evolutionarily crisis period for a variety of organisms already adapted to detoxify As(III) in the previously prevailed reducing atmosphere. These organisms could detoxify As(III) but not As(V), leading to the purifying selection in favor of the life forms that had already evolved strategies to handle inorganic As(V). They evolved biochemical circuitry as the adaptation that can reduce intracellular As(V) to more toxic As(III), for which they already had mechanisms for cytosolic removal, leading toward the emergence of three families of small As(V) reductase enzymes for detoxifying As(V) (Mukhopadhyay and Rosen, 2002). Thus, As appeared to play an important factor during the evolution of early life forms. Contemporary As selective pressure comes from natural sources such as volcanic activities and anthropogenic activities exemplified by mining coal burning and other human activities.

Rosen (1999) hypothesized that the origin and diversification of As resistance genes involved series of gradual steps conferring incremental survival advantages resulting in the present day *ars* operon. Therefore, the *ars* operon experiences positive selection in relation to the survival and therefore the reproductive fitness of the possessing life forms. It was assumed that shortly after the evolution of the *arsB* gene, *arsR* joined to form the *arsRB* operon (Gihring et al., 2003). Following the development of *arsRB* operon, it was thought that *arsC* was added to confer resistance to both As(III) and As(V). Finally, it might be hypothesized that the evolution of *arsRDABC* operon due to the addition of *arsDA* genes to the pre-existing *arsRBC* operon might have conferred elevated As resistance in gram positive bacteria (Rosen, 1999).

It has been suggested that the horizontal gene transfer events of *ars* genes may be common in nature (Wang et al., 2016). There still remain many unanswered questions relating to the nature of geological environments that first led to the origin and progressive diversification of As resistance operon. Whether the proteins conferring As resistance in microbes, evolved for the purpose of As detoxification or primarily evolved for other function and later evolutionarily adopted for As detoxification is yet to be fully understood. The insights gained from the study of molecular circuitry of the arsenic resistance bacteria improve our understanding of the flexible adaptation of microorganisms to resist As (Wang et al., 2016).

5. Intervention of As contamination by microbial adaptation – the bioremediation approach

Decontamination of arsenic from environment is of great significance to local agriculture and the population elsewhere in the As-affected area. The conventional techniques for the decontamination of As includes chemical precipitation, chemical redox reactions, ion exchange, filtration, and reverse osmosis (Malik, 2004). The disadvantage of these techniques includes less accuracy, particularly, in very low As concentration (Chaalal et al., 2005) and secondary environmental pollution due to the chemicals used in the remediation process. The cost that is involved restricts the utilization of the prevailing techniques. In recent years, bioremediation of heavy metals using microorganisms has gained attention. Microorganisms play major roles in the biochemical cycle of arsenic and can convert to different oxidation states with different solubility and mobility, therefore influencing the toxicity (Silver and Phung, 2005). Certain microorganisms in nature have evolved the needed genetic components that provide resistance mechanisms which enable them to survive and grow in an environment containing toxic levels of arsenic. The *ars* operon located on genomes of prokaryotes which confers As resistance is well characterized (Xu et al., 1998). The microbial As detoxification involves the reduction of As(V) to As (III) via cytoplasmic arsenate reductase (*arsC*) and further, As (III) is extruded by a membrane associated with *arsB* efflux pump. Other genes such as *arsR*, *arsD*, and *arsA* form part of *ars* operon along with *arsB* and *arsC* in most prokaryotes (Rosen, 2002). Considering the threat of As to human health, the future challenge is to remove this toxic metalloid from our habitable ecological niche. The myriad arrays of As resistant adaptations in contemporary life forms are the evolutionary tools for the sustainable environmental As decontamination. Furthermore, deeper investigations for linking As resistant properties of organisms with respect to their life history and the environmental issues leading toward As decontamination are essential for successful bioremediation.

References

Acharyya S.K., Chakraborty P., Lahiri S., Raymahashay B.C., Guha S., Bhowmik A. Arsenic poisoning in the Ganges delta. Nature 401, 545–546, 1999.

Afkar E., Lisak J., Saltikov C., Basu P., Oremland R.S., Stolz J.F. The respiratory arsenate reductase from *Bacillus selenitireducens* strain MLS10. FEMS Microbiol. Lett. 226, 107–112, 2003.

Ahmann D., Roberts A.L., Krumholz L.R., Morel F.M. Microbe grows by reducing arsenic. Nature 371, 750–750, 1994.

Ajees A.A., Yang J., Rosen B.P. The ArsD As(III) metallochaperone. Biometals 24, 391–399, 2011.

Ajees A.A., Marapakala K., Packianathan C., Sankaran B., Rosen B.P. Structure of an As(III) S-adenosylmethionine methyltransferase: insights into the mechanism of arsenic biotransformation. Biochemistry 51, 5476–5485, 2012.

Anderson G.L., Williams J., Hille R. The purification and characterization of arsenite oxidase from *Alcaligenes faecalis*, a molybdenum-containing hydroxylase. J. Biol. Chem. 267, 23674–23682, 1992.

Bentley R., Chasteen T.G. Microbial methylation of metalloids: arsenic, antimony, and bismuth. Microbiol. Mol. Biol. Rev. 66, 250–271, 2002.

Bhattacharjee H., Li J., Ksenzenko M.Y., Rosen B.P. Role of cysteinyl residues in metalloactivation of the oxyanion-translocating ArsA ATPase. J. Biol. Chem. 270, 11245–11250, 1995.

Bhattacharjee H., Mukhopadhyay R., Thiyagarajan S., Rosen B.P. Aquaglyceroporins: ancient channels for metalloids. J. Biol. 7, 33–39, 2008.

Bhattacharjee H., Rosen B.P. Arsenic metabolism in prokaryotic and eukaryotic microbes. *In*: Nies D.H., Silver S. (eds.). Molecular Microbiology of Heavy Metals. Springer, Berlin, pp. 371–406, 2007.

Bowell R.J., Parshley J. Arsenic cycling in the mining environment. Characterization of waste, chemistry, and treatment and disposal, proceedings and summary report on U.S. EPA workshop on managing arsenic risks to the environment, Denver, Colorado, 2001.

Boyle R.W., Jonasson I.R. The geochemistry of As and its use as an indicator element in geochemical prospecting. J. Geochem. Explor. 2, 251–296, 1973.

Canfield D.E., Rosing M.T., Bjerrum C. Early anaerobic metabolisms. Philos. Trans. R. Soc. B 361, 819–834, 2006.

Carlin A., Shi W., Dey S., Rosen B.P. The *ars* operon of *Escherichia coli* confers arsenical and antimonial resistance. J. Bacteriol. 177, 981–986, 1995.

Caussy D. Case studies of the impact of understanding bioavailability: arsenic. Ecotoxicol. Environ. Saf. 56, 164–173, 2003.

Chaalal O., Zekri A.Y., Islam R. Uptake of heavy metals by microorganisms: an experimental approach. Energ. Source 27, 87–100, 2005.

Chakraborti D., Das B., Rahman M.M., Chowdhury U.K., Biswas B., Goswami A.B., Nayak B., Pal A., Sengupta M.K., Ahamed S., Hossain A., Basu G., Roychowdhury T., Das D. Status of groundwater arsenic contamination in the state of West Bengal, India: a 20-year study report. Mol. Nutr. Food. Res. 53, 542–551, 2009.

Chen Y., Rosen B.P. Metalloregulatory properties of the ArsD repressor. J. Biol. Chem. 272, 14257–14262, 1997.

Chou W.C., Hawkins A.L., Barrett J.F., Griffin C.A., Dang C.V. Arsenic inhibition of telomerase transcription leads to genetic instability. J. Clin. Invest. 108, 1541–1547, 2001.

Cullen W.R., Reimer K.J. Environmental arsenic chemistry. Chem. Rev. 89, 713–764, 1989.

Davies P.C., Benner S.A., Cleland C.E., Lineweaver C.H., McKay C.P., Wolfe-Simon F. Signatures of a shadow biosphere. Astrobiology 9, 241–249, 2009.

Dembitsky V.M., Levitsky D.O. Arsenolipids. Prog. Lipid. Res. 43, 403–448, 2004.

Dey S., Dou D., Rosen B.P. ATP-dependent arsenite transport in everted membrane vesicles of *Escherichia coli*. J. Biol. Chem. 269, 25442–25446, 1994.

Duval S., Ducluzeau A.L., Nitschke W., Schoepp-Cothenet B. Enzyme phylogenies as markers for the oxidation state of the environment: the case of respiratory arsenate reductase and related enzymes. BMC Evol. Biol. 8, 206–219, 2008.

Fendorf S., Herbel M.J., Tufano K.J., Kocar B.D. Biogeochemical processes controlling the cycling of arsenic in soils and sediments. *In*: Violante A., Huang P.M., Gadd G.M. (eds.). Biophysico-Chemical Processes of Heavy Metals and Metalloids in Soil Environments. Wiley, Hoboken, pp. 313–338, 2008.

Fendorf S., Michael H.A., van Geen A. Spatial and temporal variations of groundwater arsenic in South and Southeast Asia. Science 328, 1123–1127, 2010.

Gihring T.M., Bond P.L., Peters S.C., Banfield J.F. Arsenic resistance in the archaeon "*Ferroplasma acidarmanus*": new insights into the structure and evolution of the *ars* genes. Extremophiles 7, 123–130, 2003.

Gladysheva T.B., Oden K.L., Rosen B.P. Properties of the arsenate reductase of plasmid R773. Biochemistry 33, 7288–7293, 1994.

Green H.H. Description of a bacterium which oxidizes arsenite to arsenate, and one of which reduces arsenate to arsenite, isolated from a cattle-dipping tank. S. Afr. J. Sci. 14, 465–467, 1918.

IPCS. Environmental Health Criteria 224. Arsenic and Arsenic Compounds, 2nd Ed. World Health Organization, Geneva. 2001.

Ji G., Garber E.A.E., Armes L.G., Chen C.M., Fuchs J.A., Silver S. Arsenate reductase of *Staphylococcus aureus* plasmid pI258. Biochemistry 33, 7294–7299, 1994.

Kargel J.S., Lewis J.S. The composition and early evolution of Earth. Icarus 105, 1–25, 1993.

Krafft T., Macy J.M. Purification and characterization of the respiratory arsenate reductase of *Chrysiogenes arsenatis*. Eur. J. Biochem. 255, 647–653, 1998.

Kulp T.R., Hoeft S.E., Asao M., Madigan M.T., Hollibaugh J.T., Fisher J.C., Stolz J.F., Culbertson C.W., Miller L.G., Oremland R.S. Arsenic(III) fuels anoxygenic photosynthesis in hot spring biofilms from Mono Lake, California. Science 321, 967–970, 2008.

Kulp T.R., Hoeft S.E., Miller L.G., Saltikov C., Murphy J.N., Han S., Lanoil B., Oremland R.S. Dissimilatory arsenate and sulfate reduction in sediments of two hypersaline, arsenic-rich soda lakes: Mono and Searles Lakes, California. Appl. Environ. Microbiol. 72, 6514–6526, 2006.

Lebrun E., Brugna M., Baymann F., Muller D., Lievremont D., Lett M.C., Nitschke W. Arsenite oxidase, an ancient bioenergetic enzyme. Mol. Biol. Evol. 20, 686–693, 2003.

Lin Y.F., Walmsley A.R., Rosen B.P. An arsenic metallochaperone for an arsenic detoxification pump. Proc. Natl. Acad. Sci. 103, 15617–15622, 2006.

Lin Y.F., Yang J., Rosen B.P. ArsD residues Cys12, Cys13, and Cys18 form an As(III)-binding site required for arsenic metallochaperone activity. J. Biol. Chem. 282, 16783–16791, 2007.

Ma L.Q., Komar K.M., Tu C., Zhang W., Cai Y., Kennelley E.D. A fern that hyperaccumulates arsenic. Nature 409, 579–579, 2001.

Malasarn D., Saltikov C.W., Campbell K.M., Santini J.M., Hering J.G., Newman D.K. arrA is a reliable marker for As(V) respiration. Science 306, 455–455, 2004.

Malik A. Metal bioremediation through growing cell. Environ. Int. 30, 261–278, 2004.

Mandal B.K., Suzuki K.T. Arsenic round the world: a review. Talanta 58, 201–235, 2002.

Meng Y.L., Liu Z., Rosen B.P. As(III) and Sb(III) uptake by GlpF and efflux by ArsB in *Escherichia coli*. J. Biol. Chem. 279, 18334–18341, 2004.

Messens J., Silver S. Arsenate reduction: thiol cascade chemistry with convergent evolution. J. Mol. Biol. 362, 1–17, 2006.

Mukhopadhyay R., Rosen B.P. Arsenate reductases in prokaryotes and eukaryotes. Environ. Health. Perspect. 110, 745–748, 2002.

Mukhopadhyay R., Rosen B.P., Phung L.T., Silver S. Microbial arsenic: from geocycles to genes and enzymes. FEMS Microbiol. Rev. 26, 311–325, 2002.

Newman D.K., Ahmann D., Morel F.M.M. A brief review of microbial arsenate respiration. Geomicrobiol. J. 15, 255–268, 1998.

Nickson R., McArthur J., Burgess W., Ahmed K.M., Ravenscroft P., Rahmanñ M. Arsenic poisoning of Bangladesh groundwater. Nature 395, 338–338, 1998.

Oden K.L., Gladysheva T.B., Rosen B.P. Arsenate reduction mediated by the plasmid-encoded ArsC protein is coupled to glutathione. Mol. Microbiol. 12, 301–306, 1994.

Oremland R.S., Saltikov C.W., Wolfe-Simon F., Stolz J.F. Arsenic in the evolution of Earth and extraterrestrial ecosystems. Geomicrobiol. J. 26, 522–536, 2009.

Oremland R.S., Stolz J.F. Arsenic, microbes and contaminated aquifers. Trends. Microbiol. 13, 45–49, 2005.

Oremland R.S., Stolz J.F. The ecology of arsenic. Science 300, 939–944, 2003.

Qin J., Rosen B.P., Zhang Y., Wang G., Franke S., Rensing C. Arsenic detoxification and evolution of trimethylarsine gas by a microbial arsenite S-adenosylmethionine methyltransferase. Proc. Natl. Acad. Sci. 103, 2075–2080, 2006.

Ravenscroft P., Brammer H., Richards K. Arsenic Pollution: A Global Synthesis. John Wiley & Sons, West Sussex, UK, 2009.

Rensing C., Ghosh M., Rosen B. Families of soft-metal-ion-transporting ATPases. Bacteriology 181, 5891–5897, 1999.

Richey C., Chovanec P., Hoeft S.E., Oremland R.S., Basu P., Stolz J.F. Respiratory arsenate reductase as a bidirectional enzyme. Biochem. Biophys. Res. Commun. 382, 298–302, 2009.

Rosen B.P. Biochemistry of arsenic detoxification. FEBS Lett. 529, 86–92, 2002.

Rosen B.P. Families of arsenic transporters. Trends. Microbiol. 7, 207–212, 1999.

Rosen B.P., Ajees A.A., McDermott T.R. Life and death with arsenic. BioEssays 33, 350–357, 2011.

Ruan X., Bhattacharjee H., Rosen B.P. Cys-113 and Cys-422 form a high affinity metalloid binding site in the ArsA ATPase. J. Biol. Chem. 281, 9925–9934, 2006.

Saltikov C.W., Newman D.K. Genetic identification of a respiratory arsenate reductase. Proc. Natl. Acad. Sci. 100, 10983–10988, 2003.

Shi H., Shi X., Liu KJ. Oxidative mechanism of arsenic toxicity and carcinogenesis. Mol. Cell. Biochem. 255, 67–78, 2004.

Shi W., Dong J., Scott R.A., Ksenzenko M.Y., Rosen B.P. The role of arsenic-thiol interactions in metalloregulation of the *ars* operon. J. Biol. Chem. 271, 9291–9297, 1996.

Shi W., Wu J., Rosen B.P. Identification of a putative metal binding site in a new family of metalloregulatory proteins. J. Biol. Chem. 269, 19826–19829, 1994.

Silver S., Phung T. Genes and enzymes involved in bacterial oxidation and reduction of inorganic arsenic. Appl. Environ. Microbiol. 71, 599–608, 2005.

Smedley P.L., Kinniburgh D.G. A review of the source, behaviour and distribution of arsenic in natural waters. Appl. Geochem. 17, 517–568, 2002.

Smith E., Naidu R., Alston A.M. Arsenic in the soil environment: a review. Adv. Agron. 64, 149–195, 1998.

Stolz J.F. Overview of microbial arsenic metabolism and resistance. *In:* Santini J.M., Ward S.A. (eds.). The Metabolism of Arsenite. CRC Press, London, UK, pp. 55–60, 2012.

Stolz J.F., Basu P., Oremland R.S. Microbial arsenic metabolism: new twists on an old poison. Microbe 5, 53–59, 2010.

Stolz J.F., Basu P., Santini J.M., Oremland R.S. Arsenic and selenium in microbial metabolism. Annu. Rev. Microbiol. 60, 107–130, 2006.

Sun W., Sierra-Alvarez R., Hsu I., Rowlette P., Field J.A. Anoxic oxidation of arsenite linked to chemolithotrophic denitrification in continuous bioreactors. Biotechnol. Bioeng. 105, 909–917, 2010a.

Sun W., Sierra-Alvarez R., Milner L., Field J.A. Anaerobic oxidation of arsenite linked to chlorate reduction. Appl. Environ. Microbiol. 76, 6804–6811, 2010b.

Tapio S., Grosche B. Arsenic in the aetiology of cancer. Mutat. Res.-Gen. Tox. En. 612, 215–246, 2006.

van Lis R., Nitschke W., Duval S., Schoepp-Cothenet B. Arsenic as bioenergetic substrates. Biochim. Biophys. Acta. 1827, 176–188, 2013.

Wang L., Zhuang X., Zhuang G., Jinga C. Arsenic resistance strategy in *Pantoea* sp. IMH: Organization, function and evolution of *ars* genes. Sci. Rep. 6, 39195, 2016.

Wang T.S., Hsu T.Y., Chung C.H., Wang A.S.S., Bau D.T., Jan K.Y. Arsenite induces oxidative DNA adducts and DNA-protein cross-links in mammalian cells. Free. Radic. Biol. Med. 31, 321–330, 2001.

Wenzel W.W. Arsenic. *In:* Alloway B.J. (ed.), Heavy Metals in Soils: Trace Metals and Metalloids in Soils and Their Bioavailability. Springer, Netherland, Dordrecht, pp. 241–281, 2013

Wolfe-Simon F., Switzer B.J., Kulp T.R., Gordon G.W., Hoeft S.E., Pett-Ridge J., Stolz J.F., Webb S.M., Weber P.K., Davies P.C., Anbar A.D., Oremland R.S. A bacterium that can grow by using arsenic instead of phosphorus. Science 332, 1163–1166, 2010.

Wu J., Rosen B.P. Metalloregulated expression of the *ars* operon. J. Biol. Chem. 268, 52–58, 1993.

Wu J., Rosen B.P. The ArsR protein is a trans-acting regulatory protein. Mol. Microbiol. 5, 1331–1336, 1991.

Xu C., Rosen B.P. Metalloregulation of soft metal resistance pumps. *In:* Sarkar B. (ed.). Metals and Genetics. Plenum, New York, pp. 5–19, 1999.

Xu C., Zhou T.Q., Kuroda M., Rosen B.P. Metalloid resistance mechanisms in prokaryotes. J. Biochem. 123, 16–23, 1998.

Yang H.C., Fu H.L., Lin Y.F., Rosen B.P. Pathways of arsenic uptake and efflux. Curr. Top. Membr. 69, 325–358, 2012.

Yang J., Rawat S., Stemmler T.L., Rosen B.P. Arsenic binding and transfer by the ArsD As(III) metallochaperone. Biochemistry 49, 3658–3666, 2010.

Yoshida T., Yamauchi H., Fan Sun G. Chronic health effects in people exposed to arsenic via the drinking water: dose-response relationships in review. Toxicol. Appl. Pharm. 198, 243–252, 2004.

Zargar K., Conrad A., Bernick D.L., Lowe T.M., Stolc V., Hoeft S., Oremland R.S., Stolz J., Saltikov C.W. ArxA, a new clade of arsenite oxidase within the DMSO reductase family of molybdenum oxidoreductases. Environ. Microbiol. 14, 1635–1645, 2012.

Zargar K., Hoeft S., Oremland R., Saltikov C.W. Identification of a novel arsenite oxidase gene, arxA, in the haloalkaliphilic, arsenite-oxidizing bacterium *Alkalilimnicola ehrlichii* strain MLHE-1. J. Bacteriol. 192, 3755–3762, 2010.

Zhao F.J., Dunham S.J., McGrath S.P. Arsenic hyperaccumulation by different fern species. New Phytol. 156, 27–31, 2002.

Zhao F.J., McGrath S.P., Meharg A.A. Arsenic as a food chain contaminant: mechanisms of plant uptake and metabolism and mitigation strategies. Annu. Rev. Plant Biol. 61, 535–559, 2010.

Zhou T., Radaev S., Rosen B.P., Gatti D.L. Conformational changes in four regions of the *Escherichia coli* ArsA ATPase link ATP hydrolysis to ion translocation. J. Biol. Chem. 276, 30414–30422, 2001.

Zhou T., Radaev S., Rosen B.P., Gatti D.L. Structure of the ArsA ATPase: the catalytic subunit of a heavy metal resistance pump. EMBO J. 19, 4838–4845, 2000.

Zhu Y.G., Yoshinaga M., Zhao F.J., Rosen B.P. Earth abides arsenic biotransformations. Annu. Rev. Earth. Planet. Sci. 42, 443–467, 2014.

PART II
STRATEGIES OF BIOREMEDIATION

CHAPTER 5

Bioaccumulation and Biosorption of Heavy Metals

Ana Belén Segretin, Josefina Plaza Cazón* and Edgardo R. Donati

1. Introduction

Metals can enter the environment through different pathways, of which, anthropogenic sources play a significant role in increasing metal concentrations. Wastewaters from industries are often disposed directly into the rivers or other nearby water sources without any treatment procedure. This entails a high ecological risk as these water sources are usually considered to be the basis for drinking water (Vijayaraghavan and Balasubramanian, 2015). Several remediation techniques to remove metal ions from aqueous solutions are available ranging from traditional physicochemical methods to emerging biotechnological techniques as biosorption and bioaccumulation (Chojnacka, 2010). The main advantages of these biological methods are low operating costs, selectivity for specific metal remediation, minimization of the volume of chemical and biological sludge, and high efficiency in detoxifying very dilute effluents. Both processes mentioned involve interactions and concentration of toxic metals either on living (bioaccumulation) or non-living (biosorption) biomass. Two of the most notable differences between both processes are their kinetics and values of activation energy (Chojnacka, 2010; Chojnacka et al., 2005). While biosorption is a fast process independent of the presence of specific nutrients, bioaccumulation is slow and nutrient dependent. In biosorption, there is no danger of toxicity by sorbate to the sorbent but in bioaccumulation such danger exists (Chojnacka, 2010). Biosorption, also known as passive metal uptake, is the metabolism-independent uptake of metals by non-living biomass. Mechanisms of cell surface sorption are based

Centro de Investigación y Desarrollo en Fermentaciones Industriales, CINDEFI (CCT La Plata-CONICET, UNLP), Facultad de Ciencias Exactas, 50 y 115, (1900) La Plata. Argentina.
* Corresponding author: joplaca@hotmail.com

on physicochemical interactions between metal and functional groups of the cell wall (Veglio et al., 2003; Schiewer and Volesky, 2000). Conversely, bioaccumulation comprises intracellular metal accumulation which occurs in two-stages: the first, identical to biosorption which is fast and the subsequent is slower and includes transport of sorbate inside the living cells, most frequently using active transport systems. Bioaccumulation is a non-equilibrium process, more complex than biosorption itself. Several living organisms (yeast, fungi, algae, and bacteria) have been reported with the ability to capture large quantities of heavy metals (Wang and Chen, 2009). Particularly, algae have been extensively studied for heavy metal removal from wastewater due to their ubiquitous occurrence in nature. A number of microalgal strains (*Chlorella vulgaris*, *Chlorella fuscas*, *Spirogyra* species, *Spirulina* sp., *Chaetophora elegans*, *Cladophora fascicularis*, *Cladophora* sp., and *Enteromorpha* sp.), potentially suitable for heavy metal removal in aqueous solution, were used in several studies showing varying removal efficiencies (Andrade et al., 2005; Chojnacka et al., 2005; Gupta and Rastogi, 2008; Deng et al., 2007; Vogel et al., 2010; Zbikowski et al., 2007). However, comparison between the processes of bioaccumulation and biosorption of heavy metals by algae cells is rarely found on literature (Flouty and Estephane, 2012).

2. Bioaccumulation and biosorption

2.1 Bioaccumulation

As it has been already mentioned, bioaccumulation involves the cultivation of an organism in the presence of a sorbate. Cells can offer binding sites both on the cell surface and inside the cell. As part of the sorbate is transported inside the cell, binding sites present on the surface are released; therefore, additional amount of sorbate can be bound there according to the course of the equilibrium biosorption dependence. Also, the concentration of the biomass eventually increases which enables to bind even more sorbate (Chojnacka, 2010). After entering into the cell, the metal ions are compartmentalized into different subcellular organelles (e.g., mitochondria, vacuole, etc.). Vijver et al. (2004) summarizes some metal ion accumulation strategies especially internal compartmentalization strategies. Metal accumulation strategies for essential and non-essential metal ions may be different. For essential metals, limiting metal uptake or strategies with active excretion, storage in an inert form or excretion of stored metal are the main strategies. For non-essential metals, excretion of the metal excess pool and internal storage without elimination are the major strategies and the metal concentration in the cells gets higher with increasing external concentrations. There are two main cellular sequestration mechanisms: the formation of distinct inclusion bodies and the binding of metals to heat-stable proteins. The first one includes three types of granules: type A, amorphous deposits of calcium phosphates, e.g., Zn; type B, mainly containing hydrogen phosphates, accumulating, e.g., Cd,

Cu, Hg, and Ag; and type C, excess iron stored in granules as haemosiderin. The latter mechanism mainly relates to a specific metal-binding protein, metallothioneins (MT), which are low molecular weight and cysteine-rich and can be induced by many substances, including heavy metal ions such as Cd, Cu, Hg, Co, Zn, etc. (Wang, 2009). When considering the operational aspects, bioremoval by growing cells is usually performed in batch systems. Simple batch contacting for a sufficient period of time results in very low residual metal concentrations (Aksu and Dönmez, 2000). Moreover, the process of bioaccumulation does not need to include a separate biomass cultivation mode, which is a positive aspect for this technique. Also, additional unit processes are reduced: harvesting, drying, processing, and storage (Chojnacka, 2010).

Bioaccumulation organisms should be selected among species which are resistant to high loads of pollutants and do not have mechanisms which protect from excessive accumulation inside the cell. Some microorganisms produce toxic substances during their growth such as bacterial toxins, cyanotoxin, aflatoxins, ochratoxin, citrinin, and many other mycotoxins in adverse environmental conditions (Deng and Wilson, 2001; Koçberber and Dönmez, 2007). Hence, while employing growing cells for bioaccumulation of metals, careful considerations and proper choice shall be made for such strains which must be non-pathogenic and do not produce toxins. Additionally, the chosen bioaccumulant should have preferably some mechanism of intracellular binding such as special proteins rich in thiol groups—metallothioneins, phytochelatins, which are synthesized as the response to the presence of toxic metal ions in their living environment as well as the ability to complex those pollutants, thus excluding them from normal metabolic processes (Chojnacka, 2010).

If bioaccumulation is to be performed under laboratory conditions, in the first stage, the biomass should be suspended in the solution containing the sorbate. However, if heterotrophic organisms (bacteria or fungi) are intended to be used, organic carbon source should be supplied to wastewater. This is a severe limitation because wastewaters which are to be treated rarely contain any organic carbon source besides the sorbate. This concerns wastewater from metallurgical industry and supplementation of organic source is not advantageous. A solution to this problem could be the use of photosynthetic organisms: either algae or aquatic plants as their nutritive requirements are rather small and they use inorganic carbon source which could be carbon dioxide from flue gases. For example, high bioaccumulation of arsenate was found under limited phosphorus source conditions for the microalgae *Scenedesmus obliquus* (Wang, 2013). At the same time, the wide spread cyanobacteria *Microcystis aeruginosa* was reported to accumulate large quantities of both, Cd and Pb (Rzymski, 2014). Even if the use of photosynthetic microorganisms could be positive as a bioremediation strategy, they can also serve as major components of the food web chain turning into potential source of toxic metals for the aquatic organisms. It is therefore suggested that the occurrence of these microorganisms in the surface water in industrial areas can lead to multiple environmental and health hazards (including the release of toxins and incorporation of toxic metals in the food chain) and therefore its blooms should be prevented. Additionally, some papers report that at higher level of pollutants, some algae activate systems which protect from excessive accumulation resulting in low bioaccumulation capacity. The process is thus useful only at low level of pollutants (Chojnacka, 2010).

Compared to algae, bacteria usually grow faster and are able to accumulate heavy metals under a wider range of external conditions (Malik, 2004). Bacillus strains were found capable of bioaccumulating Cr^{+6}, bringing down the concentration of Cr^{+6} to 0.06 mg.L^{-1}, which is permissible when the concentration of Cr^{+6} is 50 mg.L^{-1} or below (Srinath, 2002). It was also published that *Lysinibacillus* sp. strain SS11 displayed high arsenic tolerance and showed bioaccumulation capacity of 23.43 mg.L^{-1} for arsenate and 5.65 mg.L^{-1} for arsenite. Additionally, it was observed that the brake fern *Pteris vittata* is able to take up more arsenic and iron from soil in the presence of this bacterial strain than in its absence, leading to contaminant-free soil. Thus, this symbiotic system appears to be a rapid, inexpensive, and environmentally friendly bioremediation strategy for arsenic-contaminated soils (Singh, 2015).

Sometimes wastewater's physicochemical characteristics may be restrictive when choosing a microorganism capable of tolerating these conditions. For example, high salt concentrations usually interfere with microorganisms' growth and development. However, it has been reported that some halotolerant microorganisms can be used for bioaccumulation processes. The yeasts *Zygosaccharomyces rouxii* and *Saccharomyces cerevisiae* have shown increasing effects on the heavy metal tolerance and bioaccumulation capacity with rising NaCl concentrations. For both yeasts, NaCl improved Cd and Zn tolerance. Additionally, the bioaccumulation capacities of Cu, Zn, and Fe increased after the addition of NaCl. These microorganisms might be promising for heavy metal removal in high salt environment (Li, 2013). Other fungi were also reported as potential organisms for bioremediation processes. *Aspergillus versicolor* was found to be a promising bioaccumulator of heavy metal ions in wastewater effluents. Heavy metal bioaccumulation for 50 mg.L^{-1} Cr^{+6}, Ni, and 5 for Cu ions resulted in 99.89%, 30.05%, and 29.06% removal yield respectively (Taştan, 2010).

Conclusively, the great diversity of microorganisms able to capture heavy metals in a wide range of conditions might be a very promising alternative to the design of different bioremediation strategies. Although many considerations are to be taken into account, the use of microorganisms still represents an economic and efficient alternative. Furthermore, the application of bioaccumulation processes could be used not only for removing undesirable products. In the future, it may be use in the separation of valuable biomolecules from a mixture. This would enable to reduce the number of conventionally used separation steps and would enable a single-step recovery.

2.2 Biosorption

Biosorption can be defined as the passive uptake of pollutants by dead or inactive biological materials through different physicochemical mechanisms (Vijayaraghavan and Yun, 2008). Mechanisms of metal removal usually include physical adsorption, ion exchange, chelation, complexation, and micro-precipitation (Veglio et al., 2003; Abdolali et al., 2014). Since biosorption involves a variety of metabolism-independent processes taking place essentially in the cell wall, the mechanisms responsible for the metal binding differ according to the biomass type. Biosorbents that are commonly used for the removal of metal ions include algae (fresh and marine), fungi, bacteria, industrial wastes, agricultural wastes, and other polysaccharide materials. When choosing the biomass for metal biosorption, its origin is a major factor to be taken into account. Several bacterial, fungal, and yeast biomasses were found to have an excellent biosorption capacity (Vijayaraghavan and Yun, 2008) due to their cell wall composition. However, most of these biomasses were also cultivated and used to examine their biosorption potential (Wang and Chen, 2009; Vijayaraghavan and Yun, 2008). Apart from increasing costs, cultivation increases the uncertainty in maintaining a continuous supply of biomass for the process. Another alternative is to use microbial wastes generated by food/fermentation and pharmaceutical industries. The possible reuse of these wastes for another process (biosorption) might be environmentally benign and also generate additional incomes for the industries. Some studies explored the possibility of employing microbial wastes of several industries with good success in metal biosorption. This type of biomass includes *Saccharomyces cerevisiae* (Ramirez Carmona et al., 2012), *Corynebacterium glutamicum* (Vijayaraghavan et al., 2008), *Trametes versicolor* (Song et al., 2015), and *Streptomyces rimosus* (Selatnia et al., 2004). Another important heavy metal biosorbent is marine algae, otherwise known as seaweeds. Seaweeds are biological resources and are available in many parts of the world. They are present in abundance and grow at a fast pace. Thus, utilizing seaweeds as biosorbents can be beneficial to local economies. Apart from being rigid, seaweeds are proved to be an excellent biosorbent for different metal ions (Romera et al., 2007). In particular, brown seaweeds are established biosorbents due to their alginate content (Davis et al., 2003). Apart from this, their macroscopic structure offers a convenient basis for the production of biosorbent particles suitable for sorption process applications (Vijayaraghavan et al., 2005). However, it should be noted that seaweeds are not regarded as wastes; in fact, seaweeds are the only source for the production of agar, alginate, and carrageenan. Therefore, some care should be taken while selecting seaweeds for biosorption process. Several investigators used low-cost industrial and agricultural wastes for heavy metal biosorption (Crini, 2005; Mahajan and Sud, 2013; Abdolali et al., 2014). Among these wastes, crab shell (Cadogan et al., 2014), activated sludge (Hammaini et al., 2007; Sulaymon et al., 2014), and rice husk (Chuah et al., 2005) deserves particular attention. Apart from being cheap and available for sustainable biosorbent production, these wastes generated from various industrial and agricultural activities are found to possess excellent biosorption capacity and reasonable rigidity.

The performance of a biosorbent does not only depend on the chemical composition of the biosorbent and the nature of the solutes but it is also strongly influenced by operational parameters such as pH, temperature, ionic strength, co-ion concentration, sorbent size, reaction time, sorbent dosage, and initial solute concentration. Among the different operational parameters, pH is the most important one, significantly influencing the biosorbent characteristics and solution chemistry. The binding site/functional group of a biosorbent, which plays a vital role in biosorption, strongly depend on the solution pH. Most biosorbents, irrespective of its nature, were found to be influenced by solution pH.

For instance, in the case of brown seaweeds, a maximum biosorption always occurs in the range of pH 3–5 for almost all metal cations (Davis et al., 2003). This is because of the negatively charged carboxyl groups (pKa = 3.5–5.5) which are responsible for binding metal cations through the ion-exchange mechanism (Davis et al., 2003). For metal anions, strong acidic pH is often required to protonate the functional groups to increase binding capacity (Niu et al., 2007). The solution pH also affects the solution chemistry of metals. At a higher solution pH, the solubility of most metal complexes decrease due to the precipitation and this may complicate the sorption process. At a lower solution pH, most of the cationic metals exist in a stable state and are easier to be adsorbed. Overall, to enhance the biosorption capacity of the particular biosorbent, the optimum pH should be found.

The presence of co-ions strongly influences the removal capacity of the biosorbent towards a particular solute. Since biosorption is a passive process in which several chemical groups or chemical components of the same biomass play a vital role in metal biosorption, one can expect a complicated interaction in the presence of many ions. There were reports which indicated that the presence of light metal ions and anions affected the removal efficiency of biosorbents (Vijayaraghavan et al., 2011). The presence of anions can lead to the formation of complexes that have a lower affinity to the sorbent than free metal ions (Schiewer and Volesky, 2000). Several research reports identified that in a multi-component system, a strong competition prevails among metallic species in occupying the binding sites (Baig et al., 2009; Vijayaraghavan and Balasubramanian, 2010).

Important reasons for competition among species include the nature and number of binding components of biomass as well as the nature and concentration of metal ions. Each functional group has a particular preference toward some metal ions and this affinity toward metal ions generally depends on ionic radio, electronegativity, and atomic mass of metal ions (Vijayaraghavan and Balasubramanian, 2010). Furthermore, it should be noted that the initial concentration of metal usually plays a major role on the uptake capacity of the biomass in the multi-component systems. Since most biosorbents possess only limited active binding sites, a reduced biosorption capacity toward a particular ion in multi-component systems compared to single-component is expected.

Another aspect to study in biosorption field is the desorption process. Reuse potential of spent biomass is an important criterion for selection of any biosorbent. The possibility of biomass regeneration decreases the overall process cost and the dependency of the process on the continuous supply of the biomass. The success of a desorption process depends on the mode of removal mechanisms and the mechanical stability of the biomass. Considering that most biosorbents exhibit an ion-exchange mechanism for cationic heavy metal ions, a mild to strong acidic condition is sufficient for desorption. Using acids for desorption is also beneficial because acidic solutions are common wastes in almost all industries and if biosorbents are employed in industrial wastewater schemes, these acidic solutions can be used to regenerate biosorbents. However, in many cases, the integrity of biomass is affected by the acidic environment. Chemical agents as EDTA (Oyetibo et al., 2014) and $CaCl_2$ (Davis et al., 2000) have been effective and non-detrimental for the biomasses.

2.2.1 Biosorption kinetics

The biosorption kinetics plays an important role in selection and design of reactor systems as well as operations. Since heavy metal biosorption is metabolism-independent, it typically occurs rapidly, in particular for uptake of cationic metal ions. Most of cationic metal uptake takes place within the first 20–60 min followed by a relatively slow uptake process. The adsorption equilibrium for cationic heavy metal ions usually can be reached within 2–6 h (Ibrahim, 2011; Apiratikul et al., 2011; Vijayaraghavan and Yun, 2008), which is much faster than adsorption on activated carbons and metal oxide/hydroxide-typed adsorbents.

However, biosorption for uptake of anionic contaminants (e.g., hexavalent chromium) is much lower than that of cationic contaminants. Typically, reaching the biosorption equilibrium takes several hours to few days. For example, it was reported that the complete uptake of hexavalent chromium was achieved in 20 h when a chemically modified *Sargassum* sp. was used (Yang and Chen, 2008).

A few kinetic models have been employed to describe the adsorption kinetics (Mahajan and Sud, 2013). Among these models, pseudo-first order model and pseudo-second order model are mostly used to describe the adsorption kinetics. The mathematical equations of the pseudo-first- and second-order rate models are expressed as follows:

Pseudo First Order

$$\log(q_{eq} - q_t) = \log q_{eq} - \frac{-K_1}{2.303} t \qquad (1)$$

Pseudo Second Order

$$\frac{t}{q_t} = \frac{1}{K_2 q_{eq}^2} + \frac{1}{q_{eq}} \qquad (2)$$

where, q_{eq} and q_t: adsorption capacity at the equilibrium and time t respectively; K_1 and K_2 are the pseudo first and pseudo second order rate constants, respectively.

The kinetics model fitting curves and comparison of experimental and calculated q_e values can be used to determine the suitable kinetics model. In addition, the obtained correlation coefficient of R^2 values can help to decide the suitable model. The high R^2 value would indicate the suitable kinetics model to describe the adsorption kinetics.

Being the best theoretical model, the intraparticle diffusion model is also employed to describe the adsorption process:

Intraparticle Diffusion

$$q_t = K_{dif} t^{1/2} + \qquad (3)$$

where, K_{dif} is the intraparticle diffusion rate constant.

The parameter values of the above models are normally affected by many factors including the properties of the sorbent and solution and physical parameters (e.g., stirring speed and adsorbent size).

2.2.2 Biosorption equilibrium

Biosorption equilibrium is highly dependent upon the water chemistry and the nature of heavy metal ions and the biosorbents. Higher cationic metal uptake occurs when pH is higher (e.g., above 4–6). However, better removal for anionic heavy ions can be obtained at lower pH. Ionic strength plays an important role in the biosorption. Higher ionic strength would lead to lower biosorption of heavy metals due to competitive sorption between light metals (represented by ionic strength) and heavy metals for the functional groups. The biosorption isotherm models are extensively used to evaluate the maximum biosorption capacity, the concentration of treated effluent, and a few other engineering parameters. The distribution of metal ions in the bulk solution and on the biomass can be described by one or more isotherms such as Langmuir model, Freundlich model, and Dubinin-Radushkevich (D-R) model. Among them, Langmuir model and Freundlich model are the most commonly used for the description of isothermal biosorption. Langmuir model assumes that the sorption takes place onto a homogeneous surface of the sorbent and a monolayer sorption occurs on the surface. It has been successfully applied to describe many adsorption processes to evaluate the maximum adsorption capacity of a sorbate on a sorbent.

The models can be expressed by the following equations:

Langmuir Model

$$\frac{C_{eq}}{q_{eq}} = \frac{C_{eq}}{q_m} + \frac{1}{q_m b} \tag{4}$$

Freundlich Model

$$logC_{eq} = logK_f + \frac{1}{n} logC_{eq} \tag{5}$$

Dubinin-Radushkevich (D-R) Model

$$lnq_{eq} = lnq_m - \beta\,\epsilon^2 \qquad E = \frac{1}{\sqrt{-2\beta}} \tag{6}$$

where, q_{eq} sorption is the capacity at equilibrium; q_m is the maximum sorption capacity; C_{eq} is the equilibrium concentration; b is the Langmuir affinity constant; K_f and n are the Freundlich constants; β activity coefficient related to biosorption means free energy; ϵ is the Polanyi potential; E is the energy sorption.

2.2.3 Binary biosorption system

Many studies till date have been restricted to simple solutions containing a single metal and only in a limited number of cases effects of other cations or anions on the metal uptake process have been reported. The presence of other cations (co-cations) can affect the sorption of metal ions (primary cation) and may hamper the removal efficiency to some extent. It is known that metal ions often interact to give rise to effects which may be synergistic, antagonistic, or non-interactive, the results of which cannot be predicted on the basis of single metal studies. Also, wastewaters often contain more than one type of metal ion which may interfere in the removal and/or recovery of the metal ion of interest. This is the reason that studies on the effect of other cations on the uptake of the primary metal ion are so relevant (Pranik and Paknikar, 1999). The mathematic models to understand the effect of the competition in bi component biosorption systems are described by the following equations:

Competitive Langmuir

$$q_1 = \frac{q_{m1} x b_1 x C_{eq1}}{1 + b_1 x C_{eq1} + b_2 x C_{eq2}} \qquad\qquad q_2 = \frac{q_{m2} x b_2 x C_{eq2}}{1 + b_2 x C_{eq2} + b_1 x C_{eq1}} \qquad (7)$$

Non Competitive Langmuir

$$q_t = \frac{q_m x b_1 x C_1 x \left[1 + \dfrac{K}{b_1} x C_2\right]}{1 + b_1 x C_1 + b_2 x C_2 + 2 x K x C_1 x C_2} \qquad (8)$$

Jain and Snoeyink Model

$$q_1 = \frac{(q_{m1} - q_{m2}) x b_1 x C_{eq1}}{1 + b_1 x C_{eq1}} + \frac{q_{m2} x b_1 x C_{eq1}}{1 + b_1 x C_{eq1} + b_2 x C_{eq2}} \qquad q_2 = \frac{q_{m2} x b_2 x C_{eq2}}{1 + b_1 x C_{eq1} + b_2 x C_{eq2}} \qquad (9)$$

where, q_1 and q_2 are adsorption capacity at the equilibrium of metal 1 and metal 2, respectively; b_1 and b_2 are the Langmuir affinity constants of metal 1 and metal 2, respectively; q_{m1} and q_{m2} are maximum adsorption capacity of metal 1 and metal 2 respectively; C_{eq1} and C_{eq2} are the equilibrium concentration of metal 1 and metal 2 respectively; q_t is the total adsorption of both metals (metal 1 and metal 2) at the equilibrium; K is the reciprocal affinity coefficient (b^{-1}).

2.2.4 Dynamic biosorption system

In order to validate the biosorption data obtained under batch conditions, evaluation of sorption performance in a continuously operated column is necessary because the sorbent uptake capacity is more efficiently utilized than in a completely mixed system and also the contact time required to attain equilibrium is different under column operation mode. For column operation, the adsorbent is continuously in contact with fresh wastewater and consequently the concentration in the solution in contact with a given layer of the biosorbent in a column changes very slowly. A fixed-bed column is simple to operate and economically valuable for wastewater treatment. Experiments using a laboratory-scale fixed-bed column yield performance data that can be used to design a larger pilot and industrial scale plant with a high degree of accuracy (Acheampong et al., 2013).

For the successful design of a column adsorption process, it is important to predict the concentration-time profile or breakthrough curve for effluent parameters. A number of mathematical models have been developed for use in the design of continuous fixed bed sorption columns. The Bed Depth Service Time (BDST), Thomas, Yoon-Nelson, and Adams Bohart models were used in predicting the behavior of the breakthrough curve because of their effectiveness. Among the various design approaches, the BDST approach based on the Bohart-Adams model is the most widely used. It assumes that the rate of adsorption is governed by the surface reaction between the adsorbate and the unused capacity of the adsorbent. The BDST model describes the relation between the breakthrough time often called the service time of the bed and the packed-bed depth of the column. The advantage of the BDST model is that any experimental test can be reliably scaled up to other flow rates and inlet solute concentrations without further experimental test (Acheampong et al., 2013).

Thomas model

This model is one the most general and widely used in the column performance theory. It has been applied for biosorption progress where the external and internal diffusion limitations are absent. The linearized form of the model is given as (El Messaoudi et al., 2016):

$$Ln\left(\frac{C_0}{C_t} - 1\right) = \frac{K_{Th} \times m \times q_0}{F} - K_{Th} \times C_o \times \tag{10}$$

where, C_0 is the inlet dye concentration, C_t is the outlet dye concentration at time t, q_0 is the maximum solid phase concentration of solute, m is the amount of biosorbent, F is the flow rate and K_{Th} is the Thomas model constant.

Bohart-Adams model

The Bohart-Adams model is usually used for the description of the initial part of the breakthrough cure. The linearized form can be expressed as follows (El Messaoudi et al., 2016):

$$Ln\left(\frac{C_t}{C_0}\right) = K_{BA} \times C_o \times t - \frac{K_{BA} \times N_o \times z}{U_o} \tag{11}$$

where, N_0 is the biosorption capacity of bed, U is the linear velocity, K_{BA} is Bohart-Adams model constant and Z is the bed height of the column.

Yoon and Nelson Model

Yoon and Nelson developed a model to investigate the breakthrough behavior of adsorbed gases on activated carbon. The model was based on the assumption that the rate of decrease in the probability of adsorption of each adsorbate molecule is proportional to the probability of the adsorbate adsorption and the adsorbate breakthrough on the adsorbent.

$$Ln\left(\frac{C_t}{C_o - C_t}\right) = k_{YN} - \tau k_{YN} \tag{12}$$

where, k_{YN} is a constant, τ is the time required for adsorbing 50% of initial adsorbate.

Bed Depth Service Time Model (BDST)

$$t_\tau = \frac{N_0}{C_0 U} Z - \frac{1}{K_a C_0} \ln\left(\frac{C_0}{C} - 1\right) \tag{13}$$

where, N_0 is the biosorption capacity of the bed, U is the linear velocity, K_a is the constant velocity.

In conclusion, biosorption has great potential to compete with conventional technologies for the treatment of metal-contaminated waters. However, most of the work done on biosorption so far has been confined to laboratory based investigations from the fundamental research viewpoint. Only limited investigations were attempted to examine the suitability of biosorption to industrial effluents. It is well-known that comprehensive fundamental understanding of the key concepts affecting the performance of biosorption including mechanisms of biosorption involved with different types of biosorbents, the relative influence of different experimental parameters affecting biosorption, modes of operation and biosorption capacity, and other related phenomena has already been established. With this knowledge, further advances are needed to transform this highly effective technique into practical applications, in particular, scale-up of biosorption processes to various types of wastewaters in a continuous large-scale operation. Biosorption can find potential applications in areas where heavy metals need to be extracted such as laboratory effluents, mining effluents, dyes effluents, etc.

3. Case study: Metal ion adsorption by dry cyanobacterial mat

3.1 Introduction

Heavy metal pollution of waterbodies due to indiscriminate disposal of industrial and domestic wastes threatens all kinds of inhabiting organisms (De Filippis and Pallaghy, 1994). Therefore, it is necessary to alleviate heavy metal burden of wastewaters before discharging them into waterways. A number of physicochemical methods such as chemical precipitation, adsorption, solvent extraction, ion exchange, membrane separation, etc., have been commonly employed for stripping toxic metals from wastewaters (Eccles, 1999). However, these methods have several disadvantages such as incomplete metal removal, expensive equipment and monitoring system requirements, high reagent or energy requirements, and generation of toxic sludge or other waste products that require disposal. Further, they may be ineffective or extremely expensive when metal concentration in wastewater is in the range 10–100 mg L^{-1} (Mehta and Gaur, 2005).

The use of biological processes for the treatment of metal enriched wastewaters can overcome some of the limitations of physical and chemical treatments and provide a way for cost-effective removal of metals. A great deal of interest has recently been generated using different kind of inexpensive biomass for adsorbing and removing heavy metals from wastewater (Volesky and Holan, 1995). In this context, the metal sorption capacity of many microorganisms, including algae, has been known for a few decades but has received increased attention only in recent years because of its potential for application in environmental protection or recovery of strategic metals.

Cell walls of microbial biomass mainly composed of polysaccharides, proteins and lipids, offer particularly abundant metal-binding functional groups such as carboxylate, hydroxyl, sulphate, phosphate, and amino groups. The physico-chemical phenomenon of metal biosorption, based on adsorption, ion exchange, complexation, and/or microprecipitation is relatively rapid and can be reversible. The metal-uptake ability of microorganisms has been known for a long time and apart from academic interest it is of great concern for toxicological and ecological fields.

Metal sorption ability of algae varies greatly from species to species and even among strains of a single species for any metal, although this variation may also be due to variable experimental conditions in different studies. A suggestion has also been made that cells grown under different conditions vary with regard to composition of their cell wall and hence, in biosorption characteristics (Chojnacka et al., 2005). Some algae show a high affinity for sorbing a particular ion, whereas others do not show such specificity and sorb several metal ions. The affinity of various algal species for binding of ions shows different hierarchies. In general, ions with greater electronegativity and smaller ionic radii are preferentially sorbed by algae (Mehta and Gaur, 2005).

The process of heavy metal biosorption involves mechanisms such as ion exchange, complexation, electrostatic attraction, and microprecipitation. Ion exchange has been shown to be the most important mechanism for the biosorption of metal ions by algal biomass, thus this process is highly dependent on pH conditions. Sorption and removal of heavy metals by algal biosorbents is also largely dependent on initial metal concentration in the solution. Metal sorption initially increases with metal concentration in the solution becoming saturated after a certain concentration of metal.

Many algae have immense capability to sorb metals and there is considerable potential for using them to treat wastewaters. Metal sorption involves binding on the cell surface and there are numerous reports of cyanobacterial and algae that remove heavy metals from contaminated effluents. The utilization of *Spirulina platensis* was investigated in order to biosorbe Cu(II) by Al-Homaidan et al. (2014). Maximum biosorption (90.6%) was found to be for 0.05 g dose, 90 min of contact time, at 37°C and pH = 7. *Chlorella vulgaris* was also investigated for the adsorption of Cr(VI) (Indhumathi et al., 2014). The maximum adsorption obtained was at pH3, while the equilibrium attained at 120 min for all the studied concentrations.

Scenedesmus quadricauda is another algal which was sufficiently applied for the sequestration of Cr^{3+} and Cr^{6+} (Shokri et al., 2014). Maximum biosorption for Cr(III) (pH = 1–6) and Cr(VI) (pH = 1–9) was found at pH 6 and pH 1 respectively.

Both living (Bender et al., 1995; Bender and Phillips, 2004) and dead (Mehta and Gaur, 2005; Kumar et al., 2010; Chakraborty et al., 2011; Kumar and Gaur, 2011) bacterial biomass metal sorption studies have been carried out by several researchers. Bender et al. (1997) found that a cyanobacterial mat almost completely removed cadmium, lead, and chromium in 24 h from an initial concentration of 10 mg.L^{-1}. Sheoran and Bhandari (2005) reported considerable alleviation of metal burden from acid mine drainage by a microbial mat. Bioremediation has been also performed using cyanobacterial mats (consisting on a consortium of cyanobacteria/blue–green algae such as *Chlorella* sp., *Phormidium* sp., and *Oscillatoria* sp.) in the form of biological treatment to clean up chromium (VI) contaminant in surface water at pH 5.5–6.2 for both low and high levels of contamination (Shukla et al., 2012). Kumar and Gaur (2012) published that live *Phormidium bigranulatum—*dominated mat successfully removed Pb(II), Cu(II), and Cd(II) from aqueous solution. Equilibrium of metal removal was achieved within 4 h, independent of mat thickness (0.2–1.6 mm), in batch system. But metal removal percentage increased with increase in mat thickness due to enhancement of biomass which provided more metal binding sites. Although cyanobacterial mats occur in nature as stratified communities of cyanobacteria and some other bacteria, they could also be cultured on large scale and used for bioremediation processes. Furthermore, Liu et al. (2015) studied the surface properties of the cyanobacterium *Synechococcus* sp. PCC 7002 and its interaction in cadmium removal. As this genus is one of the dominant marine phytoplankton, the results presented highlight the potential role of surface sorption by phytoplankton in the cycling of metals in the ocean.

Particularly, the use of no living biomass have several advantages as low cost biosorption conditions and storage, the obtainment of stable biosorbent particles, no limitations for toxicity, no requirement for growth media and nutrients, easily desorption of biosorbed metal ions, and reusability of biomass.

Materials used on biosorption processes are likely to be cheap, stable, and to have high affinity for the pollutants. In this work, dry biomass of a wide spread photosynthetic biomat from a geothermal area of Neuquén, Argentina, was studied as a potential biosorbent material for heavy metal bioremediation. Considering former evidence of algal and cyanobacterial capacity of biosorption, the dried biomat used is considered to be a promising and stable source of biosorption material.

3.2 Materials and methods

3.2.1 Biosorbent selection

Photosynthetic biomat widely spread in the geothermal area of Domuyo in the province of Neuquén in Argentina was selected as biosorbent material. Domuyo geothermal area is emplaced in the Southern slope of the Domuyo hill, the highest peck in Patagonia with an elevation of 4.709 m.a.s.l. Domuyo is not a stratovolcano, however there are magmatic chambers near the base of the hill that control the geothermal activity of the area (Pesce, 2013). There the surface manifestations that include fumaroles, hot springs, and geysers are mostly of neutral pH and high temperatures.

In the Domuyo area of Los Tachos, the biomat selected for this work grows adhere to stones being splashed with high temperature water. Temperature and pH inside the biomat are 40.5°C and 7 respectively. This particular biomat is many centimetres thick and presents three clearly different layers at different depths. In the superficial area, active photosynthetic microorganisms are responsible for the green–blue colour of this layer. It is between 1–3 cm deep and is mainly dominated by filamentous cyanobacteria identified as *Leptolyngbya* sp. by 16s rRNA gene sequencing. An orange filamentous like layer is found under the green one. This inner layer is thicker than the green one but has the same filamentous texture. Most probably, it is part of the same cyanobacteria former growth and its reddish colour is due to catabolism products of the chlorophyll-A pigment (Gossauer et al., 1996). The lowest layer, in adherence to the surface, is thin and white with complete chlorophyll degradation. Similar biomat formations have been reported on other geothermal areas (Boomer et al., 2009). As it is a widely spread material on the area, its biosorbent capacity was tested for different heavy metals. The algal biomass was dried in oven at 50°C for 72 h to constant weigh. The dried biomass was shredded, ground in a mortar and was used for biosorption experiments at a concentration of 0.1 g/L. Figure 1 illustrates the treatment process to convert the biomass into the biosorptive material.

Figure 1. Photosynthetic biomat used as biosorbent material (a) Domuyo area of Los Tachos in the Argentinean province of Neuquén; (b) Biomat attached to a rock; (c) Biomat sample in the laboratory previous to dehydration; (d) Dry biomat used on the biosorption assays.

3.2.2 Biosorption conditions

For the biosorption assays, five different metal ions were tested, both anions and cations. Between the anions, four metals were selected: Cu^{2+} ($CuCl_2$), Ni^{2+} ($NiSO_4.6H_2O$), Pd^{2+} ($Pb(NO_3)_2$), and Cr^{3+} ($Cr_2(SO_4)_3.6H_2O$) and one anion was tested, Cr^{6+} ($K_2Cr_2O_7$). Batch systems in 250 ml Erlenmeyer flasks were done in duplicates for each metal. The concentration tested was 10 ppm of the metal of interest in 100 ml of distilled water at pH 5 for the cations and at pH2 for the anion. Negative controls without biomass were also done. All the systems were incubated at 30°C and 120 rpm agitation.

3.2.3 Biosorption kinetics studies

Samples were taken at different time intervals and were filtered by a 0.45 μm filter. Total concentration of each metal was measured by Atomic Absorption Spectroscopy (AAS).

The metal biosorption (q) by the algae and bioremoval efficiency (R) were calculated by:

$$q = \frac{(C_i - C_f)V}{M} \qquad R(\%) = \frac{(C_i - C_f)}{C_i} \times 100 \qquad (14)$$

where q is the metal adsorption (mg/g); M is the dry mass of algae (g); V is the volume of initial metal ion solution (L); R is the bioremoval efficiency (%); C_i is the initial concentration of metal in aqueous solution (mg/L); C_f is the final concentration of metal in aqueous solution (mg/L).

3.3 Results and discussion

Figure 2 shows the kinetics of metal adsorption onto the biomat obtained by batch contact time studies, namely, Pb^{2+}, Cu^{2+}, Ni^{2+}, Cr^{3+}, and Cr^{6+} ions. The plots represent the amount of metal adsorbed onto the biosorbent versus time for an initial metal concentration of 10 ppm. Kinetic parameters were calculated from AAS metal measures from samples taken at different times. Table 1 summarizes metal adsorption values (q) and bioremoval efficiency percentage (R) for each of the ions tested.

Biomat used as biosorbent material was successful for cations while the anion tested (Cr^{6+}) was practically not sorbed at all by the material in the conditions tested (3.29% bioremoval efficiency).

Despite the same experimental conditions used, it is interesting to note that the fixation capacities were also different according to the metal sorbed. Between cations, the highest bioremoval capacities were obtained for Cr^{3+} and Cu^{2+} ions (8.39 and 7.84 mg.g^{-1}), while the capacity for the other metal ions evaluated was 3.93 mg.g^{-1} for Ni^{2+} and 6.65 mg.g^{-1} for Pb^{2+}.

The order of affinity (in mg of metal per gram of biomat) is: $Cr^{3+} > Cu^{2+} > Pb^{2+} > Ni^{2+}$. Differences of metal uptake are related with the interaction between metals and the functional groups on the biomat surface plus the difference in atomic masses. Some authors reported that metal sorption increased with increasing valence and atomic number (Holan and Volesky, 1994) but our results do not follow such behavior.

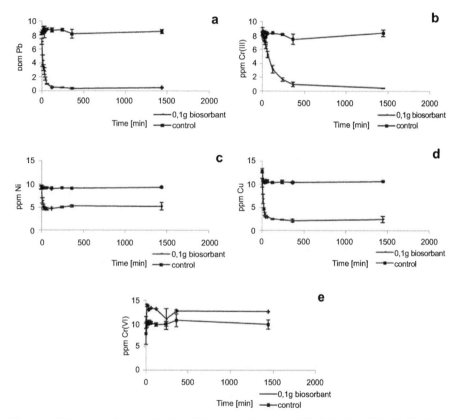

Figure 2. Biosorption kinetics for the different metals used. (a) Pb^{+2}; (b) Cr^{+3}; (c) Ni^{+2}; (d) Cu^{+2}; (e) Cr^{+6}.

Table 1. Summary of the adsorption values (q: mg of metal ion/g biosorbent) and bioremoval efficiency percentage (R) obtained for the metal ions used on the assay.

	q (mg·g⁻¹)	R (%)
Pb^{2+}	6.65	93.66
Cr^{3+}	8.39	95.09
Ni^{2+}	3.93	43.05
Cu^{2+}	7.84	76.04
Cr^{6+}	0.41	3.29

4. Conclusion

The biosorption of ions Cr^{3+}, Pb^{2+}, Cu^{2+}, Ni^{2+}, and Cr^{6+} onto a dried photosynthetic biomat from Domuyo, Neuquén in Argentina was studied in a batch system with respect to the fixed values of pH, temperature, and initial metal ion concentration. The maximum uptake capacities for Cr^{3+} and Cu^{2+} ions were the highest (8.39 and 7.84 $mg.g^{-1}$), while the capacities for the other metal ions evaluated were 3.93 $mg.g^{-1}$ for Ni^{2+} and 6.65 $mg.g^{-1}$ for Pb^{2+} ions. These results suggested that Cr^{3+} has greater affinity to the binding sites present on the surface of the biomat. Consequently, dried biomat from Domuyo is a good adsorbing agent for metals and has an especially high adsorption capacity for Cr^{3+}. This last feature could be useful as the last step on chromate remediation processes following Cr(VI) reduction reaction to completely remove Cr(III) from contaminated effluents.

References

Abdolali A., Guo W.S., Ngo H.H., Chen S.S., Nguyen N.C., Tung K.L. Typical lignocellulosic wastes and by-products for biosorption process in water and wastewater treatment: A critical review. Bioresour. Technol. 160, 57–66, 2014.

Acheampong M.A., Pakshirajan K., Annachhatre A.P., Lens P.N.L. Removal of Cu(II) by biosorption onto coconut shell in fixed-bed column systems. J. Ind. Eng. Chem. 19, 841–848, 2013.

Aksu Z., Dönmez G. The use of molasses in copper(II) containing wastewaters: effects on growth and copper(II) bioaccumulation properties of *Kluyveromyces marxianus*. Process Biochem. 36, 451–458, 2000.

Al-Homaidan A.A., Al-Houri H.J., Al-Hazzani A.A., Elgaaly G., Moubayed N.M. Biosorption of copper ions from aqueous solutions by *Spirulina platensis* biomass. Arabian J. Chem. 7, 57–62, 2014.

Anastopoulos I., Kyzas G.Z. Progress in batch biosorption of heavy metals onto algae. J. Mol. Liq. 209, 77–86, 2015.

Andrade A.D., Rollemberga M.C.E., Nóbrega J.A. Proton and metal binding capacity of the green freshwater alga *Chaetophora elegans*. Process Biochem. 40, 1931–1936, 2005.

Apiratikul R., Madacha V., Pavasant P. Kinetic and mass transfer analyses of metal biosorption by *Caulerpa lentillifera*. Desalination 278, 303–311, 2011.

Baing Shahzad K., Doan H.D., Wu J. Multicomponent isotherms for biosorption of Ni^{2+} and Zn^{2+}. Desalination 249, 429–439, 2009.

Bender J., Lee R.F., Phillips P. Uptake and transformation of metals and metalloids by microbial mats and their use in bioremediation. J. Ind. Microbiol. 14, 113–118, 1995.

Bender J., Phillips P. Microbial mats for multiple applications in aquaculture and bioremediation. Bioresour. Technol. 94, 229–238, 2004.

Bender J., Phillips P., Lee R., McNally T., Rodrıguez-Eaton S., Félix C. Rapid heavy metal removal in a continuous-flow batch reactor by microbial mat. *In Situ* and On-Site Bioremediation 3, 373–378, 1997.

Boomer S., Noll K.L., Geesey G., Dutton B.E. Formation of multilayered photosynthetic biofilms in an alkaline thermal spring in Yellowstone National Park, Wyoming. Appl. Environ. Microbiol. 75, 2464–2475, 2009.

Cadogan E.I., Lee C.H., Popuri S.R., Lin H.Y. Efficiencies of chitosan nanoparticles and crab shell particles in europium uptake from aqueous solutions through biosorption: Synthesis and characterization. Int. Biodeter. Biodegr. 95, 232–240, 2014.

Chakraborty N., Banerjee A., Pal R. Accumulation of lead by free and immobilized cyanobacteria with special reference to accumulation factor and recovery. Bioresour. Technol. 102, 4191–4195, 2011.

Chojnacka K. Biosorption and bioaccumulation—the prospects for practical applications. Environ. Int. 36, 299–307, 2010.

Chojnacka K., Chojnacki A., Gorecka H. Biosorption of Cr^{3+}, Cd^{2+} and Cu^{2+} ions by blue green algae *Spirulina* sp.: kinetics, equilibrium and the mechanism of the process. Chemosphere 59, 75–84, 2005.

Chuah T.G., Jumasiah A., Azni I., Katayon S., Thomas Choong S.Y. Rice husk as a potentially low-cost biosorbent for heavy metal and dye removal: an overview. Desalination 175, 305–316, 2005.

Crini G. Recent developments in polysaccharide-based materials used as adsorbents in wastewater treatment. Prog. Polym. Sci. 30, 38–70, 2005.

Davis T.A., Volesky B., Mucci A. A review of the biochemistry of heavy metal biosorption by brown algae. Water Res. 37, 4311–4330, 2003.

Davis T.A., Volesky B., Vieira R.H.S.F. *Sargassum* seaweed as biosorbent for heavy metals. Water Res. 34, 4270–4278, 2000.

De Filippis L.F., Pallaghy C.K. Heavy metals: sources and biological effects. *In*: Rai L.C., Gaur J.P., Soeder C.J. (eds.). Advances in Limnology Series: Algae and Water Pollution. E. Scheizerbartsche Press, Stuttgart, pp. 31–77, 1994.

Deng L., Su Y., Su H., Wang X., Zhu X. Sorption and desorption of lead (II) from wastewater by green algae *Cladophora fascicularis*. J. Hazard. Mater. 143, 220–225, 2007.

Eccles H. Treatment of metal-contaminated wastes: why select a biological process? Trends Biotechnol. 17, 462–465, 1999.

El Messaoudi N., El Khomri M., Dbik A., Bentahar S., Lacherai A., Bakiz B. Biosorption of Congo red in a fixed-bed column from aqueous solution using jujube shell: Experimental and mathematical modeling. J. Ind. Eng. Chem. 4, 3848–385, 2016.

Flouty R., Estephane G. Bioaccumulation and biosorption of copper and lead by a unicellular algae *Chlamydomonas reinhardtii* in single and binary metal systems: A comparative study. J. Environ. Manage. 111, 106–114, 2012.

Gossauer A., Engel N. Chlorophyll catabolism—structures, mechanisms, conversions. J. Photochem. Photobiol. B 32, 141–151, 1996.

Gupta V.K., Rastogi A. Biosorption of lead from aqueous solutions by green algae *Spirogyra* species: Kinetics and equilibrium studies. J. Hazard. Mater. 152, 407–414, 2008.

Hammaini A., González A., Ballester A., Blázquez M.L., Muñoz J.A. Biosorption of heavy metals by activated sludge and their desorption characteristics. J. Environ. Manage. 84, 419–426, 2007.

Holan Z.R., Volesky B. Biosorption of lead and nickel by biomass of marine algae. Biotechnol. Bioeng. 43, 1001–1009, 1994.

Ibrahim W.M. Biosorption of heavy metal ions from aqueous solution by red macroalgae. J. Hazard. Mater. 192, 1827–1835, 2011.

Indhumathi P., Shabudeen P.S., Shoba U.S., Saraswathy C.P. The removal of chromium from aqueous solution by using green micro algae. J. Chem. Pharm. Res. 6, 799–808, 2014.

Kumar D., Gaur J.P. Chemical reaction-and particle diffusion-based kinetic modeling of metal biosorption by a *Phormidium* sp.—dominated cyanobacterial mat. Bioresour. Technol. 102, 633–640, 2011.

Kumar D., Pandey L.K., Gaur J.P. Evaluation of various isotherm models, and metal sorption potential of cyanobacterial mats in single and multi-metal systems. Colloids Surf. B 81, 476–485, 2010.

Kumar D., Rai J., Gaur J.P. Removal of metal ions by *Phormidium bigranulatum* (Cyanobacteria)–dominated mat in batch and continuous flow systems. Bioresour. Technol. 104, 202–207, 2012.

Li C., Xu Y., Jiang W., Dong X., Wang D., Liu B. Effect of NaCl on the heavy metal tolerance and bioaccumulation of *Zygosaccharomyces rouxii* and *Saccharomyces cerevisiae*. Bioresour. Technol. 143, 46–52, 2013

Liu Y., Alessi D.S., Owttrim G.W., Petrash D.A., Mloszewska A.M., Lalonde S.V., Konhauser K.O. Cell surface reactivity of *Synechococcus* sp. PCC 7002: Implications for metal sorption from seawater. Geochim. Cosmochim. Acta 169, 30–44, 2015.

Mahajan G., Sud D. Application of ligno-cellulosic waste material for heavy metal ions removal from aqueous solution. J. Environ. Chem. Eng. 1, 1020–1027, 2013.

Malik A. Metal bioremediation through growing cells. Environ. Int. 30(2), 261–278, 2004.

Mehta S.K., Gaur J.P. Use of algae for removing heavy metal ions from wastewater: progress and prospects. Crit. Rev. Biotechnol. 25, 113–152, 2005.

Niu C.H., Volesky B., Cleiman D. Biosorption of arsenic(V) with acid-washed crab shells. Water Res. 41, 2473–2478, 2007.

Oyetibo G.O., Ilori M.O., Obayori O.S., Amud O.O. Equilibrium studies of cadmium biosorption by presumed non-viable bacterial strains isolated from polluted sites. Int. Biodeter. Biodegr. 91, 37–44, 2014.

Pesce A.H. The Domuyo Geothermal Area, Neuquén, Argentina. Geothermal Resources Council Transactions 37, 309–314, 2010.

Pranik P.R., Paknikar K.M. Influence of co-cations on biosorption of lead and zinc—a comparative evaluation in binary and multimetal systems. Bioresour. Technol. 70, 269–276, 1999.

Ramírez Carmona M.E., Pereira da Silva M.A., Ferreira Leite S.G., Vasco Echeverri O.H., Ocampo-López C. Packed bed redistribution system for Cr(III) and Cr(VI) biosorption by *Saccharomyces cerevisiae*. J. Taiwan Inst. Chem. Eng. 4, 428–432, 2012.

Romera E., González F., Ballester A., Blázquez M.L., Muñoz J.A. Comparative study of biosorption of heavy metals using different types of algae. Bioresour. Technol. 98, 3344–3353, 2007.

Rzymski P., Poniedzialek B., Niedzielski P., Tabaczewski P., Wiktorowicz K. Cadmium and lead toxicity and bioaccumulation in *Microcystis aeruginosa*. Front. Env. Sci. Eng. 8(3), 427–432, 2014.

Schiewer S., Volesky B. Biosorption processes for heavy metals removal. *In*: Lovley D.R. (ed.). Environmental Microbe-metal Interactions. ASM Press. Washington, D. C., pp. 329–362, 2000.

Selatnia A., Bakhti M.Z., Madani A., Kertous L., Mansouri Y. Biosorption of Cd^{2+} from aqueous solution by a NaOH-treated bacterial dead *Streptomyces rimosus* biomass. Hydrometallurgy 75, 11–24, 2004.

Sheoran A.S., Bhandari S. Treatment of mine water by a microbial mat: bench-scale experiments. Mine Water Environ. 24, 38–42, 2005.

Shokri Khoubestani R., Mirghaffari N., Farhadian O. Removal of three and hexavalent chromium from aqueous solutions using a microalgae biomass-derived biosorbent. Environ. Prog. Sustainable Energy 34, 949–956, 2015.

Shukla D., Vankar P.S., Srivastava S.K. Bioremediation of hexavalent chromium by a cyanobacterial mat. Appl. Water Sci. 2, 245–251, 2012.

Singh R., Singh S., Parihar P., Singh V.P., Prasad S.M. Arsenic contamination, consequences and remediation techniques: a review. Ecotoxicol. Environ. Saf. 112, 247–270, 2015.

Song W., Zhang M., Liang J., Han G. Removal of As(V) from wastewater by chemically modified biomass. J. Mol. Liq. 206, 262–267, 2015.

Srinath T., Verma T., Ramteke P.W., Garg S.K. Chromium (VI) biosorption and bioaccumulation by chromate resistant bacteria. Chemosphere 48(4), 427–435, 2002.

Sulaymon A.H., Yousif S.A., Al-Faize M.M. Competitive biosorption of lead mercury chromium and arsenic ions onto activated sludge in fixed bed adsorber. J. Taiwan Inst. Chem. Eng. 45, 325–337, 2014.

Taştan B.E., Ertuğrul S., Dönmez G. Effective bioremoval of reactive dye and heavy metals by *Aspergillus versicolor*. Bioresour. Technol. 101(3), 870–876, 2010.

Veglio F., Beolchini F., Prisciandaro M. Sorption of copper by olive mill residues. Water Res. 37, 4895–4903, 2003.

Vijayaraghavan K., Balasubramanian R. Is biosorption suitable for decontamination of metal bearing wastewaters? A critical review on the state-of-the-art of biosorption processes and future directions. J. Environ. Manage. 160, 238–296, 2015.

Vijayaraghavan K., Balasubramanian R. Single and binary biosorption of cerium and europium onto crab shell particles. Chem. Eng. J. 163, 337–343, 2010.

Vijayaraghavan K., Mahadevan A., Sathishkumar M., Pavagadhi S., Balasubramanian R. Biosynthesis of Au(0) from Au(III) via biosorption and bioreduction using brown marine alga *Turbinaria conoides*. Chem. Eng. J. 176, 223–227, 2011.

Vijayaraghavan K., Yun Y.-S. Bacterial biosorbents and biosorption. Biotechnol. Adv. 26, 266–291, 2008.

Vijayaraghavana K., Jeganb J., Palaniveluc K., Velana M. Biosorption of cobalt(II) and nickel(II) by seaweeds: batch and column studies. Sep. Purif. Technol. 44, 53–59, 2005.

Vijayaraghavana K., Wona S.W., Mao J., Yun Y.S. Chemical modification of *Corynebacterium glutamicum* to improve methylene blue biosorption. Chem. Eng. Prog. 145, 1–6, 2008.

Vogel M., Günther A., Rossberg A., Li B., Bernhard G., Raff J. Biosorption of U(VI) by the green algae *Chlorella vulgaris* in dependence of pH value and cell activity. Sci. Total Environ. 409, 384–395, 2010.

Volesky B., Holan Z.R. Biosorption of heavy metals. Biotechnol. Prog. 11, 235–250, 1995.

Wang J., Chen C. Biosorbents for heavy metals removal and their future. Biotechnol. Adv. 27, 195–226, 2009.

Yang L., Chen J.P. Biosorption of hexavalent chromium onto raw and chemically modified *Sargassum* sp. Bioresour. Technol. 99, 297–307, 2008.

Żbikowski R., Szefer P., Latała A. Comparison of green algae *Cladophora* sp. and *Enteromorpha* sp. as potential biomonitors of chemical elements in the southern Baltic. Sci. Total Environ. 387, 320–332, 2007.

CHAPTER 6

Heavy Metal Bioprecipitation
Use of Sulfate-Reducing Microorganisms

*Graciana Willis** and *Edgardo R. Donati*

1. Introduction

The environmental impact of heavy metal contamination due to anthropogenic activities has increased enormously in the last years. As a result of industrial activity, highly polluted waters from industrial effluents, municipal waste treatment plants, landfill leaching and mining activities among others, are released into the environment without any concern or government regulation. An inadequate disposal of these materials could cause hazardous environmental problems if their large amounts of toxic metals are mobilized and reach soil and groundwater. Unlike organic pollutants, heavy metals cannot be degraded and accumulate over time in the environment, including the food chain. Depending on their concentration and speciation, some heavy metals can be toxic and non-essential (e.g., Hg, Cd, and Pb), while others can be essential in certain amounts but they can also become toxic at higher doses (e.g., Fe, Zn, Cu, Mn, Co, Ni, and Cr) (Viera and Donati, 2004; Singh and Gadi, 2009).

To make the environment healthier for human beings, contaminated water bodies and land need to be amended to make them free from heavy metals (Dixit et al., 2015). The feasible possibilities are either their immobilization in a non-bioavailable form or their speciation into less toxic forms. Both physicochemical and biological techniques are available nowadays (Viera and Donati, 2004). Abiotic treatments include reduction, reverse osmosis, filtration, electrochemical treatment, evaporation, ion-exchange, and

Centro de Investigación y Desarrollo en Fermentaciones Industriales, CINDEFI (CCT LA PLATA-CONICET, UNLP). Facultad de Ciencias Exactas. Universidad Nacional de La Plata. Argentina.
* Corresponding author: willis.graciana@biotec.quimica.unlp.edu.ar

chemical precipitation the latter being the most widely used. In brief, it involves the addition of alkalizing chemicals (such as $CaCO_3$) to raise water pH and precipitate metals as hydroxides and carbonates (Fu and Wang, 2011). Nevertheless, all these technologies have the main disadvantages of being expensive and inefficient at low metal concentrations (less than 100 mg.L^{-1}). The generation of high volumes of mixed-metal sludge is another drawback. Those residues require careful disposal, often in specially designated landfill sites. Also, metal hydroxides are sensitive to pH variations so they are less stable than other metal precipitates such as sulfides (Hashim et al., 2011; Colin et al., 2012).

1.1 Sulfate reducing microorganisms

In recent decades and as alternative to traditional physiochemical methods, biological and more environmental-friendly techniques have been developed and are generally known as bioremediation. Bioremediation offers high specificity in the removal of heavy metals of interest while also operational flexibility when comparing with physiochemical techniques (Viera and Donati, 2004; Gadd, 2010). Among these biological approaches, bioprecipitation using sulfate-reducing microorganisms (SRM) is a viable alternative to treat heavy metal-impacted waters due to the advantages it offers over chemical precipitation. Moreover, it is an interesting option to be used in the treatment of mine-impacted waters (Johnson and Hallberg, 2005). The classification SRM is functional and comprises all microorganisms that are capable of anaerobic sulfate respiration including both bacteria and archaea groups. Since only three species of archaea are known to use sulfate respiration, SRM are still indicated as SRB (sulfate-reducing bacteria) (Barton, 2015; Hao et al., 2014). Under anaerobic conditions, these microorganisms reduce sulfate (SO_4^{2-}) to sulfide (S^{2-}), oxidizing low molecular weight organic substrates as electron donors (Sheoran et al., 2010). The process is summarized as follow:

$$2CH_2O + SO_4^{-2} \rightarrow H_2S + 2HCO_3^- \tag{1}$$

where CH_2O represents the organic substrate. Depending on the SRM and growth conditions, the electron donor can be incompletely oxidized and end products of metabolism, such as acetate, could be present (Thauer et al., 2007).

As described above, among the most used electron donors by SRM are the low molecular weight organic acids such as lactate, formate, acetate, pyruvate, and some alcohols such as ethanol and glycerol (Cao et al., 2012). In addition, glucose, fructose, and galactose are effective electron donors that can be easily degraded under anaerobic conditions by some groups of SRM (Liamleam and Annachhatre, 2007). Various types of organic wastes have been employed in recent studies as alternative electron donors and carbon sources for sulfate reduction. For example, Das et al. (2015) used different sweetmeat waste (SMW) concentrations, in terms of chemical oxygen demand (COD)/SO_4^{2-} ratios, to reactivate an exhausted-upflow packed bed bioreactor (PBR). Sulfate was removed in 99% of the bioreactor using little amounts of SMW at a high rate (1417 mg.L^{-1}.d^{-1}), indicating the high sulfate removal efficiency of the system. Similar studies using other organic wastes, other SRM consortia, or other operational conditions were recently reported (Costa et al., 2009; Dev et al., 2016; McCullough et al., 2008; Sánchez-Andrea et al., 2012b; Zhang et al., 2016b).

Hydrogen is another attractive electron donor for sulfate reduction (Liamleam and Annachhatre, 2007). Hydrogenotrophic sulfate reduction involves:

$$4\,H_2 + SO_4^{2-} + H^+ \rightarrow HS^- + 4\,H_{20} \tag{2}$$

SRM are a complex physiological group and various properties have been used in traditional classification. Based on 16S sequence analysis four groups of SRM are stablished: Gram-negative mesophilic SRM, Gram-positive spore forming SRM, thermophilic bacterial SRM, and thermophilic archaeal SRM. Gram-negative mesophilic SRM are located within the delta subdivision of the Proteobacteria and include, among others, the genera *Desulfovibrio, Desulfomicrobium, Desulfobulbus, Desulfobacter, Desulfobacterium, Desulfococcus, Desulfosarcina, Desulfomonile, Desulfonema, Desulfobotulus,* and *Desulfoarculus* (Castro et al., 2000).

The Gram-positive spore-forming SRM comprise mainly the *Desulfotomaculum* and *Desulfosporosinus* genera. Due to their metabolic versatility and their ability to persist as endospores, this group is commonly found in a wide variety of environments such as deep fresh water lakes, geothermally active areas, acidic mine-impacted sediments, etc. (Aullo et al., 2013).

The most well characterized species within the thermophilic SRM are *Thermodesulfobacterium commune* and *Thermodesulfovibrio yellowstonii.* Both bacteria were isolated from geothermally active vents in Yellowstone National Park and their optimal growth temperature are higher than those of Gram-positive spore forming SRM but lower than those of the archaeal SRM (Castro et al., 2000).

Finally, the archaeal group is characterized by optimal growth temperature above 80°C. Two species have been completely described: *Archaeoglobus fulgidus* and *A. profundus* (Castro et al., 2000).

Sulfate-reducing microorganisms play an important role in the geochemical carbon and sulfur cycles. They are widely distributed in a large variety of anaerobic marine, terrestrial, and subterrestrial ecosystems and can coexist along with another anaerobic microorganism. These interactions are particularly important in both oxic/anoxic interface and in the deeper anoxic regions (Thauer et al., 2007). In the last years, several studies on the isolation and identification by molecular approaches of SRM from these environments were reported. Falagan et al. (2014) analyzed the indigenous microbial communities of two extremely acidic and metal-rich stratified pit lakes located in the Iberian Pyrite Belt (Spain), using a combination of cultivation-based and cultured independent approach. SRM belonging to *Desulfomonile* and *Desulfosporosinus* genus were isolated from the chemocline zone. Other microorganisms that take part in the iron cycle were also found, such as *L. ferrooxidans*, *A. ferrooxidans*, and *Acidocella* sp. Similar studies performed in other extreme environments, such as acidic hot-spring sediments, the Tinto River and acid mine drainage-affected areas were reported (Alazar et al., 2010; Rowe et al., 2007; Sánchez-Andrea et al., 2012a, 2013; Willis et al., 2013).

2. Metal precipitation using SRM

The hydrogen sulfide generated as a waste product by dissimilatory sulfate-reducing bacteria (eq. 1) can be used for metal precipitation. It can react with metal ions present in contaminated waters and precipitate as insoluble metal sulfide (eq. 3)

$$H_2S + M^{+2} \rightarrow MS_{(s)} + 2H^+ \tag{3}$$

For a biotechnological point of view, heavy metal precipitation using SRM implies important applications—in addition to metal precipitation, decrease in sulfate concentration and the increase in pH (eq. 1) are also produced (Viera and Donati, 2004). Those facts are particularly important in metal-rich acid mine drainage (AMD) treatment, where pH values are low and sulfate is present at high concentrations (Bertolino et al., 2013; Ñancucheo and Johnson, 2014). One of the advantages of this treatment, in which most of the precipitates are sulfides, is the reduction of the volume of sludge that is generated as compared to hydroxides and carbonates that are usually colloidal and more voluminous. Furthermore, under anaerobic conditions, metal sulfides are more stable and insoluble than the corresponding hydroxides or carbonates. Also, the low solubility products of most metal sulfides allow the metals to be removed even at low concentrations in wastewaters (Sheoran et al., 2010).

In addition, metal sulfides can be selectively precipitated controlling the pH, which dictates the soluble species of sulfide (H_2S, HS^-, and S^{2-}) and/or the concentration of electron donor. On the other hand, the precipitated metallic sulfides can be easily recovered and reused in further industrial processes (Johnson, 2014). For example, in a recent study published by Hedrich and Johnson (2014) they used two modular pH-controlled bioreactors to remediate and selectively recover metals from an acidic mine water rich in Zn, Fe, and small amounts of As. Zn was removed as ZnS in an acidophilic sulfidogenic bioreactor with controlled pH from which the metal could be recovered. On the other hand, Fe was precipitated as schwertmannite after microbial iron oxidation. Then, a small proportion (~11%) of the schwertmannite produced was used to remove As at the initial step in the process and other chalcophilic metals (Cu, Cd, and Co) were removed (as sulfides). Cibati et al. (2013), using a SRM consortia, reported the use of biogenic H_2S and NaOH to selectively precipitate Mo (36–72%) and V from synthetic spent refinery catalyst leach liquors at pH 2.

2.1 Bioprecipitation under acidic conditions by SRM

One of the major limitations of the application of sulfate-reducing systems is the sensitivity of SRM to acidity and high heavy metal concentration. However, the growth of SRM and metal precipitation from acidic liquors (such as AMD) in the same bioreactor vessel has many advantages including the simple engineering design and the reduction of construction and operational costs.

As explained in the previous paragraph the requirement of sulfidogenic active cultures at low pH values becomes strictly necessary (Johnson, 2012). Inhibition at low pH is associated with the use of low molecular weight organic acids in culture media to isolate SRM. At low pH values, these molecules can diffuse the cell membrane in the non-dissociated form and dissociate into the cells where the pH is almost neutral. To overcome such inhibition organic acids are replaced by a non-acidic organic substrate (such as glycerol or sugars). Using this strategy, many acidophilic/acid-tolerant sulfidogenic consortia and isolates of sulfate reducers have been obtained in the last years from diverse acidic natural or human-impacted environments (Chiacchiarini et al., 2010; Falagán et al., 2014; Kimura et al., 2006; Sánchez-Andrea et al., 2015; Rowe et al., 2007; Willis et al., 2013). Moreover, a recent study reported by Ñancucheo and Johnson (2012) describes the use of a mix population of acidophilic/acid-tolerant SRM in the selective precipitation of ZnS and CuS using different continuous-flow bench-scale bioreactors from synthetic acidic mine-water. Zn was totally precipitated as ZnS at pH 4.0 whereas the selective precipitation of CuS was achieved at pH 2.2–2.5. They also observed changes in the bacterial population in response to varying operational parameters.

The use of zero-valent iron (Fe⁰) is another promising strategy to get over the inhibition of SRM at low pH values and high heavy metal concentration. When Fe⁰ interacts with water at low pH without oxygen, the oxidation of Fe⁰ to ferrous ion (Fe^{+2}) occurs at the anodic areas of the surface of Fe⁰, and H_2 released in the cathode is utilized by SRM as an electron donor for sulfate reduction (Karri et al., 2005). Consequently, the acidity of AMD is neutralized by the H_2 utilization and the alkaline production during the oxidation of Fe⁰ which results in a suitable pH for the growth of SRM. In a recent study, Bai et al. (2013) used a mix population of SRM and Fe⁰ (SRM+Fe⁰) to remove Cu from a copper-containing synthetic wastewater. The SRM+Fe⁰ system reduced sulfate twice as much as that of the SRM alone system at a loading rate of 125 mg.L^{-1}.h^{-1}. Cu removal was held above 95% in SRM+Fe^0 system at all influent Cu concentrations. A similar study but using heavy metal contaminated sediments was reported by Li et al. (2016). On the other hand, to enhance tolerance to high concentration of heavy metals and low pH, immobilized sulfate-reducing bacteria beads are used. For example, Zhang et al. (2016b) prepared novel SRM sludge beads for synthetic AMD treatment with high concentration of heavy metals. The tolerance of immobilized SRM beads to heavy metals was significantly enhanced compared with that of the suspended SRM. The bacterial population analyzed by denaturing gradient gel electrophoresis (DGGE) revealed a synergism between *D. desulfuricans* and other fermentative bacteria which plays a key role in the substrate utilization. Other studies evaluate the combined effect of metals on each other's removal using an ANOVA analysis and the effect of Fe(III) on the biotreatment of bioleaching solutions (Cao et al., 2013; Kiran et al., 2016).

2.2 Metals with more than two oxidation stages

Metals that have more than two oxidation stages can be first reduced by SRM and eventually precipitated by direct or indirect action of SRM. For example, U(VI) is reduced to U(IV) that can precipitate as oxide or carbonate. Interesting examples are the cases of Cr, As, Al, and Sb. Cr(VI) and Cr(III) are the most common forms of Cr, last one being less soluble and in consequence less toxic than Cr(VI). The use of SRM in Cr(III) bioprecipitation has been reported in many different studies (Viera et al., 2003; Perez et al., 2010; Kikot et al., 2010). However, the most common approaches followed to remove hexavalent chromium from wastewaters involve the reduction to Cr(III) followed by immobilization produced by the increase of pH (produced by sulfate reduction) to neutral values. Also, the H$_2$S generated by biological sulfate reduction can be used in a separate reduction of Cr(VI) to Cr(III). In addition, Cr(VI) can be used as an electron acceptor by SRM in the absence of sulfate (Barton et al., 2015). Sahinkaya et al. (2012) treated Cr(VI) synthetic acidic wastewater in an anaerobic baffled reactor (ABR) supplemented with ethanol. They observed both complete removal of Cr(VI) and high sulfate and COD removal rates. Other studies with different SRM population, organic substrates, and bioreactor configurations were also reported (Pagnanellia et al., 2012; Tekerlekopoulou et al., 2010; Singh et al., 2011; Cirik et al., 2013; Marquez-Reyes et al., 2013). Arsenic (As) is usually present as oxyanions: arsenate (AsO$_4^{3-}$), most abundant in aerobic environments, and arsenite (AsO$_3^{3-}$) commonly found in reducing conditions. It is a toxic metal frequently present in acid mine water and effluents (Hashim et al., 2011). The most common approach to remove As(V) is the co-precipitation or absorption into Fe oxyhydroxides (Hedrich and Johnson, 2014). As(V) could also be reduced by H$_2$S produced by SRM or used as electron terminal acceptor and then precipitated as As_2S_3. Battaglia-Brunet et al. (2012) evaluated the precipitation of As_2S_3 under different pH conditions and electron donors in sulfate-reducing reactors at low pH. The highest global arsenic removal rate was close to 2.5 $mg.L^{-1}.h^{-1}$ using glycerol as electron donor. The microbial community includes fermentative bacteria and *Desulfosporosinus*-like SRM.

The alternatives to remediate contaminated waters with Sb and Al are less reported in literature. Recent studies evaluated the use of sulfidogenic bioreactors in the treatment of contaminated sites with those elements. The strategy to be used depends on the chemistry and the oxidation states of the elements. For example, Zhang et al. (2016a) reported high removal of Sb(V) (93%) from simulated wastewater using batch SRM cultures. The mechanism followed was the same as explained for the removal of Cr(VI) and As(V). However, the removal of Al from acidic water was reported by Falagán et al. (2016). The process involves acidophilic/acid-tolerant SRB initially using protons present in the acidic water bodies causing the pH to increase to around 5 and then utilizing further protons generated as Al hydrolyses and precipitates.

2.3 Other biotechnological uses of SRM

Although their participation in bioprecipitation processes is highlighted, SRM can also participate in biosorption processes. In a pH range of 4 to 7, these microorganisms can retain heavy metals by biosorption because the components of the cell wall are in their deprotonated state. Biosorption in SRM has been reported to be independent of metabolism (sorption on cell walls) or dependent on metabolism (transport, or intracellular sequestration) (Sheoran et al., 2010). Besides their application in heavy metal bioprecipitation, the interest in SRM in the last decades has been their negative ecological and economic impact due to these microorganisms that are involved in microbially influenced corrosion (MIC) of platform structures, pipes, transmission lines, and general equipment. The production of Fe sulfides and the formation of SRM biofilms play a critical role during biocorrosion. Several studies to inhibit SRM activity and/or to avoid the attachment and consequent formation of biofilms are continuously reported (Gadd, 2010; Javed et al., 2015; Wikiel et al., 2013).

3. Case study

As it was mentioned above, interest in sulfate-reducing microorganisms that inhabit extreme environments characterized by low pH values (such as volcanic areas, hydrothermal, and acid mine drainage-affected areas) has increased enormously in recent years due to their application in many biotechnological processes. In this case study, results about metal bioprecipitation using sulfate-reducing consortia from Copahue Volcano are shown. The geothermal Caviahue-Copahue system is an extreme environment located in the north-west of Neuquén Province, Argentina. This area is characterized by a wide range of temperatures (20–90°C) and pH (< 1 to 8), and high concentrations of some metals (Chiacchiarini et al., 2010). Samples detected by negative Eh values—indicating anaerobic conditions—were collected, characterized, and used in Cr and Cd bioprecipitation. Comparison with the precipitation under acidic and neutral conditions is also included.

3.1 Methodology

Establishment of SRM enrichments

Sediment samples were taken from different hot springs with geothermal activity at Copahue volcano. They are geographically grouped in different zones: Copahue thermal centre, Las Máquinas, Las Maquinitas, Anfiteatro, and Chancho-Co (situated in the Chilean side of the cordillera). Las Máquinas (LMa), with a moderate temperature pool (38°C), low pH value (about 3), and reducing conditions (Eh about –70 mV) was chosen because it has been less affected by human activity. Las Maquinitas (LMi) is the smaller thermal spring at Copahue geothermal site. It is characterized by high temperature values (around 92°C) in the thermal ponds. In the last years, this place has been influenced by human activity because its sediments are important for health treatments. Finally, the section of thermal bath called Baño 9 (B9) is located at the Copahue Thermal Centre. This area is of interest as it has a wide range of pH (2–7) and temperature (15–90°C). Table 1 show different physicochemical parameters of the hot spring sediments collected at LMa, LMi, and B9 measured *in situ*. All samples were characterized by negative Eh values and low pH values.

Samples for acidophilic consortia were enriched in *aSRM* medium with glycerol 3 mM and Zn 4 mM at pH 3.0 (Willis et al., 2013). Nitrogen was bubbled through the media to displace oxygen to create an environment for the growth of anaerobic microorganisms. After that, the medium was sterilized by autoclaving and rapidly divided into 50 mL flasks under sterile conditions. Sediment samples were added to the flasks containing anaerobic medium. Flasks were incubated in sealed anaerobic jars at 30°C. After one month, glycerol was measured by an enzymatic colorimetric method, Zn

Table 1. Physicochemical characteristics of Copahue samples.

Site	pH	T [°C]	E_h [mV]
B9 (3)	5	67	–290
LMi	5	90	–126
LMa2	3	38	–70

concentration was measured using atomic absorption spectrophotometry and the remaining sulfate concentration using a turbidimetric method with $BaCl_2$ (APHA-AWWA-WPCF, 1992).

To obtain SRM consortia at neutral pH, samples from Copahue were enriched using Postgate B medium with lactate as the electron and carbon source, and sulfate as the terminal electron acceptor (Kikot et al., 2010). The pH was adjusted to 7.0 with NaOH 5 M. The medium was divided into 50 mL flasks, sealed with rubber stoppers and sterilized by autoclaving at 121°C for 20 minutes. After that, the flasks were opened under sterile conditions, sediment samples added and immediately closed. Flasks containing the medium and sediment samples were incubated at 40°C for 15–20 days. Growth of cultures was monitored periodically by measuring the remaining sulfate concentration using the turbidimetric method mentioned above and by observing the formation of black precipitates (FeS).

The microbial composition of both enrichments was analysed by cloning and sequencing. The procedure followed was the same that has been described by Willis et al. (2013).

Heavy metals bioprecipitation at low pH by SRM enrichments

In the preparation of the inoculum for bioprecipitation assays, 10 mL of the grown enrichments were added to 90 mL of sterile *aSRM* medium (initial pH 3) or Postgate C medium and incubated anaerobically. For acidophilic SRM inoculum, the medium was sparked with N_2 before sterilization to remove traces of dissolved oxygen. In the case of Postgate C medium, 50 µL of anaerobic solution (0.2 g ascorbic acid, 200 µL thioglycolic acid, and 10 mL distilled water) was added to obtain anoxic conditions. The concentrations of glycerol, Zn, sulfate, and pH were measured at the end of growth.

Samples for bioprecipitation tests were prepared in 10 ml flasks. Each of them was filled under sterile conditions with 9 ml of *aSRM* medium or Postgate C medium (prepared as above) at pH 3 or 7 respectively. In each case, the medium was prepared without the addition of $ZnSO_4$ or $FeSO_4$ to test the precipitation of other heavy metals. Different volumes of a 1000 mg.L^{-1} stock solution of each heavy metal were added to reach different concentration (5, 10, 25, and 50 mg.L^{-1}). The metals tested were Cr(III) and Cd. Finally, 1 mL of grown SRM enrichment cultures was added to each flask. The flasks were then sealed with rubber stoppers and incubated anaerobically at 30°C or 40°C for one month. Experiments were performed in quadruplicates. One extra tube was prepared in the same conditions and was used to determine the initial precipitation of metals due to the dissolved sulfide found in the inoculum.

Control tests without SRM inoculation were also performed with both media to distinguish the amount of heavy metals removed by biological mechanisms (bioprecipitation) from those removed by chemical precipitation. Control tests were incubated in the same conditions as for the inoculated systems. Samples for analytical determinations were previously filtered through 0.22 μm cellulosic acetate membrane to remove any possible precipitates. The initial and final concentrations of metals were determined by atomic absorption spectrophotometry. Previously, samples were diluted in 0.14 N nitric acid if it was necessary. For both enrichment cultures, bioprecipitation percentage (%BP) was calculated using the following equation:

$$\frac{([M]t = 0 - [M]t = f)}{[M]t = 0} x100 = \%BP \tag{4}$$

where $[M]_{t=0}$ is the dissolved metal concentration at initial time of abiotic controls and $[M]_{t=f}$ is the dissolved metal concentration after 30 days.

To determine the influence of different concentrations of Cr(III) and Cd on the growth of SRM consortia, glycerol and sulfate were measured at the beginning and at the end of the experiment for the acidophilic and neutrophilic SRM consortia respectively.

3.2 Results and discussion

Acidophilic and neutrophilic SRM enrichments

After 30 days of incubation, positive SRM growth was observed in both acidophilic and neutrophilic enrichments. Glycerol is used by sulfate reducing bacteria as carbon and energy source under anaerobic conditions; this carbon source is used at low pH values because it does not inhibit SRM growth. Complete reduction of glycerol was observed in all acidophilic enrichments. Sulfate concentration decreased by 1 mM and H_2S smell was detected in all cultures confirming sulfate was reduced to sulfide. Zn concentrations were also very low after one month of incubation, surely due to the precipitation as ZnS. However in SRM enrichments, at neutral pH, sulfate decreased from 3 to 0.5 g.L^{-1} and black precipitates were observed after 30 days of incubation at 40°C.

The microbial composition analyzed by cloning and sequencing revealed the presence of the sulfate reducer *"Desulfobacillus acidavidus"* as the dominant organism in the acidophilic SRM enrichments (99% 16S sequence similarity). For the enrichments established at neutral pH, the dominant sulfate reducer was found to be *Desulfotomaculum reducens* (99% 16S sequence similarity).

Very few acidophilic SRM have been reported to be present in acidophilic environments despite the fact that sulfate is usually present in elevated concentrations (Alazar et al., 2010; Falagán et al., 2014; Sánchez-Andrea et al., 2012a; Rowe et al., 2007). As it was mentioned above, one probable reason could be the pH-related toxicity of commonly used enrichment substrates as well as metabolic end products (such as sulfide and acetic acid).

Non-acidic alternative electron donors, such as glycerol have been tested to overcome this problem. Using glycerol, it was possible to establish a sulfidogenic SRM consortium being able to grow at low pH values in samples collected from different acidic hot springs at Copahue volcano. This indicates that the SRM metabolic processes were not adversely affected nor inhibited by the initial low pH. However, the presence of sulfate reducers, able to grow at neutral pH, was fully reported in literature. The members of the genus *Desulfotomaculum* were particularly found to inhabit extreme environments such as volcanic geothermal and mine-impacted areas, probably due to their ability to form endospore structures (Aullo et al., 2013).

In the next section, results about metal precipitation are described. Only one enrichment culture from the acidophilic and neutrophilic SRM was selected for the bioprecipitation test because all of them had the same microbial composition and physiological behavior.

Heavy metal precipitation by SRM enrichments

Figure 1 shows glycerol concentration by the acidophilic SRM enrichment culture with different concentrations of Cr(III) and Cd at the beginning and after one month. Glycerol consumption was high in the enrichment culture with 5, 10, and 25 mg.L^{-1} of Cr(III) and Cd. An increase in Cr(III) concentration did not significantly affect the growth and all glycerol was consumed after 30 days of incubation. However, the raise of Cd concentration to 50 mg.L^{-1} adversely affects the growth of the enrichment (glycerol concentration after 30 days of incubation was comparable to the beginning). Figure 2 shows the sulfate concentration in cultures at pH 7 under different concentrations of Cr(III) and Cd at the beginning and after one month of incubation at 40°C. Firstly, a decrease in sulfate concentration from 3.5 to 1.7 g.L^{-1} was observed in cultures with 5 mg.L^{-1} of Cd, although this decline was lower than that observed for cultures without metal.

Moreover, an increase in Cd concentrations to 50 mg.L^{-1} did not affect the growth of the SRM consortia. Also, it can be observed that there is no clear evidence of an inhibitory effect of Cr(III) increase on the growth of the SRM consortia.

The results of Cd bioprecipitation are shown in Fig. 3. Neutrophilic SRM consortium could precipitate Cd more than in abiotic controls at the lower metal concentrations (5 and 10 mg.L^{-1}). When Cd concentrations increase (25 and 50 mg.L^{-1}), no differences between inoculated and sterile systems were observed, although microorganisms were not inhibited (Fig. 2). However, acidophilic SRM showed an excellent performance for Cd precipitation at 5, 10, and 25 mg.L^{-1} and even at this pH value, the abiotic controls reached low precipitation percentages. These data correlate with the complete oxidation of glycerol observed at the same metal concentration (Fig. 1). The significantly decrease in glycerol consumption observed at 50 mg.L^{-1} of Cd was reflected

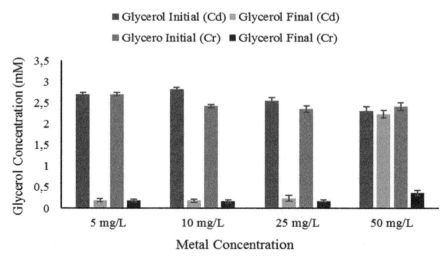

Figure 1. Glycerol concentrations at the beginning and end of the bioprecipitation experiment under different concentrations of Cd and Cr (III) for the SRM enrichment at pH 3.

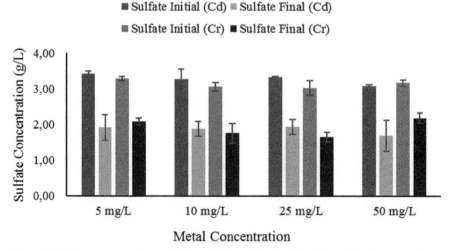

Figure 2. Sulfate concentrations at the beginning and end of the bioprecipitation experiment under different concentrations of Cd and Cr (III) for the SRM enrichment at pH 7.

in their bioprecipitation percentages (not higher than 4%), indicating that the precipitation of Cd was highly related with sulfide production. This metal is reported to be more toxic than, for example, Cr. Hao et al. (1994) found an inhibitory effect of sulfate reduction at 20 mg.L⁻¹ of Cd. These results are similar to the ones obtained in the experiments with our acidophilic SRM enrichment. However, neutrophilic SRM consortium was less inhibited at higher Cd concentration, although it was not reflected in its precipitation percentages at 25 and 50 mg.L⁻¹. An enhanced sulfate reduction under 60 mg.L⁻¹ of Cd was observed by Gonzalez-Silva et al. (2009) in a sulfate reducing granular sludge. In the case of Cr, the neutrophilic SRM consortium is only able to overcome the precipitation in abiotic controls at higher metal concentrations (50 mg.L⁻¹) since abiotic Cr precipitation is very important at pH values close to neutral (Fig. 4). Like Cd, the acidophilic microbial consortium reached high percentages of Cr precipitation at all concentrations

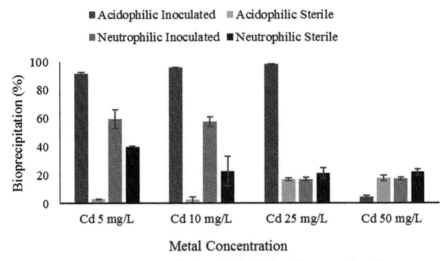

Figure 3. Cd bioprecipitation at pH 3.0 and 7.0 by SRM enrichments and sterile systems.

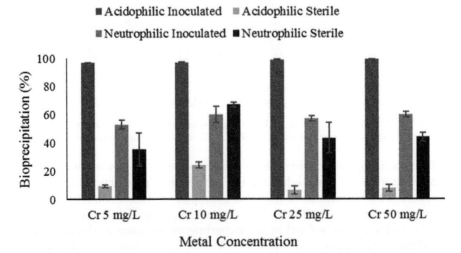

Figure 4. Cr(III) bioprecipitation at pH 3.0 and 7.0 by SRM enrichments and sterile systems.

tested even at 50 mg.L^{-1}. These data correlate with the complete glycerol consumption observed (Fig. 1).

From the results, it can be concluded that at all concentrations of Cr(III), no differences on the acidophilic consortium growth were observed and glycerol was completely consumed after one month of incubation. Cabrera et al. (2006) have shown that low concentrations of Cr(III) can improve the growth in SRM pure cultures. In addition, total Cr precipitation was achieved at all concentrations.

4. Conclusion

Bioprecipitation using sulfate-reducing microorganisms is a promising technology for the removal of heavy metals from contaminated waters. One of the difficulties to use this technology is the high sensitivity of these microorganisms to low pH values which is a common characteristic of mine-impacted waters. As it was demonstrated in this case study, the geothermal Caviahue-Copahue area, specially its acidic hot springs, is an extreme environment with a great potential for the cultivation of anaerobic microorganisms, especially sulfate reducers. The use of a non-acidic electron donor-like glycerol-allowed the cultivation of SRM at low pH values from sediment samples. However, neutrophilic SRM were also cultivated from acidic sediments, demonstrating the versatility of the microorganisms found in this area. Regarding bioprecipitation, using sulfate-reducing consortia obtained at both neutral and acidic conditions from Copahue volcano, it was possible to precipitate Cr and Cd. The precipitation values correlate with the tolerance of the consortia to the concentrations of heavy metals tested. In general, biological metal precipitation was more efficient at low pH values. All these results suggest that bioprecipitation is a promising technology that can be used for the removal of heavy metals from contaminated wastewaters.

References

Alazar D., Joseph M., Battaglia-Brunet F., Cayol J.L., Olliver B. *Desulfosporosinus acidiphilus* sp. nov.: a moderately acidophilic sulfate-reducing bacterium isolated from acid mining drainage sediments. Extremophiles 14, 305–312, 2010.

APHA-AWWA-WPCF. Métodos normalizados para el análisis de aguas potables y residuales (1st ed). Díaz de Santos, Madrid, 1992.

Aüllo T., Ranchou-Peyruse A., Ollivier B., Magot M. *Desulfotomaculum* spp. and related gram-positive sulfate reducing bacteria in deep subsurface environments. Front. Microbiol. 4, 1–12, 2013.

Bai H., Kang Y., Quan H., Han Y., Sun J., Feng Y. Bioremediation of copper-containing wastewater by sulfate-reducing bacteria coupled with iron. J. Environ. Manage. 129, 350–356, 2013.

Barton L.L., Tomei-Torres F.A., Xu H., Zocco T. Metabolism of metals and metalloids by the sulfate-reducing bacteria. *In:* Saffarini D. (ed.). Bacterial-Metal Interactions. Springer International Publishing, pp. 57–83, 2015.

Battaglia-Brunet F., Crouzet C., Burnol A., Coulon S., Morin D., Joulian C. Precipitation of arsenic sulphide from acidic water in fixed-film bioreactor. Water Res. 46, 3923–3933, 2012.

Bertolino S.M., Melgaco L.A., Quites N.C., Leao V.A. Performance evaluation of two anaerobic reactors for removing sulphate from industrial effluents. Adv. Mat. Res. 825, 491–495, 2013.

Cabrera G., Perez R., Gomez J.M., Abalos A., Cantero D. Toxic effects of dissolved heavy metals on *Desulfovibrio vulgaris* and *Desulfovibrio* sp. strains. J. Hazard. Mater. 135, 40–46, 2006.

Cao J., Li Y., Zhang G., Yang C., Cao X. Effect of Fe(III) on the biotreatment of bioleaching solutions using sulfate-reducing bacteria. Int. J. Miner. Process. 125, 27–33, 2013.

Cao J., Zhang G., Mao Z.S., Li Y., Fang Z., Yang C. Influence of electron donors on the growth and activity of sulfate-reducing bacteria. Int. J. Miner. Process. 106–109, 58–64, 2012.

Castro H., Williams N., Ogram A. Phylogeny of sulfate-reducing bacteria. FEMS Microbiol. Ecol. 31, 1–9, 2000.

Chiacchiarini P., Lavalle L., Giaveno A., Donati E. First assessment of acidophilic microorganisms from geothermal Copahue-Caviahue system. Hydometallurgy 104, 334–341, 2010.

Cibati A., Cheng K.Y., Morris C., Ginige M.P., Sahinkaya E., Pagnanelli F., Kaksonen A.H. Selective precipitation of metals from synthetic spent refinery catalyst leach liquor with biogenic H_2S produced in a lactate-fed anaerobic baffled reactor. Hydrometallurgy 139, 154–161, 2013.

Cirik K., Dursun N., Sahinkaya E., Çinar O. Effect of electron donor source on the treatment of Cr(VI)-containing textile wastewater using sulfate-reducing fluidized bed reactors (FBRs). Bioresour. Technol. 133, 414–420, 2013.

Colin V.L., Villegas L.B., Abate C.M. Indigenous microorganisms as potential bioremediators for environments contaminated with heavy metals. Int. Biodeterior. Biodegr. 69, 28–37, 2012.

Costa M.C., Santos E.S., Barros R.J., Pires C., Martins M. Wine wastes as carbon source for biological treatment of acid mine drainage. Chemosphere 75, 831–836, 2009.

Das B.K., Roy S., Dev S., Das D., Bhattacharya J. Improvement of the degradation of sulfate rich wastewater using sweetmeat waste (SMW) as nutrient supplement. J. Hazard. Mater. 300, 796–807, 2015.

Dev S., Roy S., Bhattacharya J. Understanding the performance of sulfate reducing bacteria based packed bed reactor by growth kinetics study and microbial profiling. J. Environ. Manage. 177, 101–110, 2016.

Dixit R., Waisullah A., Malaviya D., Pandiyan K., Singh U.B., Sahu A., Shukla R., Singh B.P., Rai J.P., Sharma P.K., Lade H., Paul D. Bioremediation of heavy metals from soil and aquatic environment: An overview of principles and criteria of fundamental processes. Sustainability 7, 2189–2212, 2015.

Falagán C., Sanchez-España J., Johnson D.B. New insights into the biogeochemistry of extremely acidic environments revealed by a combined cultivation-based and culture-independent study of two stratified pit lakes. FEMS Microbiol. Ecol. 87, 231–243, 2014.

Falagán C., Yusta I., Sanchez-España J., Johnson D.B. Biologically-induced precipitation of aluminium in synthetic acid mine water. Miner. Eng. http://dx.doi.org/10.1016/j.mineng.2016.09.028, 2016.

Fu F., Wang Q. Removal of heavy metal ions from wastewaters: A review. J. Environ. Manage. 92, 407–418, 2011.

Gadd G.M. Metals, minerals and microbes: geomicrobiology and bioremediation. Microbiology 156, 609–643, 2010.

Gonzalez-Silva B.M., Briones-Gallardo R., Razo-Flores E., Celis L.B. Inhibition of sulfate-reduction by iron, cadmium and sulfide in granular sludge. J. Hazard. Mater. 172, 400–407, 2009.

Hao O.J., Huang L., Chen J.M. Effects of metal additions on sulfate reduction activity in wastewaters. Toxicol. Environ. Chem. 46, 197–212, 1994.

Hao T., Xiang P., Mackey H., Chi K., Lu H., Chui H., van Loosdrecht M.C.M., Chen G.H. A review of biological sulfate conversions in wastewater treatment. Water Res. 65, 1–24, 2014.

Hashim M.A., Mukhopadhyay S., Sahu J.N., Sengupta B. Remediation technologies for heavy metal contaminated groundwater. J. Environ. Manage. 92, 2355–2388, 2011.

Hedrich S., Johnson D.B. Remediation and selective recovery of metals from acidic mine waters using novel modular bioreactors. Environ. Sci. Technol. 48, 12206–12212, 2014.

Javed M.A., Stoddart P.R., Wade S.A. Corrosion of carbon steel by sulphate reducing bacteria: Initial attachment and the role of ferrous ions. Corros. Sci. 93, 48–57, 2015.

Johnson D.B. Recent developments in microbiological approaches for securing mine wastes and for recovering metals from mine waters. Minerals 4, 279–292, 2014.

Johnson D.B. Reductive dissolution of minerals and selective recovery of metals using acidophilic iron- and sulfate-reducing acidophiles. Hydrometallurgy 127-128, 172–177, 2012.

Johnson D.B., Hallberg K.B. Acid mine drainage remediation options: a review. Sci. Total Environ. 338, 3–14, 2005.

Karri S., Alvarez R.S., Field J.A. Zero-valent iron as an electron donor for methanogenesis and sulfate reduction in anaerobic sludge. Bioresour. Technol. 92, 810–819, 2005.

Kikot P., Mignone C., Viera M., Donati E. Study of the effect of pH and dissolved heavy metals on the growth of sulfate-reducing bacteria by a fractional factorial design. Hydrometallurgy 104, 494–500, 2010.

Kimura S., Hallberg K.B., Johnson D.B. Sulfidogenesis in low pH (3.8–4.2) media by a mixed population of acidophilic bacteria. Biodegradation 17, 159–167, 2006.

Kiran M.G., Pakshirajan K., Das G. Heavy metal removal from multicomponent system by sulfate-reducing bacteria: Mechanism and cell surface characterization. J. Hazard. Mater. 324, 62–70, 2017.

Li X., Wu Y., Zhang C., Liu Y., Zeng G., Tang X., Dai L., Lan S. Immobilizing of heavy metals in sediments contaminated by nonferrous metals smelting plant sewage with sulfate reducing bacteria and micro zero valent iron. Chem. Eng. J. 306, 393–400, 2016.

Liamleam W., Annachhatre A.P. Electron donors for biological sulfate reduction. Biotechnol. Adv. 25, 452–463, 2007.

Márquez-Reyes J.M., López-Chuken U.J., Valdez-González A., Luna-Olivera H.A. Removal of chromium and lead by a sulfate-reducing consortium using peat moss as carbon source. Bioresour. Technol. 144, 128–134, 2013.

McCullough C.D., Lund E.M.A., May E.J.M. Field-scale demonstration of the potential for sewage to remediate acidic mine waters. Mine Water Environ. 27, 31–39, 2008.

Ñancucheo I., Johnson D.B. Selective removal of transition metals from acidic mine waters by novel consortia of acidophilic sulfidogenic bacteria. Microb. Biotechnol. 5, 34–44, 2012.

Ñancucheo I., Johnson D.B. Removal of sulfate from extremely acidic mine waters using low pH sulfidogenic bioreactors. Hydrometallurgy 150, 222–226, 2014.

Pagnanellia F., Cruz Viggi C., Cibati A., Uccelletti D., Toro L., Palleschi C. Biotreatment of Cr (VI) contaminated waters by sulphate reducing bacteria fed with ethanol. J. Hazard. Mater. 199-200, 186–192, 2012.

Perez R.M., Cabrera G., Gomez J.M., Abalos A., Cantero D. Combined strategy for the precipitation of heavy metals and biodegradation of petroleum in industrial wastewaters. J. Hazard. Mater. 182, 896–902, 2010.

Rowe O.F., Sanchez-España J., Hallberg K.B., Johnson D.B. Microbial communities and geochemical dynamics in an extremely acidic, metal-rich stream at an abandoned sulfide mine (Huelva, Spain) underpinned by two functional primary production systems. Environ. Microbiol. 9, 1761–1771, 2007.

Sahinkaya E., Altum M., Bektas S., Komitsas K. Bioreduction of Cr(VI) form acidic wastewaters in a sulfidogenic ABR. Miner. Eng. 32, 38–44, 2012.

Sanchez-Andrea I., Rojas-Ojeda P., Amils R., Sanz J.L. Screening of anaerobic activities in sediments of an acidic environment: Tinto River. Extremophiles 16, 829–839, 2012a.

Sanchez-Andrea I., Triana D., Sanz J.L. Bioremediation of acid mine drainage coupled with domestic wastewater treatment. Water Sci. Technol. 66, 2425–2431, 2012b.

Sanchez-Andrea I., Stams A.J.M., Amils R., Sanz J.L. Enrichment and isolation of acidophilic sulfate-reducing bacteria from Tinto River sediments. Environ. Microbiol. 5, 672–678, 2013.

Sanchez-Andrea I., Stams A.J.M., Hedrich S., Ñancucheo I., Johnson D.B. *Desulfosporosinus acididurans* sp. nov.: an acidophilic sulfate-reducing bacterium isolated from acidic sediments. Extremophiles 19, 39–47, 2015.

Sheoran A.S., Sheoran V., Choudhary R.P. Bioremediation of acid-rock drainage by sulphate-reducing prokaryotes: A review. Miner. Eng. 23, 1073–1100, 2010.

Singh N., Gadi R. Biological methods for speciation of heavy metals: different approaches. Crit. Rev. Biotechnol. 29, 307–312, 2009.

Singh R., Kumar A., Kirrolia A., Kumar R., Yadav N., Bishnoi N.R., Lohchab R.K. Removal of sulphate, COD and Cr (VI) in simulated and real wastewater by sulphate reducing bacteria enrichment in small bioreactor and FTIR study. Bioresour. Technol. 102, 677–682, 2011.

Tekerlekopoulou A.G., Tsiamis G., Dermou E., Siozios S., Bourtzis K., Vayenas D.V. The Effect of Carbon Source on Microbial Community Structure and Cr(VI) Reduction Rate. Biotechnol. Bioeng. 107, 478–487, 2010.

Thauer R.K., Stackebrandt E., Hamilton W.A. Energy metabolism and phylogenetic diversity of sulphate-reducing bacteria. *In*: Barton L., Hamilton W.A. (eds.). Sulphate-reducing Bacteria: Environmental and Engineered Systems. Cambridge University Press, pp. 1–37, 2007.

Viera M., Curutchet G., Donati E. A combined bacterial process for the reduction and immobilization of chromium. Int. Biodeterior. Biodegr. 52, 31–34, 2003.

Viera M., Donati E. Microbial processes to metal recovery from waste products. Curr. Top. Biotechnol. 1, 117–127, 2004.

Wikiel A.J., Datsenko I., Vera M., Sand W. Impact of *Desulfovibrio alaskensis* biofilms on corrosion behaviour of carbon steel in marine environment. Bioelectrochemistry 97, 52–60, 2014.

Willis Poratti G., Hedrich S., Ñancucheo I., Johnson B., Donati E.R. Microbial diversity in acidic anaerobic sediments from geothermal Caviahue-Copahue system. Adv. Mat. Res. 825, 7–10, 2013.

Zhang G., Ouyang X., Li H., Fu Z., Chen J. Bioremoval of antimony from contaminated waters by a mixed batch culture of sulfate-reducing bacteria. Int. Biodeterior. Biodegr. 115, 148–155, 2016a.

Zhang M., Wang H., Han X. Preparation of metal-resistant immobilized sulfate reducing bacteria beads for acid mine drainage treatment. Chemosphere 154, 215–223, 2016b.

CHAPTER 7

Bioleaching Strategies Applied to Sediments Contaminated with Metals
Current Knowledge and Biotechnological Potential for Remediation of Dredged Materials

Viviana Fonti,[1] *Antonio Dell'Anno*[2] *and Francesca Beolchini*[2]

1. Bioleaching vs. Sediment: Is there a possible match?

In all the aquatic ecosystems, metal contaminants are not persistently stored in the sediment but may easily enter the food web or spread in the environment due to changes in the environmental conditions (Bortone et al., 2004; Agius and Porebski, 2008; Förstner and Salomons, 2010). The management of contaminated aquatic sediments as dredged material is a modern-day issue of significant concern. Industrial/commercial ports, rivers, channels, lakes, and estuaries need to be dredged periodically to ensure navigational depth and a good capacity of drainage and flood prevention. Due to the burden of the anthropogenic activities usually allocated in these environments and the low hydrographic regime, the concentration of metal contaminants in the dredged materials is often high. Currently, the main management options for these materials are landfill disposal and confined aquatic disposal—two solutions with high costs and low environmental sustainability (Bortone et al., 2004; Agius and Porebski, 2008). In this context, bioleaching has been thought to have a potential to match the need of environment-friendly

[1] Environment and Sustainability Institute, University of Exeter, Penryn Campus, Penryn, Cornwall, TR10 9FE, United Kingdom.
[2] Department Life and Environmental Science, Università Politechnica delle Marche, Via Brecce Bianche, 60131 Ancona, Italy.

and cost-effective management options for dredged aquatic sediments contaminated with metals (White et al., 1998; Blais et al., 2001; Chen and Lin, 2004; Tabak et al., 2005). Because of the non-degradable nature of metal pollutants, the remediation strategies aimed at coping with them can be only aimed either (1) at increasing the solubility of metals to facilitate their removal (i.e., mobilization) or (2) at increasing their stability to reduce their bioavailability (i.e., immobilization; Fig. 1). Ideally, dredged sediments can decontaminated by bioleaching-based strategies, where the extraction of metals and semi-metals is catalyzed by bacteria (and/or archaea) that are able to oxidize inorganic sulfur compounds and/or Fe(II). Once cleanzed, the material could well suit the building industry or could be used for beach refill as well as for numerous other applications (Lee, 2000; Ahlf and Förstner, 2001; Barth et al., 2001; Siham et al., 2008). The possibility to exploit such Fe/S oxidizing strains for the reclamation of aquatic sediment arises in the early 50's from the implementation by Kennecott Copper Corporation of an industrial scale process of copper extraction from mine dumps that was (and still is) mediated by *Acidithiobacillus ferrooxidans*. Subsequently, that technology has been improved and further mining applications have been implemented. Today, minerals/ores containing Cu, Au, and Co are processed on industrial scale. Promising results have also been obtained with sulfide ores bearing Ni, Zn, Mo, Ga, Pb, and metals in the Pt group (Ehrlich, 2001; Lee and Pandey, 2012).

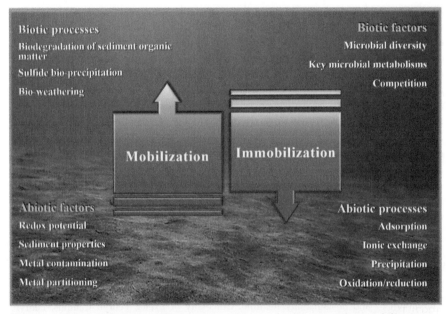

Figure 1. Abiotic and biotic influences on processes leading to either mobilization or immobilization of metal contaminants in the sediment.

In parallel with the increasing success of application of bioleaching in hydrometallurgy and mining, microorganisms involved in bioleaching have been investigated for potential application in other fields. The first record of bioleaching bacteria exploitation for non-metallurgic purposes is in 1974, the degradation of oil shale by the sulfur oxidizing bacterium *Acidithiobacillus thiooxidans* with the aim to produce kerogen, bitumen, and fuel precursors (Findley et al., 1974). Further studies have explored coal desulfurization, removal of toxic metals from contaminated environmental matrices (e.g., sludge, soil or sediment) (Tichy et al., 1998) and, in the last decade, recovery of base valuable metals from industrial and urban solid wastes. The first record of the application of bioleaching on contaminated aquatic sediments was published in 1993 by Couillard and Chartier. The research was inspired by a biological process of metal solubilization from sewage sludge, developed a few years earlier at INRS-Eau (Couillard and Chartier, 1991; Blais et al., 1992; Couillard and Mercier, 1993). Contaminated sediment shows undeniable similarities with sewage sludge but further studies have pointed out that the sediment has specific characteristics which must be taken into account (Rulkens et al., 1995; US EPA, 2005). Moreover, sediment properties can differ greatly from site to site, with significant effects on the potential application of bioleaching strategies (Mulligan et al., 2001; Bianchi et al., 2008; Fonti et al., 2013).

2. Sediment properties & contamination with metals

2.1 The sediment

The sediment is a polyphasic environmental matrix (Fig. 2) that forms by natural sedimentation of particulate materials within the water column, which in turn, originate by a variety of different natural processes: e.g., erosion, transport by rivers, deposition of organic particles, weathering of primary minerals, mineral (bio-)precipitation (e.g., carbonates, iron hydroxide, and lateral transport; Warren and Haack, 2001; Bridge, 2008). The sediment is an essential, integral, and dynamic part of any aquatic ecosystem. It provides a variety of habitats at different scales and creates favorable conditions for hosting a wide biodiversity (Torsvik et al., 2002; Lozupone and Knight, 2007; Fierer and Lennon, 2011; Wang et al., 2012). Reactions occurring in the interstitial water of sediments as well as the biogeochemical transformations occurring in them regulate the ecosystem functioning of the water body they belong to.

One of the most important sediment characteristics is its extreme heterogeneity in composition and its physicochemical properties, both at horizontal and vertical spatial scale (Salomons and Brils, 2004; Hakanson, 1992; Batley et al., 2005). Because of the intrinsic nature of sediment as a depositional material, its composition varies greatly from site to site,

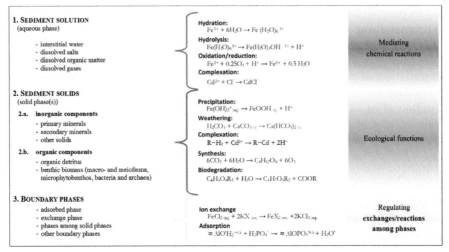

Figure 2. Physico-chemical phases in the sediment. The sediment environment can be described as consisting of three type of phases (Fig. 2): (1) the aqueous phase, or sediment interstitial water where salts, gases, and organic molecules are dissolved; (2) several solid phases that form the solid part of the sediment matrix and that fall in the following categories: (2a) minerals (inorganic component), (2b) sediment organic matter (organic component) and (2c) living organisms (biotic component); phases in the sediment are not sealed off from each other but several reactions occur among them; resulting components that allocate at the transitional interface among phases are known as (3) boundary phases. An example of boundary phase is given by the exchangeable ions that reside in the sediment exchanger phase or by the ions and molecules retained by mechanisms other than exchange and precipitation (i.e., one of the processes by which a true solid phase forms), which may be identified as residing in the adsorbed phase (Fonti et al., 2016).

depending on the anthropic activities in the water basin and surrounding areas, the hydrographical properties of the water basin, the geography of the site, and geological causes.

Sediment characteristics influence the effectiveness of treatment technologies (US EPA, 1994; Vallero, 2010). Specific chemical and physical properties are used to describe a sediment sample: grain size, grain density, water content, total organic matter (TOM), buffering capacity, carbonates, geotechnical, and agronomic properties (Batley et al., 2005). Water content, TOM content, dry bulk density, and porosity affect basic sediment characteristics such as diffusion properties and mechanical properties as well as the microbial metabolic rates (Avnimelech et al., 2001). Geochemical composition, grain size, and organic matter influence the type and the grade of sediment contamination. Grain size and geotechnical properties (e.g., plasticity, compressibility, undrained shear strength and sensitivity) influence the behavior of sediment in engineering operations, as well as large-scale treatment applications (Lee, 1982; Tessier et al., 1982; Bergamaschi et al., 1997). Freshly dredged sediment is nearly impermeable but long-term storage in the open transforms sediment into soil-like materials (Vermeulen et al., 2003). For the application of bioleaching strategies, the acid-neutralizing capacity (i.e., buffering capacity against acidification) is particularly relevant in determining metal removal efficiency from contaminated sediment. A high content in carbonates results in a high acid neutralizing capacity (eq. 1). Sediment organic matter as well contributes to neutralize protons due to protonation of functional groups (e.g., carboxyl, phenolic, amino groups) in high molecular weight organic compounds (eq. 2–3).

$$CaCO_{3(s)} + 2H^+_{(aq)} \longrightarrow Ca^{2+}_{(aq)}\ CO_{2(g)} + H_2O_{(l)} \tag{1}$$

$$R\text{-}COO^- + H^+ \longrightarrow R\text{-}COOH \tag{2}$$

$$R\text{-}NH_2 + H^+ \longrightarrow R\text{-}NH_3^+ \tag{3}$$

Al-bearing minerals contribute to the acid-neutralizing power of the sediment (eq. 4) as well despite much slower kinetics than calcite and other carbonates (White et al., 2001).

$$KAlSi_3O_{8(s)} + 4H_2O_{(l)} + 4H^+_{(aq)} \longrightarrow Al^{3+}_{(aq)} + K^+_{(aq)} + 3H_4SiO_{4(aq)} \tag{4}$$

Buffering capacity can be measured by titration of a sediment suspension with 1 M H_2SO_4 but the determination of the carbonate content provides a useful indirect estimation (Seidel et al., 2004; Löser et al., 2005; Löser et al., 2006a).

Heterogeneity, high water content, fine grain size, poor hydraulic and agronomic properties, high acid neutralizing power, mixed contamination, site-specific geochemical properties, and the other characteristics described above make the sediment a challenging matrix to be treated by bioleaching strategies.

2.2 Contamination with metals

Metals and semi-metals can enter a water body as a result of multiple processes such as atmospheric deposition, erosion of the bed-rock minerals as well as intake by a variety of anthropic activities (Salomons and Brils, 2004; Perrodin et al., 2012). In the water column, metals tend to be scavenged into the sediment due to high affinity with complex organic matter, Fe/Al/Mn oxides, sulfides, and other sediment components. As a consequence, metals typically accumulate into the sediment and reach concentrations very high than in the water column (Eggleton and Thomas, 2004; De Jonge et al., 2012). Metals and semi-metals associate with sediment solids (Fig. 2) by several mechanisms (e.g., particle surface adsorption, ion exchange, complexation with organic substances, co-precipitation, and precipitation) and distribute among various geochemical phases in a site-specific way. Metal bioavailability and toxicity are strictly dependent upon their partitioning among the geochemical phases of the sediment because the latter influences largely their mobility and speciation (Ahlf and Förstner, 2001; Chapman and Wang, 2001; Hlavay et al., 2004; Prica et al., 2010). An indirect indication of the bioavailability of metal pollutants can be gained by means of procedures of selective sequential extractions which basically describe their mobility in terms of partitioning among sediment geochemical fractions. Up to 8 main geochemical fractions can be considered: (1) the "easily exchangeable" fraction (i.e., water-soluble, non-specifically adsorbed metals), (2) the "easily mobilizable" fraction (i.e., specifically bound, surface-occluded metals; some metals bound to $CaCO_3$ and metal-organic complexes with weak bonding forces can fall in this fraction), (3) metals bound to "carbonates fraction" (an amount metals associated with acid volatile sulfides could be detected in this fraction), (4) the "organic fraction" (i.e., metals associated to functional groups of high molecular weight organic compounds), (5) the "Mn-oxide fraction", (6) the "Fe- and Al-oxide fraction" (metals in this fraction bind either amorphous and crystalline oxides), (7) the "sulfides fraction", (i.e., resistant metal sulfides), (8) the "residual fraction" which refers to metals within the crystalline lattice of primary and secondary minerals (Filgueiras et al., 2002; Gleyzes et al., 2002; Hlavay et al., 2004; Zimmerman and Weindorf, 2010). Metals in the residual fraction are considered to be very stable and are thought to reflect the natural background level of the site (relative to geological and non-anthropogenic causes). On the contrary, metals in non-residual fractions, especially those from the exchangeable to carbonates fraction can be mobilized or solubilized because of biotic and abiotic processes such as changes in ionic strength, pH, and in the oxidation/reduction potential. The consequent increase in bioavailability may lead metal pollutants to enter the food web with potential detrimental effects on ecosystem health (Gleyzes et al., 2002; Eggleton and Thomas, 2004; Toes et al., 2004; Fonti et al., 2015). One of the most applied procedures of selective sequential extraction is the three-step protocol by the European Standard Measurements and Testing Program (SMT; f.k.a. European Communities Bureau of Reference, BCR). This procedure was designed on the basis of the contributions by Salomons and Förstner (Förstner, 1993; Salomons, 1993). It divides metals into four sediment macro-fractions: (i) an "acid-soluble fraction", which comprises exchangeable and carbonate-bound fractions; (ii) a "reducible fraction", which refers to metals associated with Fe/Mn oxides; (iii) an "oxidizable fraction", with metals bound to sediment organic matter and to sulfides; and (iv) the residual fraction (Quevauviller, 2002, 1998a, 1998b). Although selective sequential extraction procedures are known to have limitations (e.g., lack of specificity in some steps and production of artefacts), such techniques are largely applied due to their high usefulness in providing considerable insights about the environmental behavior of metals (Eggleton and Thomas, 2004; Toes et al., 2004; Bacon and Davidson, 2008).

Unlike organic pollutants, metals and semi-metals cannot be degraded by biological or chemical processes but they can just be transformed. For this reason, many remediation strategies attempt to increase metal solubility (i.e., mobilization) or increase their stability and reduce bioavailability (i.e., immobilization; Tabak et al., 2005; van Hullebusch et al., 2005).

3. Mechanisms and microorganisms in bioleaching

3.1 Overview of microorganisms

In bioleaching, the metabolic products (i.e., protons, S-oxidation intermediates, and Fe^{3+} ions) of acidophilic Fe/S oxidizing bacteria and archaea are responsible for the solubilization of metals and semi-metals through acid/oxidative attack (Schippers, 2004). Fe/S oxidizers constitute a non-phylogenetic group of strains that share useful characteristics in bio-hydrometallurgical applications such as: (i) tolerance to high concentrations of metals and semi-metals; (ii) tolerance to extremely acidic conditions, even pH values lower than 1.5; (iii) the ability to acidify their environment and/or to increase the oxidation/reduction potential; (iv) the ease of handling in real application. Fe/S oxidizing bacteria are distributed among α- β- γ-Proteobacteria (*Acidithiobacillus, Acidiphilium, Acidiferrobacter, Ferrovum*), Nitrospirae (*Leptospirillum*), Firmicutes (*Alicyclobacillus, Sulfobacillus*), and Actinobacteria (*Ferrimicrobium, Acidimicrobium, Ferrithrix*), while Fe/S oxidizing archaea belong mostly to Crenarchaeota (*Sulfolobus, Acidianus, Metallosphaera, Sulfurisphaera*), although two acidophilic iron(II)-oxidizers are affiliated with Euryarchaeota (*Ferroplasma acidiphilum* and *Ferroplasma acidarmanus*). Most strains support their growth by oxidation of sulfur compounds, like S^0 and reduced inorganic sulfur compounds (RISCs), and/or Fe(II) (i.e., dissimilatory metabolism, chemolithotrophy). However, many strains are mixotrophic and need a source of organic carbon. Some acidophilic obligate heterotrophs may support bioleaching or have even been demonstrated to oxidize sulfur compounds (Dopson and Johnson, 2012; Hedrich et al., 2011; Rohwerder et al., 2003; Vera et al., 2013). The dissimilatory metabolism is often specialized in the use of certain S-compounds (i.e., thiosulfate, S^0) or Fe(II) as electron donors. However, some species, like *Acidithiobacillus ferrooxidans*, can live aerobically by using Fe(II), S^0, and RISCs as electron donors. Similarly, *Sulfobacillus thermosulfidooxidans* and *S. acidophilus* can live either on the oxidation of Fe and S, although optimal growth occurs under mixotrophic conditions. The large majority of S-oxidizing acidophiles are obligate aerobes (e.g., *At. thiooxidans* and *At. caldus*), while some strains (e.g., *At. ferrooxidans* and *Acidiferrobacter thiooxydans*) have been found to use Fe(III) as an alternative electron acceptor and grow in anoxic conditions (Mangold et al., 2011; Valdés et al., 2008). *At. ferrooxidans* and *At. ferrivorans* as well can live anaerobically through the oxidation of hydrogen (or $S^=$) coupled with Fe(III) reduction (Johnson, 2012; Ohmura et al., 2002; Osorio et al., 2013; Hallberg et al., 2010).

Fe/S oxidizers can be clustered on the basis of the temperature range in which they grow as: (i) mesophiles (below 40°C), (ii) moderate thermophiles (about 45°C or more), and (iii) thermophiles (about 70°C or more). However, such temperature thresholds are not very stringent. For example, *At. caldus* and *Acidimicrobium ferrooxidans* are active within the range of 25–55°C. The group of mesophiles is constituted exclusively by Gram-negative strains within Eubacteria domain and *Acidithiobacillus*, *Thiobacillus*, and *Leptospirillum* are the main genera. Mesophilic Fe-oxidizers are mainly affiliated to *Leptospirillum* genus (*L. ferrooxidans*, *L. ferriphilum*). Moderate thermophiles are exclusively bacteria with S-oxidizers belonging to Firmicutes (*Sulfobacillus* and *Alicyclobacillus*) and Fe-oxidizers belonging to Actinobacteria (*Ferrimicrobium*, *Acidimicrobium*, *Ferrithrix*). The thermophiles group is represented by prokaryotes belonging to the Archaea domain with the S-oxidizers affiliated mainly to the genera *Sulfolobus* and *Metallosphaera* and Fe-oxidizers with the genus *Ferroplasma*. No strictly psychrophilic sulfur-oxidizing strains have been identified, yet, although the Fe- and S-oxidizing proteobacterium *Acidithiobacillus ferrivorans* and the recently isolated *Ferrovummyxofaciens* are psychrotolerant, growing in the range between 4–35°C (Dopson and Johnson, 2012; Johnson et al., 2014).

S and Fe oxidation occurs also at circumneutral pH's. Such processes play an important role in biogeochemical cycles of circumneutral environments with high concentrations of S and Fe (e.g., oxic/anoxic interface zones) and contribute to a large extent to seafloor weathering processes. Many neutrophilic Fe-oxidizing strains are chemolithotrophic (e.g., *Galionella ferruginea*, *Sphaerotilus natans* and *Lepthotrix ochracea*) but heterotrophic bacteria are also known (Fleming et al., 2011; Gridneva et al., 2011; Hedrich et al., 2011). Within neutrophilic S-oxidizers, four metabolisms can be identified: (i) obligate chemolithoautotrophy on the oxidation of RISCs and S^0 with assimilation of CO_2 (e.g., *Halothiobacillus neapolitanus*, *Thermithiobacillus* spp. and *Thiomicrospira* spp.), (ii) facultative chemolithoautotrophy on RISCs (e.g., *Thiomonas* spp. and *Starkeya novella*), (iii) mixotrophy on RISCs as electron donors and organic compounds as C source (e.g., *Beggiatoa* spp. and *Thiotrix* spp.) and (iv) photoautotrophy under anaerobic conditions (e.g., genera *Thiocystis*, *Chlorobiaceae*, *Chromatiaceae* and *Rhodospirillaceae*) (Sklodowska and Matlakowska, 2007). However, they appear to be unlikely in industrial applications. Sklodowska and Matlakowska (2007) provided evidence that Cu can be extracted from tailings or ores with high yields (i.e., from 77% to 93%) but this would require up to 4-months in a real process.

3.2 Microorganisms in sediment bioleaching

Sediment bioleaching is usually investigated as the bio-augmentation with selected microorganisms or consortia with specific key biogeochemical functions. In a few studies, the presence of indigenous lithotrophic S-oxidizers was detected in river dredged materials, biostimulated by the addiction of S^0 (Seidel et al., 2000; Chen and Lin, 2001b; Tsai et al., 2003a; Löser et al., 2006a). The most applied strains in sediment bioleaching are *Acidithiobacillus ferrooxidans,* *At. thiooxidans, Thiobacillus thioparus, Leptospirillum ferrooxidans, L. ferriphilum.* *At. ferrooxidans, At. thiooxidans* and, *L. ferrooxidans* can together trigger both Fe oxidation and S oxidation during various phases of bioleaching (Schippers, 2004; Vera et al., 2013; Gan et al., 2015). Chen and co-authors have introduced *T. thioparus* in the studies of sediment bioleaching, because compared to other S-oxidizers it appears to adapt better to sediments and to oxidize S^0 at higher pH values. Moreover, the sediment pre-acidification step can be less extreme or even avoided in the presence of *T. thioparus* (Chen and Lin, 2000b, 2001a). Other strains find a larger use in other bioleaching applications. Nevertheless, at present, there is no important study that investigates the exploitation of archaea or acidophilic gram positive bacteria (e.g., moderate thermophilic Fe/S oxidizers, like *Sulfobacillus thermosulfidooxidans*) in the bioleaching applications on contaminated aquatic sediments.

A variety of heterotrophic acidophiles have been isolated from the same environment as Fe/S oxidizers. Some have been already reported to favor bioleaching processes (Johnson, 1998; Vera et al., 2013). According to Fournier and co-authors (1998), pure cultures of *At. ferrooxidans* would fail in iron oxidization and in environment acidification during bioleaching of wastewater sludge. Co-culture with Fe-oxidizers and heterotrophic acidophiles have displayed more efficient mineral leaching. In acid mine drainage ecosystems, heterotrophic Fe-reducing bacteria (e.g., *Acidiphilium cryptum*) can reduce ferric ions coming from Fe/S oxidizing bacteria metabolism and together feed a cycle of ferric iron supply for metal leaching reactions (Johnson, 1998; Küsel et al., 1999). Nevertheless, deeper investigations have provided evidences that *A. cryptum* does not directly affect metal and semi-metals removal during sediment bioleaching, at least with high content of marine polluted sediment (Beolchini et al., 2009, 2013; Fonti et al., 2013). Zhu et al. (2013) also suggest that consortia of autotrophic and heterotrophic strains (i.e., *At. thiooxidans* with *Pseudomonas aeruginosa*) can improve metal solubilization in a 5% river sediment bioleaching experiment and avoid the sediment pre-acidification step.

Filamentous fungi such as *Aspercillus niger* or *Penicillium chrysogenum* can also bioleach (semi-)metals from solids (i.e., "fungal leaching" or "heterotrophic leaching") by the bio-production of organic acids with metal chelating properties such as citric, oxalic, and gluconic acids (Gadd, 2010). Metals are solubilized by forming water soluble complexes (complexolysis) (Bosecker, 1997; Burgstaller and Schinner, 1993). Metal complexation can also occur with functional groups on the cell wall surface (e.g., carboxyl, carbonyl, amine, amide, hydroxyl, and phosphate groups) (Baldrian, 2003). In addition, carboxylic acids by fungi can attack the mineral surface and lead to a release of associated metals (acidolysis) (Gadd, 2007). Fungal bioleaching has been mostly investigated for metal extraction from low grade ores, mine tailings, and metal-rich industrial waste. Experiments of sediment bioleaching mediated by fungi have given lower extraction yields as compared to bioleaching with Fe/S oxidizing bacteria (Sabra et al., 2011, 2012).

3.3 *Metal sulfide oxidation mechanisms*

Due to its ability to use both sulfur and iron, *At. ferrooxidans* has been extensively used as a model microorganism in the study of mechanisms of metal bioleaching. In the past, the solubilization of metal sulfides by Fe/S oxidizing bacteria was described as a process based on two independent mechanisms: (1) a "direct mechanism", in which microorganisms would oxidize metal sulfides by attaching the mineral surface and dissolving metals without a soluble electron shuttle and (2) an "indirect mechanism", in which the dissolution of sulfides occurs by the leaching agent produced by Fe/S oxidizers (Sand et al., 2001). Several scientific evidences have demonstrated that a direct mechanism does not exist and the "indirect mechanism" has been singled-out as the sole occurring mechanism. Metal sulfides are dissolved via electron shuttle Fe(II)/Fe(III). Fe(III) ions are responsible for a chemical oxidation of metal sulfides and the resulting Fe(II) ions are (re)oxidized by cells (Donati et al., 1988; Edwards et al., 2000; Sand et al., 2001; Tributsch, 2001; Rohwerder et al., 2003; Schippers, 2004). However, a microbe attachment to the ore does occur and enhances the rate of mineral dissolution, so both "contact" and "non-contact" mechanisms do occur: Fe/S oxidizing bacteria approaches mineral surface by creating a biofilm, whereas the majority of cells attach to the sulfide surface, planktonic bacterial cells remain floating in the bulk solution. The attachment is predominantly mediated by the extracellular polymeric substances that create slime and fill the space between cell wall and mineral surface (Sand et al., 1995, 2001; Rawlings, 2002; Rohwerder et al., 2003; Rohwerder and Sand, 2007).

Due to their electronic configuration, metal sulfides FeS_2 (pyrite), MoS_2 (molybdenite), and WS_2 (tungstenite) are not sensitive to proton attack but can be solubilized only by a combination of proton and oxidative attack (i.e., acid insoluble; *"thiosulfate pathway"*). On the contrary, other metal sulfides such as As_2S_3 (orpiment), As_4S_4 (realgar), $CuFeS_2$ (chalcopyrite), FeS (troilite), Fe_7S_8 (pyrrhotite), MnS_2 (hauerite), PbS (galena), and ZnS (sphalerite) are acid-soluble and can be solubilized just by proton attack (i.e., acid soluble; *"polysulfide pathway"*). Details about the two pathways are given in Box 1. According to Sand and Schippers works, S^0 can accumulate in the "polysulfide pathway" as well as S^0 and various polythionates in the "thiosulfate pathway", if there are no acidophilic S-oxidizing prokaryotes (Schippers et al., 1996; Schippers and Sand, 1999). Primary microorganisms involved in current biomining operations are Fe-oxidizing prokaryotes but S-oxidizers play a critical role by: (i) removing sulfur rich layers on the mineral surface that can hinder metal dissolution and (ii) generating sulfuric acid that maintains the acidic conditions (pH 1–3) required by the iron-

---

Box 1. Mechanisms metal sulfide (bio-)oxidation

The sole accepted mechanism of metal sulfide oxidation is the "indirect" one (i.e., "the non-enzymatic metal sulfide oxidation by Fe(III) ions combined with the enzymatic (re)oxidation of the resulting Fe(III) ions" Rohwerder et al., 2003). Depending on the mineralogy of metal sulfides (i.e., acid soluble vs. acid insoluble) and the geochemical conditions in the environment (e.g., pH, oxidants availability), different reduced inorganic sulfur compounds (RISCs) accumulate. Resulting geochemical pathways are named with the first soluble sulfur intermediate formed.

I. Thiosulfate pathway (acid insoluble metal sulfides):

Metal sulfides FeS_2 (pyrite), MoS_2 (molybdenite) and WS_2 (tungestenite) are not are sensitive to proton attack (acid insoluble sulfides) but can be solubilized only by a combination of proton and oxidative attack.

Overall reactions:

$$FeS_2 + 6Fe^{3+} + 3H_2O \longrightarrow S_2O_3^{2-} + 7Fe^{2+} + 6H^+ \tag{5}$$

$$S_2O_3^{2-} + 8Fe^{3+} + 5H_2O \longrightarrow 2SO_4^{2-} + 8Fe^{2+} + 10H^+ \tag{6}$$

Main oxidation products: 90% sulfate and about 1 to 2% polythionates.

Details:

$$FeS_2 + 6Fe^{3+} + 3H_2O \longrightarrow S_2O_3^{2-} + 7Fe^{2+} + 6H^+ \tag{7}$$

$$2S_2O_3^{2-} + 2Fe^{3+} + 5H_2O \longrightarrow S_4O_6^{2-} + 2Fe^{2+} \tag{8}$$

$$S_4O_6^{2-} + H_2O \longrightarrow HS_3O_3^- + SO_4^{2-} + H^+ \tag{9}$$

At pH 2 main side products of equation (7) are sulfate and elemental sulfur (Schippers et al., 1996; Schippers and Jørgensen, 2001). At circumneutral pH Fe^{3+} ions are not in the solution phase but thiosulfate (eq. 8) can still form if O_2 or MnO_2 are present; apart from thiosulfate, sulfite, sulfate trithionate, tetrathionate and other polythionates can be detected.

Tetrathionate formation from thiosulfate oxidation (eq. 8) occurs in a wide pH range, although with different oxidants; contemporaneously, thiosulfate decompose to sulfite and elemental sulfur but, at least at pH 2 this reaction is very slower, thus, tetrathionate is the main product from thiosulfate reactions.

$HS_2O_3^-$ (disulfane-monosulfonic acid) rapid decompose in trithionate (eq. 10) and side products pentathionate, elemental sulfur and sulfite.

$$S_3O_3^{2-} + 1.5O_2 \longrightarrow S_3O_6^{2-} \tag{10}$$

Equation (11) can occur with Fe^{3+} as alternative oxidant.

As tetrathionate (eq. 10), trithionate hydrolyzes to thiosulfate and sulfate (eq. 11):

$$S_3O_6^{2-} + H_2O \longrightarrow S_2O_3^{2-} + SO_4^{2-} + H^+ \tag{11}$$

This series of reactions results in a cyclic degradation of thiosulfate to sulfate (Schippers et al., 1996; Schippers and Sand, 1999; Schippers, 2004).

II. Polysulfide pathway (acid-soluble metal sulfides):

Metal sulfides, like As_2S_3 (orpiment), As_4S_4 (realgar), $CuFeS_2$ (chalcopyrite), FeS (troilite), Fe_7S_8 (Pyrrhotite), MnS_2 (hauerite), PbS (galena) and ZnS (sphalerite), are acid-soluble and can be solubilized just by proton attack (Tributsch and Bennett, 1981; Schippers and Sand, 1999; Tributsch, 2001; Rohwerder and Sand, 2007).

Overall reactions:

$$MS + Fe^{3+} + H^+ \longrightarrow M^{2+} + 0.5H_2S_n + Fe^{2+} \quad n \geq 2 \tag{12}$$

Box 1 contd. ...

... Box 1 contd.

$$0.5H_2S_n + Fe^{3+} \longrightarrow 0.125S_8 + Fe^{2+} + H^+ \tag{13}$$

Main oxidation products: elemental sulfur up to 99%.
Details:

$$MS + H^+ \longrightarrow M^{2+} + H_2S \tag{14}$$

$$H_2S + Fe^{3+} \longrightarrow H_2S^{*+} + Fe^{2+} \tag{15}$$

The radical cation H_2S^{*+} dissociates to radical HS^* and disulfide formation is favored:

$$H_2S^{*+} + H_2O \longrightarrow H_3O^+ + HS^* \tag{16}$$

$$2HS^* \longrightarrow HS_2^- + H^+ \tag{17}$$

Disulfide can be further oxidized to by HS^* or Fe^{3+} and then it can dimerize to tetrasulfide or react with HS^* to trisulfide, with reactions similar to equations from (15) to (17). Chain elongation of polysulfides (H_2S_n) may proceed by analogous reactions. In acid conditions, polysulfides decompose to rings of elemental sulfur, mainly S_8:

$$HS_9^- \longrightarrow HS^- + S_8 \tag{18}$$

These series of reactions determine the formation of elemental sulfur, with yields more than 90%. Side products may be thiosulfate (Schippers et al., 1996; Schippers and Sand, 1999; Schippers, 2004).

---

oxidizers (Dopson and Johnson, 2012; Vera et al., 2013). Protons are generated in both the pathways but in the first phases of "polysulfide pathway", H^+ are consumed and the presence of S-oxidizers is needed to regenerate them and stabilize pH values (in particular, S^0 is inert to abiotic oxidation in acidic environments but it can be oxidized only by microorganisms). Acidophilic Fe-oxidizers control the oxidative conditions (ORP ca. 400–750 mV, for KCl/AgCl reference electrode), which is determined mainly by the Fe(III)/Fe(II) ratio in leaching solutions, even though S^0 bio-oxidation can determine a decrease in free electrons and an increase in ORP as well.

Since bioleaching consists of a combination of proton attack and oxidation processes, sediments with high content of metal sulfides and other metal reduced forms are supposed to be suitable for clean-up strategies based on leaching bacteria. However, sediments are depositional materials of heterogeneous nature and they may contain compounds with high acid neutralizing power (see par. 4.1), even at very high concentrations, or other metabolic products of Fe/S oxidizers may be consumed by many compounds in the sediment and do not contribute to metal dissolution; organic molecules and Fe/Mn oxyhydroxides commonly form surface coatings on other types of mineral substrates (e.g., clays and carbonates) and thus metal sulfide and metals in acid soluble fractions are not exposed (McBride et al., 1997; Warren and Haack, 2001; Yin et al., 2002). Depending on the nature of the sediment, treatment strategies based on bioleaching could require time consuming processes or bioleaching could not be the best suited biological approach (Beolchini et al., 2013).

4. Two decades of sediment bioleaching experimentation

In bioleaching of metals from sediments, the limiting step is represented by the bio-production rate of H^+ and Fe^{3+} because a great amount of the leaching agents can be consumed by chemical reaction with sediment components. The acidophilic F/S oxidizing microorganisms are the driving force: the more S^0 is microbially oxidized in sulfuric acid or Fe^{2+} bio-oxidized in Fe^{3+}, the more (semi-)metals dissolve and pH and ORP reach optimal values for keeping metals stable in the solution phase. A fair number of scientific papers have investigated scientific factors and operational parameters that can influence the application of bioleaching on dredged aquatic sediments for their remediation. However, the large majority of the studies have used a *trial and error* approach and have carried on sediment samples coming from freshwater systems, while only in a few cases marine sediments have been taken into account and no study investigates sediments coming from transitional water ecosystems. Zn, Pb, Cu, Cd, Cr, Ni, Mn, As, Co, and Hg (in order of investigation frequency) have been the main target (semi-) metals. The large majority of the researches have been based on flask-scale experimentations (experimental volume 50–250 ml) or small batch bioreactors (experimental volume up to 5 L), although few authors have performed pilot scale studies for feasibility of investigations (Löser et al., 2001, 2006a, 2007; Seidel et al., 2004, 2006a, 2006b).

4.1 Important factors influencing sediment bioleaching processes

4.1.1 Growth substrate and nutrients

Substrate choice, optimization of their concentrations, and of the application methods are hot-points in planning a bioleaching treatment. In scientific literature, S^0 and $FeSO_4$ are the most applied growth substrata but S^0 finds the largest application. Organic substances such as glucose, sucrose, or yeast extract can be also added when heterotrophs or mixotrophs are involved (Beolchini et al., 2009, 2013).

In the case of S^0, the range of investigation in sediment bioleaching experiments is between 0.1 and 20 g/L. S^0 is usually added as orthorhombic powder, although during bioleaching, a high amount of sulfur powder is not utilized. However, sulfur powder residues are difficult to be removed from sediment after treatment and are likely to cause re-acidification. S^0 pastilles represent a good source of S since they can guarantee high metal removal efficiencies and can be easily removed and reused to treat further sediment batches (Chen et al., 2003a). Biogenic waste S^0 is produced by *Acidithiobacillus* strains in wet desulfurization plants for the purification of waste gases (e.g., in paper mills). It represents a valid and very low cost energy substrate for bioleaching processes, thanks to its hydrophilic properties (Chen et al., 2003b; Seidel et al., 2006b; Fang et al., 2009b, 2013).

There are evidences that in sediment bioleaching the rate of acidification is positively affected by the amount of S^0 available as well as by the rate of sulfate bio-production (Chen and Lin, 2001a, 2001b; Tsai et al., 2003a, 2003b; Löser et al., 2005), although contrasting results were also found (Löser et al., 2005; Chen and Lin, 2001). In the range between 0.5 and 7.0 g/L, the concentration of S^0 does not significantly affect Zn bioleaching, while it has a positive significant effect on Cu and Cr solubilization up to 3.0 and 5 g/L respectively (Fang et al., 2009a). In batch reactor processes with S-oxidizing strains (e.g., *T. thioparus*, *At. thiooxidans* and indigenous strains), Chen and Lin found that the best S^0 concentration was 3 g/L or 5 g/L, with 1% and 2% (w/v) sediment concentration respectively (Chen and Lin, 2001b, 2004). Fang and co-authors used 3 g/L S^0 in suspension systems with sediment 10–12% (Fang et al., 2009a, 2009b, 2013). To obtain the highest metal solubilization yields from a marine sediment sample, Beolchini et al. (2009) increased the concentration of S^0 to 5 g/L. In a solid bed bioreactor treatment (1000 Kg treated sediment; see par. 5.2), Seidel and co-authors optimized the concentration of S^0 as a function of the sediment amount to be treated; in particular the best option was 20 g/Kg S^0 (Seidel et al., 1998, 2004; Löser et al., 2001).

Tsai et al. (2003a,b) obtained high removal of Ni, Zn, and Cu from Ell Ren river's (Taiwan) dredged materials by using thiosulfate as S source, in a range between 2.15 and 5.16 S equivalent g/L. Cr, Co, and Pb solubilization reached lower levels.

Fewer studies have used Fe(II) as an energy source for bioleaching strains. Fe is usually added as $FeSO_4$, with concentrations between 0.3 and 16 g/L. Only Kim et al. (2005) provided iron at high concentration as $FeCl_2$. When bioleaching strategies are applied with sediment samples rich in acid consuming substances (like those from seaports), the presence of S^0 can have non-significant effects on metal solubilization. In those cases, Fe^{2+} becomes a key element to have sufficiently low pH values, at least at high solid contents, however, depending on the sediment properties. If no iron is added, pH could reach values around 7 as a consequence of the buffering capacity due to the sediment (Fonti et al., 2013).

Studies comparing S^0 and Fe^{2+} as energy source for microorganisms in sediment bioleaching have given discordant results. Sabra et al. (2011) have found that S^0 10 g/kg (dry sediment) leads to the best solubilization yields for Mn, Cu, Cd, and Zn from fresh water sediments. Chen et al. (2003a) observed that a source of 1 g/L ferric ions can improve significantly the bioleaching efficiencies of Pb and Cr when S-oxidizing bacteria are bioaugmented (i.e., *At. thiooxidans* and *T. thioparus*). On the contrary, Fe^{2+} appears to be the best choice for bioleaching marine sediment samples (Beolchini et al., 2013; Fonti et al., 2013).

It has been recently reported that pyrite addition can improve the acidification potential of sediments during bioleaching treatments in stirrer tank reactors (Gan et al., 2015). Pyrite (FeS_2) dissolution occurs by oxidative attack and in acid conditions that leads to the release of Fe^{2+} and RISCs ("Thiosulfate pathway"; Box 1) and, thus, represents a potential growth substrate in metal bioleaching from solid matrices. Since S^0 is known to favor the acidification in the early stages of sediment bioleaching, it has been hypothesized that the addition of sulfur and pyrite in equal amounts can determine a more efficient acidification (Bas et al., 2013; Gan et al., 2015).

Sources of inorganic N and P in a C:N:P ratio is equal to 100:10:1 or 106:12:1 considering that C content in the sediment can enhance the microbial activity (Morgan and Watkinson, 1992; Enriquez et al., 1993). Chartier and Couillard (1997) have tested the effect of N (as $(NH_4)_2SO_4$) and P (as K_2HO_4) at various concentrations of the solubilization efficiencies of Zn, Pb, and Cu from a river sediment sample by bioleaching but have found no significant effects.

4.1.2 pH

pH is one of the most important variables that influence (semi-)metal bioleaching from solids (Rohwerder et al., 2003). A low pH value is essential for the activity of the majority of Fe/S oxidizing bacteria and archaea (Tuovinen and Kelly, 1973; Johnson, 1998; Schrenk et al., 1998) and is needed for metals to be stable in the solution phase (Blais et al., 1992; Sreekrishnan et al., 1993; Chen and Lin, 2000a, 2001a). pH influences significantly the rates and the efficiencies of metal solubilization (Sauvé et al., 2000; Chen and Lin, 2001a; Kumar and Nagendran, 2007).

Since low pH values are needed both for the activity of bioleaching strains and for the (semi)metal stability in the solution phase, different strategies are applied to favor pH decrease, for example: (i) introducing a stage of strain adaptation, (ii) the neutralization of the acid consuming substances into the sediment by an acidification stage, (iii) the use of key substrates (like S^0 and Fe^{2+}) and optimization of their concentrations as a function of sediment buffering capacity.

Although pH is a very important factor, the relationship between pH values and (semi-)metal solubilization efficiencies from sediment is not linear and metal-specific, as shown quantitatively by Chen and Lin (2001). If no precipitation occurs, the solubilization of Zn, Mn, Cu, and Pb is very significantly favored by low pH values, while Cd, Ni, and Cr are less affected by pH. Other factors affect (semi-)metal solubilization efficiencies in an element-specific way (par. 4.1.5 and section 5).

4.1.3 Sediment content and geochemical features of the sediment

The rate of slurry acidification as well as the rate of metal solubilization decrease with the increase in sediment content (Chen and Lin, 2000b, 2001a; Tsai et al., 2003b). However, the effect of the sediment content in a bioleaching treatment is strictly related to the mineralogical composition of the sediment, its physico-chemical properties, the speciation of metal contaminants, as well as their partitioning among the geochemical fractions of the sediment. The presence of sediment can inhibit the activity of bioleaching bacteria even at low content and to the presence of acid neutralizing components can lower metal removal efficiencies. The effect of sediment content as a dependent variable has a metal-specific effect and is correlated to pH values in the system (Chen and Lin, 2001a, 2001b). However, the degree of acidification depends not only on the rate of S and Fe bio-oxidation. The products of Fe/S oxidizer metabolism (H^+, Fe^{3+} and other metabolic products) react with several components in the sediment. Consequently, physico-chemical and mineralogical properties of the sediment, especially its acid neutralizing capacity highly influences the kinetics and efficiency of (semi-)metal solubilization. Important sediment characteristics are: grain size, oxidative conditions, mineralogical composition, salinity, TOM content and composition and content in carbonates.

The scientific literature provides only indirect insights on how and how much the geochemical features of the sediment influence (semi-)metal solubilization efficiency during a bioleaching treatment. Some studies have addressed the effect of specific sediment properties but a very few works have compared sediment samples with different characteristics, while in the majority of the case the geochemical properties of the sediment have been ignored. Sediment grain size influences the sediment-acid interaction (Löser et al., 2006c; Guven and Akinci, 2013). Löser et al. (2006c) have compared the efficiency of metal solubilization by bioleaching in 6 different freshwater sediment samples and have found that the acidification is influenced by the oxidation state and the carbonate content of the sediment. Tsai et al. (2003b) needed 360 g/kg S^0 to bioleach metals successfully from a sediment sample with a carbonate content equal to 11.2%. Löser et al. (2006c) have also demonstrated that the acid neutralizing capacity of sediment depends not only on its composition of pre-treatment—the same sediment sample showed a much higher buffer capacity than sediment if freshly dredged and anoxic and an easier acidification when disposed at open and consequently oxidized.

The rate and the extent to which a metal is solubilized is highly related to its speciation (Chartier et al., 2001; Chen and Lin, 2009; Fang et al., 2011; Sabra et al., 2011); as consequence, the partitioning of a particular metal among the geochemical fractions of the sediment is an important constraint that affects the efficiency of a bioleaching treatment. Chartier et al. (2001) studied metal bioleaching with three different freshwater sediments (from a lake, a canal and a river) in suspended systems, with a first effort to link the leachability of Cu, Cr, Cd, Ni, Pb, and Zn with the site-specificity of the sediment.

Metals that are weakly absorbed on mineral surface (i.e., exchangeable fraction) and those bound to carbonates are easily solubilized at low pH by ion exchange reactions (i.e., protonation) and dissolution reactions respectively (Gleyzes et al., 2002). Some studies have shown evidences that a release of (semi)-metals from the reducible fraction (i.e., Fe and Mn oxides (Chartier et al., 2001; Chen and Lin, 2004; Fang et al., 2011; Fonti et al., 2013)). However, Tsai et al. (2003a) have found that Mn-oxides became an important binding pool after bioleaching (especially when S0 available for bioleaching strains is limited). With very low pH values, hydrous Fe(III) oxides can partially dissolve by protonation at the edge of the lattice structure and cause a release of (semi-)metals from the reducible fraction (Filgueiras et al., 2002). Metals in the oxidizable fractions appear to solubilize only in highly oxidative conditions (Chartier et al., 2001); acidic non-soluble metal sulfides can be solubilized only in the presence of Fe^{3+} (Box 1), which produces very high ORP values. Some bioleaching studies have reported metal solubilization from the residual fraction. This could be partially due to known intrinsic biases in the procedure (Usero et al., 1998; Gleyzes et al., 2002; Hlavay et al., 2004) and the dissolution of small amounts of metals through edge dissolution of some acid-sensitive minerals.

Although there are evidences that metal partitioning influences significantly the solubilization efficiencies from the sediment by bioleaching, (semi-)metals behave according to metal specific patterns due to their specific chemistry (McBride et al., 1997; Sauvé et al., 2000; Fonti et al., 2013). Zn is known to be one of the most mobile metals; it is highly sensitive to abiotically/ biotically induced environmental changes and its solubilization is more influenced by ORP and pH conditions than to the presence of bioleaching strains (Chartier and Couillard, 1997; Reddy and DeLaune, 2004; Yao et al., 2012; Fonti et al., 2015). Fang et al. (2009a) observed that Zn was unaffected by the concentration of S^0. Ni, Cd, and Cu can be bioleached with very high efficiencies but they can also be poorly dissolved (Blais et al., 2001; Chartier et al., 2001). Cu removal yields are particularly affected by its partitioning among the geochemical fractions of the sediment. Cd is adsorbed to high weight organic molecules via outer-sphere complexation, while Pb and Cu via inner-sphere complexation and hence it is affected by the ionic strength; behavior of Cd, during a bioleaching treatment, is also highly affected by competition with Ca divalent cation for DOM complexation (McBride et al., 1997; Kinniburgh et al., 1999; Sauvé et al., 2000; Guo et al., 2006). Pb bioleaching efficiency is highly affected by precipitation phenomena since it can easily speciate in highly insoluble compounds ($PbSO_4$; $Pb_5(PO_4)_3Cl$; $Pb_5(PO_4)_3OH$) even at low pHs. The scientific literature very often reports a congruent and very low solubilization for Pb (Chartier et al., 2001; Chen and

Lin, 2004; Sabra et al., 2011; Fonti et al., 2013; Gan et al., 2015). To get around this problem, Chartier et al. (2001) made the sediment residuals to react with NaCl after bioleaching with the aim to favor the complexation of solubilized Pb with chloride ions and Pb removal efficiency. Such problems could also be limited by using $FeCl_2$ (Kim et al., 2005), if sulfate sources are not in the sediment which is quite a rare case. In the marine sediment, Cd, Pb, and Zn can speciate largely as carbonates and can be easily solubilized with the acid conditions of a bioleaching treatment (Fonti et al., 2013).

4.1.4 Temperature

The effect of temperature on bioleaching was extensively investigated in the field of metal recovery from low-grade ores and mineral concentrates—in general, the higher the temperature, the faster the leaching process is. This is related to the positive effect of temperature on sulfur oxidization by sulfur oxidizing strains and to the consequent rapid decrease in pH. Since main microorganisms applied in bioleaching are mesophilic bacteria, too high temperatures are inhibiting. The optimum leaching activity with mesophilic strains or indigenous sulfur oxidizing strains rounds 30–35°C, at 50°C, it is almost completely inhibited (Bosecker, 1997; Krebs et al., 1997; Rawlings, 2005). In bioleaching with ferrous iron, the temperature optimum for metal solubilization was found to be 42°C (Blais et al., 1993). Some bioleaching studies have confirmed such results in sediment bioleaching applications (Tsai et al., 2003). However, at a large scale of application, heat that is generated because of the exothermic reactions occurring in a bioreactor (e.g., solid-bed bioleaching) is accumulated and the temperature rises. The largest temperature fluctuations appear to occur in the first days of treatment and to cause delay in sulfur oxidation (Seidel et al., 2004; Löser et al., 2005; Löser et al., 2006b).

4.1.5 Oxygen

Bioleaching is a process requiring oxygen. In the presence of sediment, the oxidation of S^0 requires larger amounts of oxygen (Blais et al., 2001; Seidel et al., 2004). Although S^0 oxidization and Fe^{2+} oxidization still occur under conditions of strong oxygen limitation, a low oxygen supply results in a delay (or temporal suppression) in acidification, sulfate production, and metal solubilization (Rawlings, 2005; Löser et al., 2006b; Seidel et al., 2006a). O_2 consumption can increase even more if there is significant biodegradation of the organic compounds by indigenous heterotrophic microorganisms, especially in the first stage of a bioleaching process without a pre-acidification stage (Seidel et al., 2004). In this context, the bio-availability of sediment organic matter plays a main role in regulating oxygen concentrations.

4.2 Integration of the effects of the main factors in sediment bioleaching

A direct comparison of results gained in different scientific researches is quite complicated since the analytical techniques are not always comparable or because of important differences in the experimental plans. An additional important source of experimental differences is given by eventual sediment pre-treatment stages. Beolchini and co-authors decreased sediment salinity by a washing with deionized water to simulate a real sediment pre-treatment process (Beolchini et al., 2009, 2013; Fonti et al., 2013). Some authors perform bioleaching experiments with autoclaved sediment aliquots. Such a practice influences deeply the possibility to compare results: it acts as a mild thermal-pre-treatment that changes (semi-)metal partitioning and modify deeply the geochemical characteristics of the sediment. Other authors perform a conditioning pre-treatment to improve the permeability of sediment to oxygen and water or to simulate the characteristics of aged dredged sediments (Löser et al., 2001; Seidel et al., 2004). Last, but not the least, a statistical analysis of the hypotheses investigated and of the experimental factors is sometimes missing or very poor. Some of the oldest scientific contributions, and surprisingly even some of the more recent works, have not included proper control treatments in the experimental plan. In other cases, abiotic controls were established in an inappropriate way. If we think about metal extraction yields (Me%) as given by the sum of (i) a chemical solubilization (i.e., due to the initial pH, sediment characteristics, and metal properties; Me%$_{chemical}$) and (ii) a biologically mediated solubilization (i.e., due to the environmental conditions established by bioleaching microorganisms; Me%$_{biological}$, eq. 19):

$$Me\% = Me\%_{chemical} + Me\%_{biological} \tag{19}$$

Several controls are needed in order to differentiate the biological effects from the purely chemical ones. Moreover, metal extraction yields in bioleaching experiments should be calculated according to eq. (20), in order to provide robust estimates of the real extent of metal solubilization:

$$Me\% = \frac{Me_{solution}}{Me_{solution} + (Me_{sediment} * W/Vol)} \tag{20}$$

where $Me_{solution}$ is the concentration of the metal in the solution phase ($\mu g/L$), $Me_{sediment}$ is the concentration of the metal that remains in the sediment ($\mu g/g$), Vol is the experimental volume used (L), and W is the amount of sediment used (g, dry weight).

In sediment bioleaching, low pH values and a high ORP can favor the release of (semi-)metals associated with organic compounds. Characteristics of the organic matter, like the ratio between fulvic and humic acids, the composition in terms of functional groups (e.g., carboxyl,phenolic, alcoholic and carbonyl) and the aging state, can deeply influence the dissociation of Pb, Zn, Cr, Cd and Cu from the organic components in the sediment (McBride et al., 1997; Yin et al., 2002; Guo et al., 2006) and modify the partitioning of metals and semi-metals as well as its acid neutralizing power (Kinniburgh et al., 1999; Sauvé et al., 2000; Yin et al., 2002). That means that the chemical-mineralogical characteristics of sediment will affect substantially the feasibility of a process of sediment bioremediation based on bioleaching. A metal polluted sediment that is also rich in carbonates will exhibit a high acid-consuming capacity, hence even if metals associated with carbonate can be easily removed by bioleaching, the process would require excessive amounts of the leaching agent and would be uneconomical.

4.3 Scale-up experimentations

The effectiveness and feasibility of bioleaching as a bioremediation strategy for dredged sediments have been investigated in a 2000 L pilot scale solid-bed plant (Löser et al., 2001, 2006a, 2007; Seidel et al., 2004, 2006a, 2006b). 0.5 kg/L of contaminated dredged material were piled in the solid-bed and percolated with sulfuric acid generated by biostimulation of indigenous S-oxidizers (S⁰ 20 g/Kg of sediment) with high solubilization levels of Zn, Cd, and Ni (60–90%). Solid-bed bioreactor processes require a pre-conditioning step to increase sediment permeability to water and air. The conversion of freshly dredged sediments (nearly impermeable) into soil-like permeable material occurs spontaneously when the sediment is stored for several years in the open (Vermeulen et al., 2003) or can be obtained in few months by planting sediment with helophytes (Löser et al., 2001). Reed canary grass (*Phalaris arundinacea*) has been found to be particularly suitable for conditioning sediments (Zehnsdorf et al., 2013). The only example of pilot-scale stirred bioreactor for sediment bioleaching reported removal of yields of 80–100% for Zn and Cd, 44–70% for Cu, 14–33% for Pb and 6–20% for Ni and Cr (Blais et al., 2001). However, these yields were obtained with a much lower solid content (30 g/L) compared to the capacity of a solid-bed bioreactor.

5. Tips for bioleaching applications with contaminated aquatic sediments

On the basis of our analysis of the state of the art, we have found evidences that:

- The development of a sediment bioleaching treatment in real scale requires the assessment of the contamination and of the geochemical features of the dredged sediment to be treated.

- Since (semi-)metals in sulfide ores can be solubilized by Fe/S oxidizing bacteria/archaea, aquatic sediment with high content of sulfides are supposed to be suitable for clean-up strategies based on bioleaching. Nevertheless, sediment turn oxidized easily after dredging and this can determine significant modifications in the efficacy of bioleaching strategies.

- The main role of leaching bacteria consists in generating and re-generating leaching agents, mainly Fe(III) ions and protons. Since a "direct mechanism" does not exist and metal solubilization is mediated by metabolic products, there is no need to limit the provision of growth substrata. Nevertheless, a microbe pre-adaptation step could improve bioleaching rates.

- The main geochemical fractions of the sediment to be interested in the release of metals are expected to be the acid soluble one (i.e., metals associated to the sediment as exchangeable ions, metals that speciate as carbonates and metals in acid soluble sulfides) and the oxidizable one (i.e., metal sulfide and metals associated to the sediment organic matter). So dredged materials with metal contaminants mainly in these fractions may be suitable for a bioleaching treatment. However, this has been only partially confirmed by our analysis: metals in the reducible fraction (i.e., associated with Fe/Mn oxides) and the residual may dissolve during bioleaching but these two fractions would still be the most important binding pool after treatment.

- Due to sediment's general properties, strategies aimed at the maintenance of low pH values are needed. This could involve one or more of the following options: (1) a pre-acidification step by acid addition; (2) exploitation of S-oxidizing strains that are active at pH values almost neutral; (3) biostimulation and enrichment of indigenous S-oxidizers in the sediment (detected by CARD-FISH and/or by PCR-based techniques); (4) addition of Fe^{3+} ions; (5) exploitation of allochthonous Fe-oxidizing strains with Fe^{2+} as an energy source to stimulate Fe^{3+} ions bio-production.

- If the sediment is either (1) rich in carbonates, (2) has high acid neutralizing capacity, (3) has Pb and Cr as main contaminants, or (4) parameter control strategies require high amounts of S^0, Fe or H+, then remediation/management options other than bioleaching should be considered.

6. Conclusions: remarks and future challenges

Till date, most scientific papers about the application of bioleaching strategies for the remediation of contaminated aquatic sediments have been based on a trial-error approach, with chemical and biological processes often viewed as "black boxes". On the contrary, the development of a sound management strategy for contaminated aquatic sediments based on the exploitation of bioleaching technique requires a deep understanding of the main geochemical characteristics of the sediment, the typology, concentrations and partitioning of metal contaminants, the chemical and biological mechanisms involved, and the identification of the most appropriate microorganism. All of these aspects should be carefully taken into account before claiming bioleaching as a suitable strategy for the remediation of dredged sediments.

Although a bioleaching-based strategy may offer several advantages for the management of dredged sediments contaminated with metals, potential disadvantages should be taken into account (Table 1). One of the main limitation relays in the assumption that acidophilic microbes that are largely used in mining industry are also effective for sediment reclamation purposes.

Table 1. Advantages and disadvantages of bioleaching as a sediment bioremediation strategy.

	Advantages	Disadvantages
Biological strategy	Microbes catalyze a loop regeneration of lixiviant agents; reduction of energy and chemical consumption (Chen and Lin, 2009; Löser et al., 2007).	Operating conditions must allow microorganisms' life and activity; Sediment properties may inhibit bioleaching strains; a pre-adaptation step could be needed.
Remediation strategy	It allows the removal/decrease of metal contaminants; effectiveness demonstrated for sediments of different origins (e.g., river, lakes, seaports). A decrease of hydrocarbon concentrations have been also observed (Beolchini et al., 2009).	Not effective for refractory organic contaminants. Removal yields may vary considerably on sediment properties (Chartier et al., 2001; Fonti et al., 2013; Löser et al., 2006c). Selective for some metals.
Management of treatment of wastewaters	Generation of complex acidic leachates, rich in metal sulphates, that are suitable for bio-precipitation treatment (Fang et al., 2011).	High volumes to be treated or to be carefully disposed of (Fang et al., 2011, 2009a); volumes vary depending on reactor configuration.
Management of treated sediment	In case of solid-bed plant, sediment maintains a soil-like structure, that facilitate reuse (Löser et al., 2007; Seidel et al., 2004).	Further treatments for metal removal could be required after bioleaching (increase in costs). pH of sediment residues needs to be neutralized.
Sustainability	Low energy and chemical consumption, compared to other treatment strategies (Chen and Lin, 2009; Löser et al., 2007). Large societal acceptance.	Its sustainability, that is assumed to be better than other treatment technologies, still needs to be demonstrated.

So far, only a few acidophilic microbial strains have been tested for metal bioleaching from contaminated sediments. At the same time, geochemical variables play a major role in determining the effectiveness of the process. In view of a real and full scale application of bioleaching as a sediment remediation strategy, a quantitative determination of main geochemical constraints in a "biokinetic test" approach is urgently needed. Future studies should also quantify environmental impacts in order to develop a sustainable process.

References

Agius S.J., Porebski L. Towards the assessment and management of contaminated dredged materials. Integr. Environ. Assess. Manag. 4, 255–260, 2008.

Ahlf W., Förstner U. Managing contaminated sediments. J. Soils Sediments 1, 30–36, 2001.

Avnimelech Y., Avnimelech Y., Ritvo G., Ritvo G., Meijer L.E., Meijer L.E., Kochba M., Kochba M. Water content, organic carbon and dry bulk density in ooded sediments. Water 25, 25–33, 2001.

Bacon J.R., Davidson C.M. Is there a future for sequential chemical extraction? Analyst 133, 25–46, 2008.

Barth E., Sass B., Polaczyk A., Landy R. Evaluation of risk from using poultry litter to remediate and reuse contaminated estuarine sediments. Remediat. J. 11, 35–45, 2001.

Baldrian P. Interactions of heavy metals with white-rot fungi. Enzyme Microb. Technol. 32, 78–91, 2003.

Bas A.D., Deveci H., Yazici E.Y. Bioleaching of copper from low grade scrap TV circuit boards using mesophilic bacteria. Hydrometallurgy 138, 65–70, 2013.

Batley G.E., Stahl R.G., Babut M.P., Bott T.L., Clark J.R., Field L.J., Ho K.T., Mount D.R., Swartz R.C., Tessier A. Scientific underpinnings of sediment quality guidelines. In: Wenning R., Batley G., Ingersoll C., Moore D. (eds.). Use of Sediment Quality Guidelines and Related Tools for the Assessment of Contaminated Sediments. SETAC Press, Pensacola, FL, 2005.

Beolchini F., Dell'Anno A., De Propris L., Ubaldini S., Cerrone F., Danovaro R. Auto- and heterotrophic acidophilic bacteria enhance the bioremediation efficiency of sediments contaminated by heavy metals. Chemosphere 74, 1321–1326, 2009.

Beolchini F., Fonti V., Rocchetti L., Saraceni G., Pietrangeli B., Dell'Anno A. Chemical and biological strategies for the mobilisation of metals/semi-metals in contaminated dredged sediments: experimental analysis and environmental impact assessment. Chem. Ecol. 29, 415–426, 2013.

Bergamaschi B.A., Tsamakis E., Keil R.G., Eglinton T.I., Montluçon D.B., Hedges J.I. The effect of grain size and surface area on organic matter lignin and carbohydrate concentration, and molecular compositions in Peru Margin sediments. Geochim. Cosmochim. Acta 61, 1247–1260, 1997.

Bianchi V., Masciandaro G., Giraldi D., Ceccanti B., Iannelli R. Enhanced heavy metal phytoextraction from marine dredged sediments comparing conventional chelating agents (citric acid and EDTA) with humic substances. Water. Air. Soil Pollut. 193, 323–333, 2008.

Blais J.F., Auclair J.C., Tyagi R.D. Cooperation between two *Thiobacillus* strains for heavy-metal removal from municipal sludge. Can. J. Microbiol. 38, 181–187, 1992.

Blais J.F., Mercier G., Chartier M. Decontamination of sediments experimentally polluted with toxic metals from chemical and biological leaching. Can. J. Chem. Eng. 79, 931–940, 2001.

Blais J.F., Tyagi R.D., Auclair J.C. Bioleaching of metals from sewage sludge: effects of temperature. Water Res. 27, 111–120, 1993.

Bortone G., Arevalo E., Deibel I., Detzner H., De Propris L., Elskens F., Giordano A., Hakstege P., Hamer K., Harmsen J., Hauge A., Palumbo L., van Veen J. Sediment and dredged material treatment synthesis of the sednet work package. Outcomes 4, 225–232, 2004.

Bosecker K. Bioleaching: metal solubilization by microorganisms. FEMS Microbiol. Rev. 20, 591–604, 1997.

Bridge J. Earth surface processes, landforms and sediment deposits. Cambridge University Press. 2008.

Burgstaller W., Schinner F. Leaching of metals with fungi. J. Biotechnol. 27, 91–116, 1993.

Chapman P.M., Wang F. Assessing sediment contamination in estuaries. Environ. Toxicol. Chem. 20, 3–22, 2001.

Chartier M., Couillard D. Biological processes: the effect of initial pH, percentage inoculum and nutrient enrichment on the solubilization of sediment bound metals. Water. Air. Soil Pollut. 96, 249–267, 1997.

Chartier M., Mercier G., Blais J.F. Partitioning of trace metals before and after biological removal of metals from sediments. Water Res. 35, 1435–1444, 2001.

Chen S.Y., Chiu Y.C., Chang P.L., Lin J.G. Assessment of recoverable forms of sulfur particles used in bioleaching of contaminated sediments. Water Res. 37, 450–458, 2003a.

Chen S.Y., Lin J.G., Lee C.Y. Effects of ferric ion on bioleaching of heavy metals from contaminated sediment. Water Sci. Technol. 48, 151–158, 2003b.

Chen S.Y., Lin J.G. Factors affecting bioleaching of metal contaminated sediment with sulfur-oxidizing bacteria. Water Sci. Technol. 41, 263–270, 2000a.

Chen S.Y., Lin J.G. Influence of solid content on bioleaching of heavy metals from contaminated sediment by *Thiobacillus* spp. J. Chem. Technol. Biotechnol. Biotechnol. 75, 649–656, 2000b.

Chen S.Y., Lin J.G. Bioleaching of heavy metals from sediment : significance of pH. Chemosphere 44, 1093–1102, 2001a.

Chen S.Y., Lin J.G. Effect of substrate concentration on bioleaching of metal-contaminated sediment. J. Hazard. Mater. 82, 77–89, 2001b.

Chen S.Y., Lin J.G. Bioleaching of heavy metals from contaminated sediment by indigenous sulfur-oxidizing bacteria in an air-lift bioreactor: effects of sulfur concentration. Water Res. 38, 3205–3214, 2004.

Chen S.Y., Lin J.G. Enhancement of metal bioleaching from contaminated sediment using silver ion. J. Hazard. Mater. 161, 893–899, 2009.

Chen Z.W., Jiang C.Y., She Q., Liu S.J., Zhou P.J. Key role of cysteine residues in catalysis and subcellular localization of sulfur oxygenase-reductase of *Acidianustengchongensis*. Appl. Environ. Microbiol. 71, 621–628, 2005.

Couillard D., Chartier M. Removal of metals from aerobic sludges by biological solubilization in batch reactors. J. Biotechnol. 20, 163–180, 1991.

Couillard D., Mercier G. Removal of metals and fate of N and P in the bacterial leaching of aerobically digested sewage sludge. Water Res. 27, 1227–1235, 1993.

De Jonge M., Teuchies J., Meire P., Blust R., Bervoets L. The impact of increased oxygen conditions on metal-contaminated sediments part I: effects on redox status, sediment geochemistry and metal bioavailability. Water Res. 46, 2205–2214, 2012.

Donati E.R., Porro S.I., Tedesco P.H. Direct and indirect mechanisms in the bacterial leaching of covellite. Biotechnol. Lett. 889–894, 1988.

Dopson M., Johnson D.B. Biodiversity, metabolism and applications of acidophilic sulfur-metabolizing microorganisms. Environ. Microbiol. 14, 2620–2631, 2012.

Edwards K.J., Bond P.L., Druschel G.K., McGuire M.M., Hamers R.J., Banfield J.F. Geochemical and biological aspects of sulfide mineral dissolution: lessons from Iron Mountain, California. Chem. Geol. 169, 383–397, 2000.

Eggleton J., Thomas K.V. A review of factors affecting the release and bioavailability of contaminants during sediment disturbance events. Environ. Int. 30, 973–980, 2004.

Enriquez S., Duarte C.M., Sand-Jensen K. Patterns in decomposition rates among photosynthetic organisms: the importance of detritus C:N:P content. Oecologia 94, 457–471, 1993.

Ehrlich H.L. Past, present and future of biohydrometallurgy. Hydrometallurgy 59, 127–134, 2001.

Fang D., Liu X., Zhang R., Deng W., Zhou L. Removal of contaminating metals from soil by sulfur-based bioleaching and biogenic sulfide-based precipitation. Geomicrobiol. J. 30, 473–478, 2013.

Fang D., Zhang R., Zhou L., Li J. A combination of bioleaching and bioprecipitation for deep removal of contaminating metals from dredged sediment. J. Hazard. Mater. 192, 226–233, 2011.

Fang D., Zhao L., Yang Z.Q., Shan H.X., Gao Y., Yang Q. Effect of sulphur concentration on bioleaching of heavy metals from contaminated dredged sediments. Environ. Technol. 30, 1241–1248, 2009a.

Fang D., Zhao L., Zhou L.X., Shan H.X. Effects of sulfur forms on heavy metals bioleaching from contaminated sediments. J. Environ. Sci. Health. A. Tox. Hazard. Subst. Environ. Eng. 44, 714–721, 2009b.

Fierer N., Lennon J.T. The generation and maintenance of diversity in microbial communities. Am. J. Bot. 98, 439–48, 2011.

Filgueiras A.V., Lavilla I., Bendicho C. Chemical sequential extraction for metal partitioning in environmental solid samples. J. Environ. Monit. 4, 823–857, 2002.

Findley J., Appleman M.D., Yen T.F. Degradation of oil shale by sulfur-oxidizing bacteria degradation of oil shale by sulfur-oxidizing bacteria 28, 1974.

Fonti V., Beolchini F., Rocchetti L., Dell'Anno A. Bioremediation of contaminated marine sediments can enhance metal mobility due to changes of bacterial diversity. Water Res. 68, 637–650, 2015.

Fonti V., Dell'Anno A., Beolchini F. Influence of biogeochemical interactions on metal bioleaching performance in contaminated marine sediment. Water Res. 47, 5139–5152, 2013.

Fonti V., Dell'Anno A., Beolchini F. Does bioleaching represent a biotechnological strategy for remediation of contaminated sediments? Sci. Total Environ. 563–564, 302–319, 2016.

Förstner U. Metal speciation-general concepts and applications. Int. J. Environ. Anal. Chem. 51, 5–23, 1993.

Förstner U., Salomons W. Sediment research, management and policy. J. Soils Sediments 10, 1440–1452, 2010.

Fournier D., Lemieux R., Couillard D. Essential interactions between *Thiobacillus ferrooxidans* and heterotrophic microorganisms during a wastewater sludge bioleaching process. Environ. Pollut. 101, 303–309, 1998.

Gadd G.M. Geomycology: biogeochemical transformations of rocks, minerals, metals and radionuclides by fungi, bioweathering and bioremediation. Mycol. Res. 111, 3–49, 2007.

Gadd G.M. Metals, minerals and microbes: geomicrobiology and bioremediation. Microbiology 156, 609–643, 2010.

Gan M., Zhou S., Li M., Zhu J., Liu X., Chai L. Bioleaching of multiple heavy metals from contaminated sediment by mesophile consortium. Environ. Sci. Pollut. Res. Int. 22, 5807–5816, 2015.

Gleyzes C., Tellier S., Astruc M. Fractionation studies of trace elements in contaminated soils and sediments: a review of sequential extraction procedures. TrAC Trends Anal. Chem. 21, 451–467, 2002.

Guo X., Zhang S., Shan X.Q., Luo L.E.I., Pei Z., Zhu Y.G., Li T., Xie Y.N., Gault A. Characterization of Pb, Cu, and Cd adsorption on particulate organic matter in soil. Environ. Toxicol. Chem. 25, 2366–73, 2006.

Guven D.E., Akinci G. Effect of sediment size on bioleaching of heavy metals from contaminated sediments of Izmir Inner Bay. J. Environ. Sci. 25, 1784–1794, 2013.

Hakanson L. Sediment variability. In: Sediment Toxicity Assessment. Lewis Publishers, Boca Raton, FL, 1992.

Hallberg K.B., González-Toril E., Johnson D.B. *Acidithiobacillus ferrivorans* sp. nov.; facultatively anaerobic, psychrotolerant iron-, and sulfur-oxidizing acidophiles isolated from metal mine-impacted environments. Extremophiles 14, 9–19, 2010.

Hedrich S., Schlömann M., Johnson D.B. The iron-oxidizing proteobacteria. Microbiology 157, 1551–1564, 2011.

Hlavay J., Prohaska T., Weisz M., Wenzel W.W., Stingeder G.J. Determination of trace elements bound to soils and sediment fractions. IUPAC Technical Report. Pure Appl. Chem. 76, 415–442, 2004.

Johnson D.B. Biodiversity and ecology of acidophilic microorganisms. FEMS Microbiol. Ecol. 27, 307–317, 1998.

Johnson D.B., Kanao T., Hedrich S. Redox transformations of iron at extremely low pH: fundamental and applied aspects. Front. Microbiol. 3, 96, 2012.

Johnson D.B., Hallberg K.B., Hedrich S. Uncovering a microbial enigma: isolation and characterization of the streamer-generating, iron-oxidizing, acidophilic bacterium "*Ferrovummyxofaciens*". Appl. Environ. Microbiol. 80, 672–680, 2014.

Jones D.S., Albrecht H.L., Dawson K.S., Schaperdoth I., Freeman K.H., Pi Y., Pearson A., Macalady J.L. Community genomic analysis of an extremely acidophilic sulfur-oxidizing biofilm. ISME J. 6, 158–70, 2012.

Kim S.D., Bae J.E., Park H.S., Cha D.K. Bioleaching of cadmium and nickel from synthetic sediments by *Acidithiobacillus ferrooxidans*. Environ. Geochem. Health 27, 229–235, 2005.

Kinniburgh D.G., van Riemsdijk W.H., Koopal L.K., Borkovec M., Benedetti M.F., Avena M.J. Ion binding to natural organic matter: competition, heterogeneity, stoichiometry and thermodynamic consistency. Colloids Surfaces A Physicochem. Eng. Asp. 151, 147–166, 1999.

Krebs W., Brombacher C., Bosshard P.P., Bachofen R., Brandl H. Microbial recovery of metals from solids. FEMS Microbiol. Rev. 20, 605–617, 1997.

Kumar R.N., Nagendran R. Influence of initial pH on bioleaching of heavy metals from contaminated soil employing indigenous *Acidithiobacillus thiooxidans*. Chemosphere 66, 1775–1781, 2007.

Küsel K., Dorsch T., Acker G., Stackebrandt E. Microbial reduction of Fe(III) in acidic sediments: isolation of *Acidiphilium cryptum* JF-5 capable of coupling the reduction of Fe(III) to the oxidation of glucose. Appl. Environ. Microbiol. 65, 3633–3640, 1999.

Lee C.R. Reclamation and Beneficial Use of Contaminated Dredged Material: Implementation Guidance for Select Options. No. ERDC-TN-DOER-C12. ARMY ENGINEER WATERWAYS EXPERIMENT STATION VICKSBURG MS ENVIRONMENTAL LAB, 2000.

Lee J.C., Pandey B. Bio-processing of solid wastes and secondary resources for metal extraction—A review. Waste Manag. 32, 3–18, 2012.

Löser C., Seidel H., Hoffmann P., Zehnsdorf A. Remediation of heavy metal-contaminated sediments by solid-bed bioleaching. Environ. Geol. 40, 643–650, 2001.

Löser C., Zehnsdorf A., Görsch K., Seidel H. Bioleaching of heavy metal polluted sediment: kinetics of leaching and microbial sulfur oxidation. Eng. Life Sci. 5, 535–549, 2005.

Löser C., Zehnsdorf A., Görsch K., Seidel H. Remediation of heavy metal polluted sediment in the solid bed: comparison of abiotic and microbial leaching. Chemosphere 65, 9–16, 2006a.

Löser C., Zehnsdorf A., Görsch K., Seidel H. Bioleaching of heavy metal polluted sediment: influence of temperature and oxygen (Part 1). Eng. Life Sci. 6, 355–363, 2006b.

Löser C., Zehnsdorf A., Hoffmann P., Seidel H. Bioleaching of heavy metal polluted sediment: influence of sediment properties (Part 2). Eng. Life Sci. 6, 364–371, 2006c.

Löser C., Zehnsdorf A., Hoffmann P., Seidel H. Remediation of heavy metal polluted sediment by suspension and solid-bed leaching: estimate of metal removal efficiency. Chemosphere 66, 1699–1705, 2007.

Lozupone C.A., Knight R. Global patterns in bacterial diversity. Proc. Natl. Acad. Sci. U. S. A. 104, 11436–11440, 2007.

Mangold S., Valdés J., Holmes D.S., Dopson M. Sulfur metabolism in the extreme acidophile *Acidithiobacillus caldus*. Front. Microbiol. 2, 17, 2011.

McBride M., Sauvé, S., Hendershot W. Solubility control of Cu , Zn , Cd and Pb in contaminated soils. Eur. Journal of Soil Sci. 48, 337–346, 1997.

Morgan P., Watkinson R.J. Factors limiting the supply and efficiency of nutrient and oxygen supplements for the *in situ* biotreatment of contaminated soil and groundwater. Water Res. 26, 73–78, 1992.

Mulligan C.N., Yong R.N., Gibbs B.F. An evaluation of technologies for the heavy metal remediation of dredged sediments. J. Hazard. Mater. 85, 145–63, 2001.

Perrodin Y., Donguy G., Bazin C., Volatier L., Durrieu, C., Bony S., Devaux A., Abdelghafour M., Moretto R. Ecotoxicological risk assessment linked to infilling quarries with treated dredged seaport sediments. Sci. Total Environ. 431, 375–384, 2012.

Prica M., Dalmacija B., Dalmacija M., Agbaba J., Krcmar D., Trickovic J., Karlovic E. Changes in metal availability during sediment oxidation and the correlation with the immobilization potential. Ecotoxicol. Environ. Saf. 73, 1370–1377, 2010.

Quevauviller P. Operationally defined extraction procedures for soil and sediment analysis. I. Standardization. Trends Anal. Chem. 17, 289–298, 1998a.

Quevauviller P. Operationally defined extraction procedures for soil and sediment analysis. II. Certified reference materials. Trends Anal. Chem. 17, 632–642, 1998b.

Quevauviller P. Operationally-defined extraction procedures for soil and sediment analysis. Part 3: New CRMs for trace-element extractable contents. Trends Anal. Chem. 21, 774–785, 2002.

Rawlings D.E. Heavy metal mining using microbes. Annu. Rev. Microbiol. 56, 65–91, 2002.

Rawlings D.E. Characteristics and adaptability of iron- and sulfur-oxidizing microorganisms used for the recovery of metals from minerals and their concentrates. Microb. Cell Fact. 4, 13, 2005.

Reddy K.R., DeLaune R.D. Biogeochemistry of wetlands: science and applications. CRC Press, Boca Raton, FL, 2004.

Rohwerder T., Gehrke T., Kinzler K., Sand W. Bioleaching review part A: Progress in bioleaching: fundamentals and mechanisms of bacterial metal sulfide oxidation. Appl. Microbiol. Biotechnol. 63, 239–248, 2003.

Rohwerder T., Sand W. Oxidation of inorganic sulfur compounds in acidophilic prokaryotes. Eng. Life Sci. 7, 301–309, 2007.

Rulkens W., Grotenhuis J.T., Tichy R. Methods for cleaning contaminated soils and sediments. *In*: Förstner U., Salomons W., Mader P. (eds.). Heavy Metals. Springer-Verlag, Berlin Heidelberg, pp. 165–191, 1995.

Sabra N., Dubourguier H.C., Hamieh T. Fungal leaching of heavy metals from sediments dredged from the Deûle canal, France. Adv. Chem. Eng. Sci. 2, 1–8, 2012.

Sabra N., Dubourguier H., Duval M., Hamieh T. Study of canal sediments contaminated with heavy metals: fungal versus bacterial bioleaching techniques. Environ. Technol. 32, 1307–1324, 2011.

Salomons W. Adoption of common schemes for single and sequential extractions of trace metal in soils and sediments. Int. J. Environ. Anal. Chem. 51, 3–4, 1993.

Salomons W., Brils J. Contaminated sediments in European River Basins. EVKI-CT-2001-20002, Key Action 1.4. 1 Abatement of water pollution from contaminated land, landfills and sediments. TNO Den Helder/The Netherlands, 2004.

Sand W., Gehrke T., Jozsa P., Schippers A. (Bio)chemistry of bacterial leaching—direct vs. indirect bioleaching. Hydrometallurgy 59, 159–175, 2001.

Sand W., Gerke T., Hallmann R., Schippers A. Sulfur chemistry, biofilm, and the (in)direct attack mechanism—a critical evaluation of bacterial leaching. Appl. Microbiol. Biotech Nol. 43, 961–966, 1995.

Sauvé S., Hendershot W., Allen H.E. Critical review solid-solution partitioning of metals in contaminated soils: dependence on pH, total metal burden, and organic matter. Environmetal Sci. Technol. 34, 1125–1131, 2000.

Schippers A. Biogeochemistry of metal sulfide oxidation in mining environments, sediments, and soils. Geol. Soc. Am. Spec. Pap. 379, 49–62, 2004.

Schippers A., Jozsa P.G., Sand W. Sulfur chemistry in bacterial leaching of pyrite. Appl. Environ. Microbiol. 62, 3424–3431, 1996.

Schippers A., Sand W. Bacterial leaching of metal sulfides proceeds by two indirect mechanisms via thiosulfate or via polysulfides and sulfur. Appl. Environ. Microbiol. 65, 319–321, 1999.

Schrenk M.O., Edwards K.J., Goodman R.M., Hamers R.J., Banfield J.F. Distribution of *Thiobacillus ferrooxidans* and *Leptospirillum ferrooxidans*: implications for generation of acid mine drainage. Science 279, 1519–1522, 1998.

Seidel H., Görsch K., Schümichen A. Effect of oxygen limitation on solid-bed bioleaching of heavy metals from contaminated sediments. Chemosphere 65, 102–109, 2006a.

Seidel H., Löser C., Zehnsdorf A., Hoffmann P., Schmerold R. Bioremediation process for sediments contaminated by heavy metals: feasibility study on a pilot scale. Environ. Sci. Technol. 38, 1582–1588, 2004.

Seidel H., Ondruschka J., Morgenstem P., Stottmeister U. Bioleaching of heavy metals from contaminated aquatic sediments using indigenous sulfur-oxidizing bacteria: A feasibility study. Water Sci. Technol. 37, 387–394, 1998.

Seidel H., Ondruschka J., Morgenstern P., Wennrich R., Hoffmann P. Bioleaching of heavy metal-contaminated sediments by indigenous *Thiobacillus* spp.: metal solubilization and sulfur oxidation in the presence of surfactants. Appl. Microbiol. Biotechnol. 54, 854–7, 2000.

Seidel H., Wennrich R., Hoffmann P., Löser C. Effect of different types of elemental sulfur on bioleaching of heavy metals from contaminated sediments. Chemosphere 62, 1444–1453, 2006b.

Siham K., Fabrice B., Edine A.N., Patrick D. Marine dredged sediments as new materials resource for road construction. Waste Manag. 28, 919–928, 2008.

Sreekrishnan T.R., Tyagi R.D., Blais J.F., Campbell P.G.C. Kinetics of heavy metal bioleaching from sewage sludge—I. Effects of process parameters. Water Res. 27, 1641–1651, 1993.

Tabak H.H., Lens P., Hullebusch E.D., Dejonghe W. Developments in bioremediation of soils and sediments polluted with metals and radionuclides–1. Microbial processes and mechanisms affecting bioremediation of metal contamination and influencing metal toxicity and transport. Rev. Environ. Sci. Bio/Technology 4, 115–156, 2005.

Tessier A., Campbell P.G.C., Bisson M. Particulate trace metal speciation in stream sediments and relationships with grain size: implications for geochemical exploration. J. Geochemical Explor. 16, 77–104, 1982.

Tichy R., Rulkens W.H., Grotenhius J.T.C., Nydl V., Cuypers C., Fajtl J. Bioleaching of metals from soils or sediments. Water Sci. Technol. 37, 119–127, 1998.

Toes A.C.M., Maas B.A., Geelhoed J.S., Kuenen J.G., Muyzer G. Interactions between microorganisms and heavy metals. *In*: Verstratete W. (ed.). Proceedings of European Symposium on Environmental Biotechnology. Taylor & Francis Group plc, London, pp. 55–59, 2004.

Torsvik V., Øvreås L., Thingstad T.F. Prokaryotic diversity—magnitude, dynamics, and controlling factors. Science 296, 1064–1066, 2002.

Tributsch H. Direct versus indirect bioleaching. Hydrometallurgy 59, 177–185, 2001.

Tsai L.J., Yu K.C., Chen S.F., Kung P.Y., Chang C.Y., Lin C.H. Partitioning variation of heavy metals in contaminated river sediment via bioleaching: effect of sulfur added to total solids ratio. Water Res. 37, 4623–4630, 2003a.

Tsai L.J., Yu K.C., Chen S.F., Kung P.Y. Effect of temperature on removal of heavy metals from contaminated river sediments via bioleaching. Water Res. 37, 2449–2457, 2003b.

Tuovinen O.H., Kelly D.P. Studies on the growth of *Thiobacillus ferrooxidans*. Arch. Mikrobiol. 88, 285–298, 1973.

US EPA. ARCS Remediation Guidance Document. EPA 905-B94-003. Chicago Ill., Great Lakes National Program Office, 1994.

US EPA. Contaminated sediment remediation guidance for hazardous waste sites. Office of Solid Waste and Emergency Response: Washington, DC, 2005.

Usero J., Gamero M., Morillo J., Gracia I. Comparative study of three sequential extraction procedures for metals in marine sediments. Environ. Int. 24, 487–496, 1998.

Vallero D. Environmental biotechnology: a biosystems approach (1st ed.). Academic Press, Amsterdam, 2010.

van Hullebusch E.D., Utomo S., Zandvoort M.H., Piet P.N. Comparison of three sequential extraction procedures to describe metal fractionation in anaerobic granular sludges. Talanta 65, 549–558, 2005.

Vera M., Schippers A., Sand W. Progress in bioleaching: fundamentals and mechanisms of bacterial metal sulfide oxidation—part A. Appl. Microbiol. Biotechnol. 97, 7529–7541, 2013.

Vermeulen J., Grotenhius T., Joziasse J., Rulkens W. Ripening of clayey dredged sediments during temporary upland disposal. J. Soils Sediments 3, 49–59, 2003.

Wang Y., Sheng H.F., He Y., Wu J.Y., Jiang Y.X., Tam N.F.Y., Zhou H.W. Comparison of the levels of bacterial diversity in freshwater, intertidal wetland, and marine sediments by using millions of illumina tags. Appl. Environ. Microbiol. 78, 8264–8271, 2012.

Warren L.A., Haack E.A. Biogeochemical controls on metal behaviour in freshwater environments. Earth-Science Rev. 54, 261–320, 2001.

White A.F., Bullen T.D., Schulz M.S., Blum A.E., Huntington T.G., Peters N.E. Differential rates of feldspar weathering in granitic regoliths. Geochim. Cosmochim. Acta 65, 847–869, 2001.

White C., Sharman A.K., Gadd G.M. An integrated microbial process for the bioremediation of soil contaminated with toxic metals. Nat. Biotechnol. 16, 572–575, 1998.

Yao J., Li W.-B., Kong Q., Xia F., Shen D.S. Effect of weathering on the mobility of zinc in municipal solid waste incinerator bottom ash. Fuel 93, 99–104, 2012.

Yin Y., Impellitteri C.A., You S.J., Allen H.E. The importance of organic matter distribution and extract soil:solution ratio on the desorption of heavy metals from soils. Sci. Total Environ. 287, 107–119, 2002.

Zehnsdorf A., Seidel H., Hoffmann P., Schlenker U., Müller R.A. Conditioning of sediment polluted with heavy metals using plants as a preliminary stage of the bioremediation process: a large-scale study. J. Soils Sediments 13, 1106–1112, 2013.

Zhu J., Zhang J., Li Q., Han T., Xie J., Hu Y., Chai L. Phylogenetic analysis of bacterial community composition in sediment contaminated with multiple heavy metals from the Xiangjiang River in China. Mar. Pollut. Bull. 70, 134–139, 2013.

Zimmerman A.J., Weindorf D.C. Heavy metal and trace metal analysis in soil by sequential extraction: a review of procedures. Int. J. Anal. Chem. 2010, 2010.

CHAPTER 8

Innovative Biomining
Metal Recovery from Valuable Residues

*Camila Castro** and *Edgardo R. Donati*

1. Introduction

Metals are useful for several and different purposes; they have industrial applications and also they are present in multiples devices in the daily life (Donati et al., 2016). Nowadays, metal-bearing residues are being produced in huge amounts and this quantity is increasing due to increase in population as well as the diversification of the applications of metals. For example, in the case of copper, even when it has been partially displaced for new engineering materials in some applications, the current production is 40-fold higher than it was a century ago although the world's population only increased from 1 to about 7 billion in the same period of time (Donati et al., 2016). Numerous industries, for instance, electroplating, metal-finishing, electronic, steel and nonferrous processes, petrochemical and pharmaceutical, and the used electronic/household goods discharge a variety of heavy metals such as Zn, Cd, Cr, Cu, Ni, Pb, V, Mo, Co, etc. (Lee and Pandey, 2012). Atmospheric deposition is a major mechanism for metal input in terrestrial ecosystems. Unlike organic compounds which in general are easy to degrade, heavy metals cannot be decomposed. Metals have different reactivity and toxicity based on the nature of particular metal, concentration and speciation. In the present scenario, heavy metal pollution is a serious concern due to their harmful nature especially when they meet or exceeded the regulatory limits. An inadequate disposal of metal-bearing residues could cause serious problems because this kind of hazardous waste contaminates the sites and

CINDEFI (CCT LA PLATA–CONICET, UNLP), Facultad de Ciencias Exactas (UNLP), 47 y 115, (1900) La Plata, Buenos Aires, Argentina.
* Corresponding author: castro.camila@biotec.quimica.unlp.edu.ar

often degrades surroundings of human habitation including air, surface, and ground water if not treated properly. In contrast, these kinds of wastes contain valuable metals, precious metals, and rare elements in quantities even higher than some ores. The application of a suitable extraction process to recover some of the metals present in these wastes may be an appropriate approach to mitigate their toxicity.

However, metals are precious raw materials to the economy of a country and need to be secured for sustainable production of key components of various products such as low carbon energy technologies, automobiles, and electronic and biomedical devices (Nancharaiah et al., 2016). Metals used to support our way of life are currently obtained through mining of primary sources and processing very large quantities of rocks and metalliferous minerals which are finite and unequally distributed in the world. Nowadays, the average grade of the deposits that are found is decreasing and this trend is expected to continue as a result of urbanization, increasing standards of living, and pollution explosion (Dunbar, 2017). The availability and supply of metals greatly influence the economy of a country by affecting manufacturing, export, and job creation (Nancharaiah et al., 2016). Metal wastes (e.g., computers, printed circuit boards, electronic devices, batteries) as well as seawater, seafloor, mine tailings, and wastewaters could be alternative secondary resources for recovering metals. Recycling of secondary raw materials is critical not only to supplement the secured supply of metals and materials thereby reducing the demand on the limited natural/mineral resources, but also to reduce the environmental degradation due to disposal (Erüst et al., 2013).

There are different technologies for recovery of metals from metal-bearing residues, including pyrometallurgy, hydrometallurgy, pyro-hydrometallurgy, and biohydrometallurgy (Cui and Zhang, 2008; Ilyas and Lee, 2014). Precisely, there are two alternatives for the bioprocessing of metals from metal-bearing residues, bioleaching, and biosorption. Microorganisms use metals for structural and/or catalytic functions. This metal/microbes interaction provides possibility or promotes selective or non-selective recovery of metals. Compared to traditional technologies, bio-extracting techniques have been generally perceived as a much more environmentally benign approach, involving operational flexibility, low costs with less energy consumption, and smaller carbon footprints (Ilyas and Lee, 2014; Johnson, 2014). However, in contrast to conventional processes, these bioprocesses are relatively slow. In the last decades, considerable efforts have been made to develop bioprocess for recovery of metals from wastes and by-products generated from metallurgical and industrial processes and manmade resources. This chapter is focused on the potential applications of microbial biotechnologies that allow recycling and reuse of valuable metals from metal-bearing residues.

2. Bioleaching of metal-bearing residues

Biomining is an applied biotechnology for mobilization of metal cations from insoluble materials such as ores and concentrates by biological oxidation and complexation processes. The general term, covers both bioleaching and biooxidation techniques, although the microbial action in both cases is the same. During a bioleaching process the valuable metal is directly solubilized, while the latter term refers to situations where microorganisms are used to remove minerals that occlude target metals which are solubilized in a second process (Donati et al., 2016; Johnson, 2014; Vera et al., 2013). Today, biomining is a well-established technology; a variety of full-scale biomining operations significantly contribute to the metals mined worldwide (Brierley and Brierley, 2013; Donati et al., 2016). But the potential of biomining is yet to be explored in the case of metal recovery from alternative resources.

2.1 Mechanism of bioleaching

2.1.1 Mechanisms of autotrophs

A generalized reaction describes the biological oxidation of a mineral involved in leaching:

$$MS_{(s)} + 2O_{2(g)} \rightarrow MSO_{4(aq)} \tag{1}$$

where M is the bivalent metal that is solubilized from the mineral by action of microbial metabolites.

This reaction is catalyzed by acidophilic microorganisms capable of oxidizing iron(II) and/or sulfur-compounds. Iron-oxidizing microorganisms generate ferric iron by the following equation:

$$2Fe^{2+}_{(aq)} + 0.5O_{2(g)} + 2H^{+}_{(aq)} \rightarrow 2Fe^{3+}_{(aq)} + H_2O \tag{2}$$

Sulfur oxidizing microorganisms oxidize reduced forms of sulfur to sulfuric acid. Most relevant is the oxidation of elemental sulfur, the overall reaction may be written as:

$$S_{8(s)} + H_2O + 1.5O_{2(g)} \rightarrow H_2SO_{4(aq)} \tag{3}$$

In addition, protons keep the pH low and thus provide an acidic environment needed for the growth of acidophiles and also maintaining the dissolved metals in solution.

Generally, there are two possible mechanisms from a physical point of view: contact and non-contact leaching. In the first case, cells are attached to the surface of the mineral through different interactions and most of them can grow and generate biofilms and all the bioleaching processes occur within the microenvironment of the biofilm. In the last case, cells are not in direct contact with the surfaces and the biooxidation processes take place mainly in the solution while the chemical dissolution of sulfides occur on the surface. These terms may be useful for the description of the physical status of cells involved in bioleaching but they do not tell us anything about the underlying chemical mechanisms of biological metal sulfide dissolution.

Sand and co-workers proposed two different metal sulfide oxidation mechanisms based on the existence of two different groups of metal sulfides: the thiosulfate and the polysulfide pathways (Vera et al., 2013). The oxidative dissolution for acid-insoluble metal sulfides (such as pyrite, molybdenite, and tugstenite) proceeds exclusively through several steps of oxidative attack of ferric iron ions; where thiosulfate is the main sulfur intermediate released to the solution and in turn can be biotically or abiotically oxidized to sulfate (Donati et al., 2016). The thiosulfate mechanism can be simplified by the following equations in the case of pyrite:

$$FeS_{2(s)} + 6Fe^{3+}_{(aq)} + 3H_2O \rightarrow S_2O_3^{2-}_{(aq)} + 7Fe^{2+}_{(aq)} + 6H^+_{(aq)} \tag{4}$$

$$S_2O_3^{2-}_{(aq)} + 8Fe^{3+}_{(aq)} + 5H_2O \rightarrow 2SO_4^{2-}_{(aq)} + 8Fe^{2+}_{(aq)} + 10H^+_{(aq)} \tag{5}$$

However, the polysulfide pathway is applicable to the oxidation of acid-soluble metal sulfides (such as galena, sphalerite, arsenopyrite, and chalcopyrite). In this mechanism, metal sulfides can be oxidized by ferric iron ions and/or by protons. Here, the main sulfur intermediate is polysulfide and consequently elemental sulfur. It can be represented by:

$$MS_{(s)} + Fe^{3+}_{(aq)} + H^+_{(aq)} \rightarrow M^{2+}_{(aq)} + 0.5H_2S_{n(aq)} + Fe^{2+}_{(aq)} \; (n \geq 2) \tag{6}$$

$$0.5H_2S_{n(aq)} + Fe^{3+}_{(aq)} \rightarrow 0.125S_{8(s)} + Fe^{2+}_{(aq)} + H^+_{(aq)} \tag{7}$$

Consequently, in both pathways the main role of acidophilic microorganisms consists of the regeneration of ferric iron ions (the most important oxidant) and protons. More details of both mechanisms can be found in the literature (Schippers and Sand, 1999; Vera et al., 2013).

In some residues, metals are present in zerovalent state. In these cases ferric iron ions (produced by iron-oxidizing microorganisms) can chemically oxidize metals into more soluble forms and reduce to ferrous iron. The following equation shows the reaction in the case of copper:

$$Cu^0_{(s)} + Fe^{3+}_{(aq)} \rightarrow Cu^{2+}_{(aq)} + Fe^{2+}_{(aq)} \tag{8}$$

Microorganisms regenerate ferric iron and sulfuric acid maintaining the acid conditions and metals in solution. Also, it is reasonable that zerovalent metals can be extracted to some extent by proton attack.

2.1.2 Mechanisms of heterotrophs

Besides acidophilic species, there are also some heterotrophic species used for the recovery of valuable metals from solid wastes (Faramarzi et al., 2004). However, heterotrophic leaching of metals from waste residues has been described to a lesser extent. The metal leaching using heterotrophic microorganisms is due to the production/secretion of certain organic acids and complexing compounds, thus supplying protons and forming chelates with metal ions (Lee and Pandey, 2012; Pathak et al., 2009). The acids produced by heterotrophic microorganisms such as lactic acid, oxalic acid, citric acid, and gluconic acid contribute to acidolysis that involves the protonation of oxygen atoms in the metal compound. The protonated oxygen then combines with water resulting in the metal oxide being detached from the solid surface and being solubilized (Ilyas and Lee, 2014). Also, these acids contribute in creating a low pH environment which enhances the bioleaching of metals.

Organic acids are powerful natural chelating agents that form complex with the metals from the material to be bioleached. Complexolysis mechanism consist in the formation of bonds between metal ions and ligands stronger than the lattice bonds between metal ions and solid particles, thus the metal will be successfully solubilized from the solid particles. Cyanogenic microorganisms produce cyanide by oxidative decarboxylation of glycine (eq. 9), which could form cyanide complex with metal ions such as Au, Ag, Pd, and Pt (eqs. 10 and 11) (Cui and Zhang, 2008; Ilyas and Lee, 2014).

$$NH_2CH_2COOH \rightarrow CN^- + CO_2 + 4H^+ \tag{9}$$

$$4Au + 8CN^- \rightarrow 4Au(CN)_2^- + 4e^- \tag{10}$$

$$O_2 + 2H_2O + 4e^- \rightarrow 4OH^- \tag{11}$$

2.2 Microorganisms

Different types of microorganisms play important roles in the bioleaching process. Bacteria and archaea commonly used in bioleaching processes are able to oxidize iron and/or inorganic sulfur compounds, acidophiles, many times autotrophic, and they can tolerate elevated concentrations of metals (Johnson, 2014; Schippers, 2007). Generally, acidophilic microorganisms can be categorized broadly as mesophiles and thermophiles according to their optimum temperature for growth. The most often described mesophiles, which can grow in the temperature range 20–40°C and are used in bioleaching process, are species such as *Acidithiobacillus* sp., *Leptospirillum* sp., *Ferromicrobium* sp., and *Acidiphilum* sp. (Johnson, 2014; Lee and Pandey, 2012; Pathak et al., 2009; Schippers, 2007). *Sulfobacillus* sp., *Ferroplasma* sp. and *Acidiplasma* sp. are moderate thermophiles dominate bioleaching process at ~ 50°C (Donati et al., 2016; Lee and Pandey, 2012). The extreme thermophiles which can be used up to 70°C include archaeal species mainly belonging to the *Acidianus, Sulfolobus,* and *Metallosphaera* genera (Donati et al., 2016; Pathak et al., 2009; Wheaton et al., 2015).

Bacterial species such as *Acetobacter, Acidophilum, Arthrobacter, Bacillus, Chromobacterium, Pseudomonas,* and *Trichoderma* and fungi such as the genus *Penicillum, Aspergillus,* and *Fusarium* have been reported to carry out bioleaching process (Lee and Pandey, 2012; Pathak et al., 2009).

2.2.1 Bioleaching by bacteria and archaea

Bioleaching has been considered to be a key technology for the recovery of metals such as Al, Ni, Zn, Cu, Cd, and Cr from fly ashes, electronic scraps, spent batteries, waste slag, and spent petroleum catalysis (Bakhtiari et al., 2008; Bas et al., 2013; Funari et al., 2017; Ilyas and Lee, 2014; Kaksonen et al., 2016; Mishra and Rhee, 2014; Pathak et al., 2009; Pradhan et al., 2009; Vestola et al., 2010).

Bioleaching of fly ashes using iron and sulfur oxidizing bacteria had been resulted in significant recoveries of metals. Since waste sample lacks iron and sulfur, both should be added to allow bacterial activity. Mixed cultures of iron and sulfur oxidizing microorganisms exhibited better tolerance and metal leaching ability than pure cultures (Ishigaki et al., 2005). Funari and coworkers (2017) treated samples from an Italian incinerator of municipal solid waste by chemical leaching and bioleaching using a mixed culture of acidophilic bacteria. The two leaching methods resulted in satisfactory yields (> 85%); however bioleaching showed a significant selectivity for toxic elements and lanthanides and also halved the use of mineral acid as a consequence of the bio-production of acid which improves metal solubility. The toxic nature of metal wastes might inhibit bacterial growth as well as leaching of metals. The proper pre-adaptation of bacteria prior to bioleaching can help to overcome the detrimental effects of toxic waste. Recently, Ramanathan and coworkers (2016) reported the isolation and for the first time also investigated the use of autochthonous metal-tolerant and alkaliphilic bacteria for bioleaching a municipal solid waste incineration fly ash. Genetic characterization of the strains revealed a dominance of Firmicutes with significant pH or fly ash tolerance; recovering about 52% Cu from the waste and rendering the process to be more economical and environmental friendly. The recovery of harmful metals from electronic scrap has become an important research goal. Two and multi-stage methods, which consist of adding electronic scrap in different concentration during the bioleaching process, have some advantages. These strategies allow microorganisms to grow prior to the introduction of the waste, reducing the inhibitory effects of toxins to microbes (Brandl et al., 2001; Lee and Pandey, 2012; Liang et al., 2010; Xiang et al., 2010). Liang and coworkers (2010) demonstrated that the addition of electronic scrap with lower level at initial stage and with the higher one in the later phase improved metals recovery and minimized inhibition of bacterial growth at the same time.

The application of thermophilic microorganisms in bioprocesses to recover metals from metal-bearing residues has been poorly studied. Moderately thermophilic bacteria, including *Sulfobacillus thermosulfidooxidans*, *Sulfobacillus acidophilus*, and *Thermoplasm acidophilum*, have been used in bioleaching of electronic waste recovering approximately 80% of Ni, Cu, Zn, and Al (Ilyas et al., 2010; Ilyas et al., 2013). High temperatures promote greater metal recovery and faster leaching kinetics. They demonstrated that bioleaching by a mixed culture of *S. thermosulfidooxidans* and *T. acidophilus* decreased the leaching rate of copper, whereas the simultaneous use of *S. thermosulfidooxidans* and *S. acidophilus* showed an increase in the rate of leaching, demonstrating that enhancing bioleaching by the use of consortia requires partnering of appropriate organisms.

The cyanogenic bacteria were also used to solubilize precious metals including Au, Pt and Ag, etc. from electronic wastes. The bacterial cyanide generation has the potential of replacing chemicals to leach gold under alkaline conditions, making the metal recovery much easier and reducing the transportation charges of chemicals (Ilyas and Lee, 2014). The growth and bioleaching performance of *Chromobacterium violaceum*, *Pseudomona fluorescens*, and *Pseudomona plecoglossicida* were evaluated in the presence of various Au, Ag, and Pt containing electronic wastes (Brandl et al., 2008; Brandl and Faramarzi, 2006). The *C. violaceum* was found to be the most effective cyanogen for mobilizing gold from the printed circuit boards (68.5%) as compared to *P. fluorescens*. Recoveries of Ag and Pt from the electronic wastes were not very encouraging due to their resistance and toxic effects on the microbes. Tay and coworkers (2013) demonstrated that lixiviant metabolism in *C. violaceum* can be engineered to enhance cyanide production; a decoupling of cyanogenesis from quorum control results in a significant increase in cyanide production, and correspondingly, an increase in Au recovery from electronic waste. In a recent work, Işıldar and coworkers (2016) performed an experiment of an effective strategy for the recovery of Cu and Au from discarded printed circuit boards in a two-step bioleaching process. In the first step, chemolithotrophic acidophilic *Acidithiobacillus ferrivorans* and *Acidithiobacillus thiooxidans* were used. In the subsequent bioleaching step, the Au complexing agent produced by cyanide-producing heterotrophic *P. fluorescens* and *Pseudomona putida* was used to treat the Cu leached in the first step. Using a two-step approach, Cu and Au were removed from printed circuit board with an efficiency of 98.4% and 44.0%, respectively.

Various types of batteries, such as Li-ion and Li-Cd are used for different electronic devices. Spent batteries are sources of valuable metals, like Co, Li, Mn, etc. Most bioleaching studies about bioleaching of spent batteries use pure or mixed cultures of acidophilic bacteria in the presence of energy sources such as elemental sulfur and ferrous iron (Ijadi Bajestani et al., 2014; Mishra et al., 2008; Velgosova et al., 2013; Xin et al., 2016). Xin and coworkers (2009) found that the highest release of Li (80%) occurred at the lowest pH of 1.54 with elemental sulfur as an energy source. In contrast, the highest release of Co (90%) occurred at higher pH. They suggest that acid dissolution was the main mechanism for Li leaching independent of energy matters types, whereas ferrous iron ions catalyzed reduction of Co along with the acid dissolution. In another work, the recovery of Li, Co, Ni and Mn from spent electric vehicle Li-ion batteries by *A. thiooxidans* was studied (Xin et al., 2016). The non-contact mechanism accounted for Li extraction, whereas a contact mechanism between the cathodes material and the cells was necessary for efficient mobilization of Co, Ni, and Mn. These studies indicate that the dissolution mechanism depends of the metal species. Zeng and coworkers (2012) investigated the influence of Cu on Co bioleaching from spent Li-ion batteries by *A. ferrooxidans*. Almost all Co (99.9%) went into solution after being bioleached for 6 days in the presence of 0.75 g/L Cu ions, while only 43.1% of cobalt dissolution was obtained after 10 days without Cu. They proposed a mechanism in which $LiCoO_2$ underwent a cationic interchange with Cu ions to form $CuCo_2O_4$ on the surface of the sample which could be easy dissolved by ferric iron.

Large quantities of solid catalysts are routinely used in many chemical industries especially in petroleum refining and petrochemical industries. The catalyst contains metals such as Al, V, Mo, Fe, Sn, Sb, Co, and Ni. This waste needs to be pretreated after bioleaching, for instance, washed with acetone to remove the residual organic oil and hydrocarbons that could be toxic for bacteria (Mishra et al., 2007). Some studies examined the potential application of thermophilic archaea in bioleaching of spent catalysts (Bharadwaj and Ting, 2013; Gerayeli et al., 2013; Kim et al., 2014). Nearly 100% extraction was achieved for Fe, Ni, and Mo, and 67% for Al using the thermophilic archaea *A. brierleyi* (Bharadwaj and Ting, 2013). Recoveries ranging from 94 to 97% and 54 to 59% for Ni and Al respectively were reported in a bioleaching study carried out using *S. metallicus* (Kim et al., 2014).

2.2.2 Bioleaching by fungi

The most important species of fungi capable of bioleaching are *Aspergillus niger* and *Penicillium simplicissimum* because of their ability to excrete abundant amounts of organic acids such as oxalic acid, citric acid, and maleic acid that selectively dissolve some metals (Erüst et al., 2013; Ilyas and Lee, 2014). *A. niger* is one of the most widely used fungi in bioleaching (Xu and Ting, 2009; Amiri et al., 2012). Almost complete dissolution of Cu, Pb, Sn, and Zn was noticed in a two step leaching process using a commercial gluconic acid solution produced by *A. niger* (Brandl et al., 2001). Qu and Lian (2013) reported a study about the recovery of rare earth elements (REEs) and radioactive elements from red mud by bioleaching. They tested the bioleaching efficiencies by a filamentous and acid-producing fungus identified as *Penicillium tricolor* and isolated from red mud. The maximum leaching ratios of the REEs and radioactive elements were achieved under one-step bioleaching process at 2% pulp density. However, the highest extraction yields were achieved under two-step bioleaching process at 10% (w/v) pulp density.

3. Biosorption of metals from waste residues

Biosorption is another technique that involves the use of biomass of bacteria, fungi, or algae as adsorbents for the recovery of metals. The major advantages of biosorption are its high effectiveness in reducing the metal content and the use of inexpensive biosorvents (Fu and Wang, 2011).

3.1 Mechanisms of biosorption

Biosorption is a complex process which can include different physic-chemical mechanisms such as absorption, ion exchange, complexation, chelation, and precipitation between metal ions and ligands, depending on the specific properties of the biomass (Cui and Zhang, 2008; Fomina and Gadd, 2014; Robalds et al., 2016). The mechanism of biosorption is very intricate and the main factors that affect the biosorption processes are characteristics and concentration of the target metals, environmental conditions (e.g., pH and temperature, competing ions, contact time), type of biosorvent, and biomass concentration (Vijayaraghavan and Yun, 2008; Wang and Chen, 2009).

This process is critically regulated by the chemical groups present on the microbial cell wal, since it is the first cellular component in contact with metal ions. Differences in metal uptake are due to the properties of each microorganism such as cell wall structure, nature of functional groups (e.g., carboxyl, phosphoril, hydroxyl and tiol moieties), and surface area (Ilyas and Lee, 2014).

One of the important steps in the development of biosorption-based technologies is desorption of the loaded biosorbent which enables re-use of the biomass and recovery and/or containment of sorbates. It is desirable that the desorbing agent does not significantly damage or degrade the biomass and in some cases there might be loss of efficiency of the biomass.

3.2 Biosorbents

A wide variety of active and inactive organisms have been employed as biosorbents to sequester heavy metal ions including microbial biomass (bacteria, archaea, cyanobacteria, filamentous fungi and yeasts, microalgae), seaweeds (macro algae), industrial wastes (fermentation and food wastes, activated and anaerobic sludges, etc.), agricultural wastes (fruit/vegetable wastes, rice straw, wheat bran, sugar beet pulp, soybean hulls, etc.), natural residues (plant residues, sawdust, tree barks, weeds, sphagnum peat moss) and other materials (chitosan, cellulose, etc.) (Fomina and Gadd, 2014).

3.2.1 Bacteria

Many bacterial species such as *Bacillus*, *Pseudomonas*, *Streptomyces*, *Escherichia*, *Micrococcus*, etc., have been tested for uptake metals such as Cr, Ni, Cu, Zn, Cd, and Pb. Some studies indicated heavy metal binding onto the surface of bacterial cell wall in a two-step process (Ilyas and Lee, 2014). The first step involves the interaction of metal ions and reactive groups on cell surface and second stage includes deposition of successive metal species in greater concentrations.

3.2.2 Fungi

Important fungal biosorbents of metals such as Cr, Co, Pb, Au, and Ag include *Aspergillus*, *Rhizopus*, *Penicillium*, *Mucor*, and *Saccharomyces*. Studies conducted with fungal biomass indicated that several types of ionizable sites influence the metal uptake efficiency of fungal cell wall including proteins, nitrogen-containing ligands on protein, and chitin or chitosan (Ilyas and Lee, 2014).

3.2.3 Algae

Algal divisions include red, green and brown seaweed; of which brown seaweeds are found to be excellent biosorbents. The performance in the removal of Pb, Cu, Cd, Zn, Ni, Cr, U, and Au has been extensively studied. The brown algae can effectively remove the extremely toxic metal ions such as Pb and Cr. The cell walls of brown algae generally contain three components: cellulose, the structural support; alginic acid, a polymer of mannuronic and guluronic acids and the corresponding salts of sodium, potassium, magnesium and calcium; and sulphated polysaccharides. As a consequence, carboxyl and sulphate are the predominant active groups in this kind of algae (Romera et al., 2007).

Green and red algae can remove heavy metal ions from aqueous solutions. However, the performance of both is far below than that of brown algae (He and Chen, 2014). The cell-wall of red algae present in sulphated polysaccharides is made of galactanes (agar and carragenates), while green algae mainly contain cellulose, and a high percentage of the cell wall are proteins bonded to polysaccharides to form glycoproteins. These compounds contain several functional groups (amino, carboxyl, sulphate, hydroxyl, etc.) which could play an important role in the biosorption process (Romera et al., 2007).

4. Future perspectives

Recovery of metals from valuable residues is an important subject not only from the point of view of recovery of valuable metals but also from the point of view of waste treatment reducing the environmental degradation due to disposal. Bioleaching and biosorption are promising routes because they offer an ecological alternative in an economically feasible manner. Considerable efforts have been made to develop methodologies for metal recovery from wastes using autotrophic and heterotrophic microorganisms. Their recovery through microbial technologies seems feasible. Bioleaching has been used for the recovery of metals from ores for many years. However, limited progresses were carried out in the development of biotechnological processes for the recovery of metals wastes; the current level of development is mostly confined to the application at the laboratory scale. Further studies should be done to advance the commercial prospects of biomining technologies. In the future, a greater focus should be on understanding microbial-metal interactions in order to develop effective bioprocesses to recover and recycle metals from valuable residues.

References

Amiri F., Mousavi S.M., Yaghmaei S., Barati M. Bioleaching kinetics of a spent refinery catalyst using *Aspargillus niger* at optimal conditions. Biochem. Eng. J. 67, 208–217, 2012.

Bakhtiari F., Atashi H., Zivdar M., Bagheri S.A.S. Continuous copper recovery from a smelter's dust in stirred tank reactors. Int. J. Miner. Process 86, 50–57, 2008.

Bas A.D., Deveci H., Yazici E.Y. Bioleaching of copper from low grade scrap TV circuit boards using mesophilic bacteria. Hydrometallurgy 138, 65–70, 2013.

Bharadwaj A., Ting Y.P. Bioleaching of spent hydrotreating catalyst by acidophilic thermophilic *Acidianus brierleyi*: leaching mechanism and effect of decoking. Bioresour. Technol. 130, 673–680, 2013.

Brandl H., Bosshard R., Wegmann M. Computer-munching microbes: Metal leaching from electronic scrap by bacteria and fungi. Hydrometallurgy 59, 319–326, 2001.

Brandl H., Faramarzi M.A. Microbe-metal-interactions for the biotechnological treatment of metal-containing solid waste. China Particuology 4, 93–97, 2006.

Brandl H., Lehman S., Faramarzi M.A., Martinelli D. Biomobilization of silver, gold and platinum from solid waste materials by HCN-forming microorganisms. Hydrometallurgy 94, 14–17, 2008.

Brierley C.L., Brierley J.A. Progress in bioleaching: part B: applications of microbial processes by the minerals industries. Appl. Microbiol. Biotechnol. 97, 7543–7552, 2013.

Cui J., Zhang L. Metallurgical recovery of metals from electronic waste: A review. J. Hazard. Mater. 158, 228–256, 2008.

Donati E.R., Castro C., Urbieta M.S. Thermophilic microorganisms in biomining. World J. Microbiol. Biotechnol. 32, 179–191, 2016.

Dunbar W.S. Biotechnology and the mine of tomorrow. Trends Biotecnol. 35, 79–89, 2017.

Erüst C., Akcil A., Gahan C.S., Tuncuk A., Deveci H. Biohydrometallurgy of secondary metal resources: a potential alternative approach for metal recovery. J. Chem. Technol. Biotechnol. 88, 2115–2132, 2013.

Faramarzi M.A., Stagars M., Pensini E., Krebs W., Brandl H. Metal solubilization from metal-containing solid materials by cyanogenic *Chromobacterium violaceum*. J. Biotechnol. 113, 321–326, 2004.

Fomina M., Gadd G.M. Biosorption: current perspectives on concept, definition and application. Bioresour. Technol. 160, 3–14, 2014.

Fu F., Wang Q. Removal of heavy metal ions from wastewaters: A review. J. Environ. Manage. 92, 407–418, 2011.

Funari V., Mäkinen J., Salminen J., Braga R., Dinelli E., Revitzer H. Metal removal from Municipal Solid Waste Incineration fly ash: A comparison between chemical leaching and bioleaching. Waste Manage. 60, 397–406, 2017.

Gerayeli F., Ghojavand F., Mousavi S.M., Yaghmaei S., Amiri F. Screening and optimization of effective parameters in biological extraction of heavy metals from refinery spent catalyst using thermophilic bacterium. Sep. Purif. Technol. 118, 151–161, 2013.

He J., Chen J.P. A comprehensive review on biosorption of heavy metals by algal biomass: Materials, performances, chemistry, and modeling simulation tools. Bioresour. Technol. 160, 67–78, 2014.

Ijadi Bajestani M., Mousavi S.M., Shojaosadati S.A. Bioleaching of heavy metals from spent household batteries using *Acidithiobacillus ferrooxidans*: Statistical evaluation and optimization. Separation and Purification Technology 132, 309–316, 2014.

Ilyas S., Ruan C., Bhatti H.N., Ghauri M.A., Anwar M.A. Colum bioleaching of metals from electronic scrap. Hydrometallurgy 101, 135–140, 2010.

Ilyas S., Lee J.-c., Chi R. Bioleaching of metals from electronic scrap and its potential for commercial exploitation. Hydrometallurgy 131–132, 138–143, 2013.

Ilyas S., Lee J.-c. Biometallurgical recovery of metals from waste electrical and electronic equipment: a Review. ChemBioEng. Rev. 1, 148–169, 2014.

Ishigaki T., Nakanishi A., Tateda M., Ike M., Fujita M. Bioleaching of metal from municipal waste incineration fly ash using a mixed culture of sulfur-oxidizing and iron-oxidizing bacteria. Chemosphere 60, 1087–1094, 2005.

Işıldar A., van de Vossenberg J., Rene E.R., van Hullebusch E.D., Lens P.N.L. Two-step bioleaching of copper and gold from discarded printed circuit boards (PCB). Waste Manage. 57, 149–157, 2016.

Johnson D.B. Biomining-biotechnologies for extracting and recovering metals from ores and waste materials. Curr. Opin. Biotech. 30, 24–31, 2014.

Kaksonen A.H., Särkijärvi S., Peuraniemi E., Junnikkala S. Metal biorecovery in acid solutions from a copper smelter slag. Hydrometallurgy, in press, 2017.

Kim D.-J., Srichandan H., Gahan C.S., Lee S.-W. Thermophilic bioleaching of spent petroleum refinery catalyst using *Sulfolobus metallicus*. Canadian Metallurgical Quarterly 51, 403–412, 2014.

Lee J.-c., Pandey B.D. Bio-processing of solid wastes and secondary resources for metal extraction—A review. Waste Manage. 32, 3–18, 2012.

Liang G., Mo Y., Zhou Q. Novel strategies of bioleaching metals from printed circuit boards (PCBs) in mixed cultivation of two acidophiles. Enz. Microb. Technol. 47, 322–326, 2010.

Mishra D., Kim D.J., Ralph D.E., Ahn J.G., Rhee Y.H. Bioleaching of vanadium rich spent refinery catalysts using sulfur oxidizing lithotrophs. Hydrometallurgy 88, 202–209, 2007.

Mishra D., Kim D.J., Ralph D.E., Ahn J.G., Rhee Y.H. Bioleaching of metals from spent lithium ion secondary batteries using *Acidithiobacillus ferrooxidans*. Waste Manage. 28, 333–338, 2008.

Mishra D., Rhee Y.H. Microbial leaching of metals from solid industrial wastes. J. Microbiol. 52, 1–7, 2014.

Nancharaiah Y.V., Mohan S.V., Lens P.N.L. Biological and biochemical recovery of critical and scarce metals. Trends Biotechnol. 34, 137–155, 2016.

Pathak A., Dastidar M.G., Sreekrishnan T.R. Bioleaching of heavy metals from sewage sludge: A review. J. Environ. Manage. 90, 2343–2353, 2009.

Pradhan D., Mishra D., Kim D.J., Roychaudhury G., Lee S.W. Dissolution kinetics of spent petroleum catalyst using two different acidophiles. Hydrometallurgy 99, 157–162, 2009.

Qu Y., Lian B. Bioleaching of rare earth and radioactive elements from red mud using *Penicillium tricolor* RM–10. Bioresour. Technol. 136, 16–23, 2013.

Ramanathan T., Ting Y.-P. Alkaline bioleaching of municipal solid waste incineration fly ash by autochthonous extremophiles. Chemosphere 160, 54–61, 2016.

Robalds A., Naja G.M., Klavins M. Highlighting inconsistencies regarding metal biosorption. J. Hazard. Mater. 304, 553–556, 2016.

Romera E., González F., Ballester A., Blázquez M.L., Muñoz J.A. Comparative study of biosorption of heavy metals using different types of algae. Bioresour. Technol. 98, 3344–3353, 2007.

Schippers A., Sand W. Bacterial leaching of metal sulfides proceeds by two indirect mechanisms via thiosulfate or via polysulfides and sulfur. Appl. Environ. Microbiol. 65, 139–146, 1999.

Schippers A. Microorganisms involved in bioleaching and nucleic acid-based molecular methods for their identification and quantification. *In*: Donati E., S and W. (eds.). Microbial Processing of Metal Sulfides. Springer, Heildelberg, pp. 3–33, 2007.

Tay S.B., Natarajan G., bin Abdul Rahim M.N., Tan H.T., Chung M.C.M., Ting Y.P., Yew W.S. Enhancing gold recovery from electronic waste via lixiviant metabolic engineering in *Chromobacterium violaceum*. Scientific reports 3, 2013.

Velgosova O., Kadukova J., Marcincakova J., Palfy P., Trpcevska J. Influence of H_2SO_4 and ferric iron on Cd bioleaching from spent Ni-Cd batteries. Waste Manage. 33, 456–461, 2013.

Vera M., Schippers A., Sand W. Progress in bioleaching: fundamentals and mechanisms of bacterial metal sulfide oxidation—part A. Appl. Microbiol. Biotechnol. 97, 7529–7541, 2013.

Vestola E.A., Kusenaho M.A., Narhi H.M., Touvinen O.H., Puhakka J.A., Plumb J.J., Kaksonen A.H. Acid bioleaching of solid waste materials from copper, steel and recycling industries. Hydrometallurgy 103, 74–79, 2010.

Vijayaraghavan K., Yun Y.-S. Bacterial biosorbents and biosorption. Biotechnol. Adv. 26, 266–291, 2008.

Wang J., Chen C. Biosorbents for heavy metals removal and their future. Biotechnol. Adv. 27, 195–226, 2009.

Wheaton G., Counts J., Mukherjee A., Kruh J., Kelly R. The confluence of heavy metal biooxidation and heavy metal resistance: implications for bioleaching by extreme thermoacidophiles. Minerals 5, 397–451, 2015.

Xiang Y., Wu P., Zhu N., Zhang T., Liu W., Wu J., Li P. Bioleaching of copper from waste printed circuit boards by bacterial consortium enriched from acid mine drainage. J. Hazard Mater. 184, 812–818, 2010.

Xin B., Zhang D., Zhang X., Xia Y., Wu F., Chen S., Li L. Bioleaching mechanism of Co and Li from spent lithium-ion battery by mixed culture of acidophilic sulfur-oxidizing and iron-oxidizing bacteria. Bioresour. Technol. 100, 6163–6169, 2009.

Xin Y., Guo X., Chen S., Wang J., Wu F., Xin B. Bioleaching of valuable metals Li, Co, Ni and Mn from spent electric vehicle Li-ion batteries for the purpose of recovery. J. Cleaner Prod. 116, 249–258, 2016.

Xu T.J., Ting Y.P. Fungal bioleaching of incineration fly ash: metal extraction and modeling growth kinetics. Enz. Microb. Technol. 44, 323–328, 2009.

Zeng G., Deng X., Luo S., Luo X., Zou J. A copper-catalyzed bioleaching process for enhancement of cobalt dissolution from spent lithium-ion batteries. J. Hazard. Mater. 199–200, 164–169, 2008.

CHAPTER 9

The Challenges of Remediating Metals Using Phytotechnologies

Sabrina G. Ibañéz, Ana L. Wevar Oller, Cintia E. Paisio,
Lucas G. Sosa Alderete, Paola S. González, María I. Medina
and *Elizabeth Agostini**

1. Introduction

The term phytoremediation, derived from the Greek word: phyto = plant and the Latin word: remedium = cure or restore, refers to the strategic use of plants for the treatment of contaminated environments and has been increasingly considered as a sustainable approach for such purposes. Although the use of this terminology arose during the 80's, the study of the application of plants for the treatment of polluted environments is somewhat older. It is known that hundreds of years ago plants were used to treat some residues, reduce soil erosion, protect water quality and since the 50's their potential for extracting radionuclides from soil has been analyzed. However, research work in this area has intensified in the last 30 years. Currently, the term phytoremediation applies to the use of plants (trees, shrubs, terrestrial and aquatic plants, or *in vitro* cultures derived from them) and rhizospheric microorganisms associated with the purpose of removing, containing, degrading, transforming, or rendering harmless a wide variety of environmental contaminants (organic or inorganic) present in soils, sediments, surface and deep water, air, etc. (Cunningham and Ow, 1996). This technique can be applied *in situ* or *ex situ*, and have gained great acceptance for being cost-effective, non-invasive, and being used as a complementary

Departamento de Biología Molecular. Facultad de Ciencias Exactas, Físico-Químicas y Naturales. Universidad Nacional de Río Cuarto. Ruta 36, Km 601. Río Cuarto (Córdoba), Argentina. All authors contributed equally to this book's chapter.
* Corresponding author: eagostini@exa.unrc.edu.ar

technology. Since phytoremediation depends only on solar energy, it is considered cheaper than other remediation methods such as excavation and soil washing, among others.

It is assumed that phytoremediation is especially suitable for:

- Large contaminated surface areas for which the application of other methods is economically and methodologically unviable.
- Sites with moderate to low contamination level for which a final removal process is required.
- Sites that do not require the immediate removal of contaminants, since phytoremediation is generally a long-term strategy.
- As a barrier to avoid vertical and horizontal migration of pollutants.
- Associated with other technologies, for which plants are used as cover once the site was treated with different methods.

In the last years, the traditional term "phytoremediation" has been included in the broader term "phytotechnologies" (Maestri and Marmiroli, 2011; Marmiroli et al., 2006). Phytotechnologies are comprehended as the result of interdisciplinary studies, involving knowledge of biology and plant biochemistry, soil chemistry and microbiology, ecology, environmental engineering, among other disciplines (Ali et al., 2013).

It is important to mention that plants can participate in detoxification processes directly, through contaminant incorporation and subsequent metabolization or immobilization within the plant, or indirectly, through the promotion or support of rhizospheric microorganisms that effectively carry out the detoxification. Therefore, different processes can be recognized, such as: phytoextraction, phytostabilization, phytovolatilization, phytotransformation (phytodegradation), phytofiltration, and rhizodegradation (Table 1).

As it can be seen, there are several terms that have been introduced in relation to phytotechnologies as a natural consequence of scientific progress. However, this lack of normalization of scientific terms among researchers may lead to confusion in the marketplace, i.e., it may confuse nonspecialized stakeholders who are not familiarized with these topics as it was discussed by Conesa et al. (2012).

Considering the main focus of this book, in the present chapter, phytoremediation is narrowed down to metals as pollutants and soils as environmental compartment, focusing only on phytoextraction and phytostabilization processes. In the last decades, many studies around the world have been conducted in this field and numerous plant species have been identified and examined for their abilities to uptake and accumulate different metals. In addition, some progress has been made in phytoremediation practical application and biomass production from contaminated sites, which could economically valorize in diverse forms representing an important environmental co-benefit. These and other challenges and opportunities will be briefly reviewed in the present chapter.

Table 1. Classification of the most frequently used phytotechnologies.

Phytotechnology	Application	Mechanism	Description
Phytoextraction/ Phytoaccumulation	Soil/water	Accumulation/ Hyperaccumulation	Concentration of pollutants in harvestable parts of plants (mainly aerial parts)
Phytostabilization	Soil/water	Sorption, precipitation, complexation	Reduction of contaminants' mobility and prevention of their dispersion
Phytovolatilization	Soil/water	Elimination via transpiration	Conversion of contaminants into volatile forms
Phytotransformation/ Phytodegradation	Soil/water	Transformation; biodegradation; mineralization	Partial transformation and/ or degradation of pollutants to less toxic or non-toxic by-products
Phytofiltration/ Rhizofiltration	Water	Sorption/absorption	Pollutants uptake from aquatic environments
Rhizodegradation/ Rhizoremediation/ Phytostimulation	Soil	Rhizosphere accumulation/ biodegradation	Degradation and/or stabilization by rhizospheric microorganims, estimulated by plant exudates
Phytodesalination	Soil	Reduction of salt levels	Removal of salts using halophytes

2. Phytoextraction vs. phytostabilization

As it was previously mentioned, plants of several species are capable to remediate metals present in soils and since metals are not biodegradable only some methodologies may be considered to remediate them. In this sense, the main metals phytoremediation processes are phytoextraction and phytostabilization, which prevent their migration through ecosystems (Erakhrumen, 2007). Phytoextraction uses accumulator and hyperaccumulator plants to absorb, extract, and accumulate metals into roots, stems, leaves, and inflorescences often in combination with chelating agents and other chemicals (Nwoko, 2010). However, phytostabilization is based on the conversion of pollutants to less bioavailable forms by sorption onto roots, precipitation, complexation or metal reduction in the rhizosphere, such as the case of $Cr(VI)$ to $Cr(III)$ reduction which is a more mobile and less toxic species (Nwoko, 2010; Wu et al., 2010). In Table 2, different plant species used for metals phytoremediation are shown.

Initially, phytoremediation focussed on phytoextraction, while phytostabilization received much less attention. In this context, phytoextraction projects focussed on the search of accumulator plant species or using biotechnology to increase metal uptake and metabolism (Conesa et al., 2012). As a result, new concepts have emerged such as the term hyperaccumulator that is used to describe the ability of plants to accumulate

Table 2. Phytoremediation of metals by different plant species.

Plant species	Metals	Removed concentration (mg/kg)	Plant organs	Phytoremediation mechanism	Reference
Atriplex halimus	Cd Pb	5 50	Roots	Phytoextraction	Manousaki and Kalogerakis, 2009
Prosopis laevigata	Cr(VI) Cd(II)	8,176 8,090 21,437 5,461	Shoots Roots Shoots Roots	Phytoextraction	Buendia-Gonzalez et al., 2010
Brassica juncea	Cr(VI)	1,640 4,100	Roots Shoots	Phytoextraction	Diwan et al., 2010
Spartina argentinensis	Cr(VI)	15,100	Leaves, shoots, and roots	Phytoextraction	Redondo-Gómez et al., 2011
Cicer arietinum	Pb Cr	> 700 1,823 > 3000	Shoots Roots, seeds	Phytoextraction	Dasgupta et al., 2011
Bidens triplinervia	Pb Zn	5,180 9,900	Roots	Phytostabilization	Betch et al., 2012
Senecio sp.	Pb Zn	4,250 3,870	Shoots	Phytoextraction	Betch et al., 2012
Corrigida telephiifolia	As	1,350 2,110	Roots Stems, leaves	Phytoextraction	García-Salgado et al., 2012
Sporobolus sp.	Cu	1,320	Shoots	Phytoextraction	Mkumbo et al., 2012
Grass mixture (*Festuca rubra, Cynodon dactylon, Lolium multiforum, Pennisetum* sp.)	Cu As	866 602	Stems, leaves	Phytoextraction	Zacarías et al., 2012
Tropaeolum Majus	As	825	Stems, leaves	Phytoextraction	Zacarías et al., 2012
Poa annua	Cu	742.06 3,000	Shoots Roots	Phytoextraction	Varun et al., 2012
Populus tomentosa	Cd	550	Leaves	Phytoextraction	Jun and Ling, 2012
Populus bolleana	Cd	206.4 177.50	Stems Roots	Phytoextraction	Jun and Ling, 2012

Table 2 contd. ...

...Table 2 contd.

Plant species	Metals	Removed concentration (mg/kg)	Plant organs	Phytoremediation mechanism	Reference
Populus hopeiensis	Cd	550 450 290	Leaves Stems Roots	Phytoextraction	Jun and Ling, 2012
Populus alba L. *var. Pyramidalis*	Cd Zn Cu Pb	40.76 696 48.21 41.62	Leaves Leaves Roots Roots	Phytoextraction	Hu et al., 2013
Nopalea ochenillifera	Cr(VI)	25,263 705	Roots Shoots	Phytoextraction	Adki et al., 2013
Betula pendula	Zn	245–482	Leaves	Phytoextraction	Dmuchowski et al., 2014
Canna indica L.	Pb Cr Zn Ni Cd	34.51 48.41 74.37 48.53 34.62	Leaves, stems, roots	Phytoextraction (Pb, Cr, Zn) Phytostabilization (Ni, Cd)	Subhashini et al., 2014
Pteris vittata	As(V) Cr(VI)	4,598 1,160 234 12,630	Fronds Roots Fronds Roots	Phytoextraction	de Oliveira et al., 2014
Plantago lanceolata	Cu(II)	142 964	Shoots, Roots	Phytoextraction	Andreazza et al., 2015
Bidens pilosa	Cu(II)	36 844	Shoots Roots	Phytoextraction	Andreazza et al., 2015
Alyssoides utriculata	Ni	> 1000	Leaves	Phytoextraction	Roccotiello et al., 2015
Jatropha curcas	Hg	5.983 0. 9541 2.782	Roots Shoots Leaves	Phytoextraction	Marrugo-Negrete et al., 2015
Brassica napus L.	Zn	500	Leaves	Phytoextraction	Belouchrani et al., 2016
Helianthus tuberosus	Mn Zn Ni	> 1,680 11,400 > 853 6,060 2,600	Roots Shoots Roots Shoots Shoots	Phytoextraction	Willscher et al., 2016
Coronopus didymus	Cd	867.2 864.5	Roots Shoots	Phytoextraction	Sidhu et al., 2017

more than 1000 mg/kg of Cu, Co, Cr, Ni or Pb, or more than 10,000 mg/kg of Mn or Zn (Wu et al., 2010). Other criteria such as high growth rate, high biomass production, high tolerance to metals toxic effects, bioconcentration, and translocation factors are essential to choose a hyperaccumulator plant species (Ahmadpour et al., 2015). The mechanisms and physiology of metal hyperaccumulation have been extensively reviewed in Verbruggen et al. (2009) and Sarma (2011). Regarding bioavailability, only a small fraction of metals present in soil are bioavailable for plant uptake since they can bind to soil particles or precipitate and did not remain soluble (Sheoran et al., 2011). Therefore, chelating agents or additives called "amendments" have been used to increase metals' mobility making them more available for plant uptake and enhancing metal phytoextraction (Alkorta et al., 2004). Synthetic amendments such as EDTA, organic acids (citric acid), or ion competitors (phosphate, ammonium sulfate) among others, have been used (Barbafieri et al., 2013). Although plant uptake is increased with the use of such amendments, chelant-assisted phytoextraction has been discredited because of the high leaching: plant uptake ratio of the contaminants and the persistence of chelants in the environment (Conesa et al., 2012).

Despite the fact that high metal accumulation capacity is a desiderable trait for phytoextractor plants, there is also a concern if they could provide an entry pathway to toxic elements into food chain when consumed by hervibores (Vamerali et al., 2010). Other drawback that limits potential phytoextraction application is that it's success requires the cleansing of the soil to a level that complies with environmental regulations. As a result of these disadvantages, there has been a progressive shift away from phytoextraction to phytostabilization (Conesa et al., 2012). Although this technique is effective when rapid immobilization is needed to preserve ground and surface waters (Chhotu et al., 2009), the main weakness is that it is not a permanent solution to metal contamination since metals remain in sediments, soils, or plant roots (Vangronsveld et al., 2009).

The efficiency of phytoextraction or phytostabilization is difficult to assess and depends on several abiotic and biotic factors that play an essential role in the final outcome of phytoremediation, such as the nature of contaminants, additives (if used) and environmental and soil conditions. Among them, pH and the organic matter content of soils should be considered as well as the interactions between plants and native bacteria which may have a synergistic or antagonistic effect. Furthermore, the decision whether a phytoremediation project should be encouraged or not must be based on a case-by-case study taking into account the above mentioned characteristics of the contaminated site (Gomes, 2012).

3. Improving the realism of metal-phytoremediation through different work-scales

Increasing the efficiency of phytoremediation processes is one of the challenges to solve. Selection of suitable plants, application of traditional crossing methods and genetic engineering are some of the strategies used to improve the tolerance, root/shoot biomass, root architecture and morphology, pollutant uptake, etc. Other strategies include the management of soil microorganisms, including not only rhizospheric bacteria and fungi but also endophytic microorganisms, their selection and improvement through genetic engineering (Weyens et al., 2009; Babu et al., 2013; Ali et al., 2013; Afzal et al., 2014). In addition, proper soil management and optimization of certain agronomic factors can enhance phytoremediation. Thus, initial tests at laboratory and greenhouse scale are fundamental to address in an appropriate way this highly complex subject. Nowadays, a high number of available results are based on such laboratory scale using diverse plant experimental systems (Subhashini et al., 2014; Andreazza et al., 2015). In this context, hairy roots (HRs) have been used as an interesting model system and this is in part because roots have evolved specific mechanisms to deal with pollutants since they are the first organs to have contact with them. Thus, these *in vitro* cultures allow to study the pollutants uptake mechanisms without the interference of soil matrix. Moreover, they have several advantages such as phenotype and genotype stability as well as fast and indefinite *in vitro* growth in absence of phytohormones under sterile conditions (Onno et al., 2011). In Table 3 different HRs cultures used for metals' remediation are shown.

Even though HRs are interesting biotechnological tools to reach important progress in phytoremediation research, they have some limitations. In this sense, since HRs are maintained under *in vitro* conditions, they do not fully mimic field conditions. Besides, this system only takes into account the root capacity. For these reasons, studies at microcosms level, using entire plants growing in culture media, perlite, vermiculite or pot soil are useful to imitate the real conditions as closely as possible. After that, it is very important to confirm the behavior of these plant systems on a larger scale, for example, at mesocosms and field-scale pilot level, since contaminants' bioavailability as well as climatic traits could generate greater discrepancies between laboratory and field conditions. Thus, it is necessary to design experimental approaches following a gradient of work-scales (Fig. 1).

When a project of applying phytoremediation to a real case arises, many questions appear; as if the suggested treatment would be able to significantly reduce contaminants level in the matrix to be treated, whose answers

Table 3. Metals phytoremediation by HRs obtained from different plant species.

HRs cultures	Metals	Removed concentration (mg/g dw)	Reference
Solanum nigrum	Zn	10.7	Subroto et al., 2007
Alyssum murale	Ni	24.7	Vinterhalter et al., 2008
Daucus carota	U	563	Straczek et al., 2009
Armoracia rusticana	U	8.5	Soudek et al., 2011
Brassica juncea L.	Zn, Ni	15.5 12.0	Ismail and Theodor, 2012
Nicotiana tabacum	As(V)	0.032	Talano et al., 2014
Scirpus americanus	Pb(II), Cr(III)	522 148	Alfaro-Saldaña et al., 2016

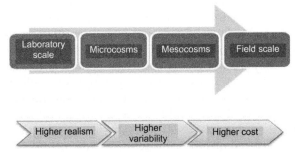

Figure 1. Gradient work-scales to improve the realism of metal phytoremediation conditions.

require a careful evaluation that necessarily involves the experimentation in microcosms and mesocosms. These systems are not only experimental devices to simply observe the behavior of the components of an ecosystem but also allow studying optimal conditions for implementing a biotechnological process. Therefore, they are interesting to make an approximation to reality since they allow evaluating bioavailability, temperature, radiation, etc. In addition, according to the site to be treated, it is required to use plants adapted to the climatic conditions of this site. These experimental systems can be implemented in laboratory chambers under controlled greenhouse conditions or can be exposed to environmental conditions. In the latter case, among other factors to be considered are seasonal changes which could profoundly alter the rate of biological activity and determine the success or failure of a project. The concept of microcosms and mesocosms is sometimes unclear and often there are discrepancies of criteria, i.e., the distinction is not clearly defined. In general, it is difficult to distinguish between them because opinions among researchers differ and there is some overlap about spatial scale (size dimensions or volume).

According to the International Union of Pure and Applied Chemistry, a mesocosm is an enclosed and essentially self-sufficient experimental environment or ecosystem that is on a larger scale than a laboratory microcosm (Duffus et al., 2007). Other definitions described them based on certain characteristics that confer more advantages to mesocosms (Amiard-Triquet, 2015). However, all researchers agree on highlighting the important advantages of these systems which allow studying a number of questions concerning ecosystems and their processes as well as they permit an improvement of the realism of field conditions. In fact, they allow taking into account both direct effects of contaminants and indirect effects from the interaction between species and the environment in a natural or reconstructed ecosystem. It is assumed that mesocosms did not cause significant artifacts and the results obtained can be extrapolated to field conditions (Tingey et al., 2008).

On the contrary, although phytotechnologies are eco-friendly alternatives, it is important to evaluate the environmental risks of their uses. In this sense, micro and mesocosms would be better options. The obtained results can be used subsequently to decide whether it is possible or not to apply phytotechnologies in a real field context, and if so, which approach provides the lowest risk in attaining the proposed goals. As it is well known, these experimental systems are not substitutes of the "real world" and the last step consists of field-scale pilot experiments which can be carried out in spatially well-confined places. It should be considered that field tests and risk analysis are important parameters to take into account when proposing to apply phytoremediation as a strategy for cleaning up of the contaminated environments. The literature provides many examples of studies related to metals' phytoextraction or phytostabilization using microcosms and mesocosms as experimental systems that have been performed in the last years (Table 4). However, it is important to note that many studies were carried out with plant species that are not useful for field applications due to their low biomass such as *Thlaspi caerulescens* and *Arabidopsis*.

In order to achieve better results, a general scheme could be adopted including four main steps:

(1) *Analysis of chemical and physical characteristics of the soil matrix*

(2) *Selection of plant species and/or treatments to be used*

(3) *Evaluation through a field-scale pilot test, including risk analysis*

(4) *Monitoring activities post phytotechnology application to evaluate ecological and economical benefits*

One of the main challenges is still the application of these phytotechnologies to larger scales. In this sense, an analysis of the actual situation in Argentina shows that there is some experience at the level of basic studies (mainly at laboratory and/or greenhouse scale) but its implementation in the field is very limited (Torri and Lavado, 2009; Branzini and Zubillaga, 2012; Zubillaga et al., 2012; Orroño et al., 2012; Velez et al., 2016). To our knowledge, only

Table 4. Summary of some metal phytoremediation studies at micro- and mesocosm scales recently performed.

Study scale	Plant species	Metal	Amendments	Highlights	Reference
Microcosm	*Thlaspi caerulescens*	Zn, Cd	N, P, K$^+$	Influence of soil microbial properties on phytoextraction process.	Epelde et al., 2010
	Brassica juncea Ricinus communis	Ni, Zn, Fe	No	Bacterial inoculation improved plant biomass production and metal accumulation.	Ma et al., 2015
	Arabidopsis halleri	Cd, Zn	No	Plants accumulated approximately 100% more Cd and 15% more Zn when grown on untreated soil containing the native microbial community compared to gamma-irradiated soil.	Muehe et al., 2015
	Typha angustifolia	Cd, Pb	No	Cd inhibits Pb uptake.	Panich-Pat et al., 2010
	Lolium Multiflorum	Pb, Zn, Cd	Saponin	Saponin enhances phytoextraction. Plants were more efficient for Zn uptake, followed by Cd and Pb.	Zhu et al., 2015
	Brassica juncea Lupinus albus	Hg, As	Ammonium thiosulphate	Thiosulphate has positive effects in increasing Hg bioavailability by plants. It has the ability to mobilize As and to promote its uptake by plants.	Petruzzelli et al., 2014
Mesocosm	*Retama sphaerocarpa Coronilla juncea Anthylli scytisoides, Bituminaria bituminosa, Piptatherum miliaceum*	Pb, Zn, Cd, Fe, Mn	Urban organic waste compost (OWC), Fermented sugar beet residue (SBR)	The application of SBR residue improved soil fertility and reduced metal phytoaccumulation in aerial parts, leading improved nutritional status and plant growth.	Kohler et al., 2014
	Brassica juncea Helianthus annuus	Hg	Ammonium thiosulphate	The addition of ammonium thiosulphate promoted Hg uptake by plants of both species.	Petruzzelli et al., 2012

few works were carried out on pilot and full field scale trial and most of them were performed using artificial wetlands in order to remove metals from aquatic environments (Maine et al., 2009; Sarandón et al., 2009; Hadad et al., 2010). Likewise, phytotechnologies' real application in South America is quite limited. It is estimated that development and subsequent implementation will increase and that progress should be linked to greater control by Regulatory Organizations and adequate Legislation with greater exigencies and vigilance activities.

Another drawback is, in most cases, the lack of an exhaustive evaluation of costs which should be site-specific and represent one of the main barriers to commercial application (Conesa et al., 2012). However, in other countries, such as USA, Canada, Italy, and Germany, important advances have been made with the use of these phytotechnologies as shown in the examples provided in Table 5.

Recently, to assess trends, a bibliometric approach using data from SciVerse Scopus, SciVerseHub, and GoogleTM Trends was performed by Koelmel and co-workers (Koelmel et al., 2015). They found that globally there is a linear increase in publications containing the word phytoremediation with China, India, and the Philippines concentrating relatively more research in this topic.

In the 90's, there were several companies in North America and Europe, mostly private, involved in phytoremediation studies including Phytotech (USA) Phytoworks (USA); Earthcare (USA), Aquaphyte Remedy (Canada), Plantechno (Italy), Piccoplant (Germany), etc. (Eapen et al., 2007; Vasavi et al., 2010). Some of them such as Ecolotree (USA), BioPlanta (Germany), and Slater (United Kingdom) are still providing treatment services for polluted environments using the biotechnological advantages offered by phytoremediation.

4. After phytoremediation: what can we do with metal-enriched plant biomass? Thinking of economically valuable strategies for sustainable development

Phytoremediation sustainability is still questioned by scientific and nonscientific stakeholders. Since the beginning, safe disposal of contaminant enriched biomass has been the main concern of phytotechnologies, thus different strategies have been explored to achieve this goal. For example, a detailed description of them is available on Sas-Nowosielska et al. (2004). In this section, the focus will be to emphasize economically valuable phyto-products obtained from metal-enriched biomass.

From a cost-benefit point of view, we could consider that if a phytoremediation treatment allows soil erosion control, carbon sequestration, recovering soil fertility, and usefulness, thus, soil productivity; treatment costs will be compensated over time, i.e., some to several years. Nevertheless,

Table 5. Recent examples of phytotechnologies applied for metals' remediation at pilot and field scales.

Work scale	Soil type	Site	Metal	Phytoremediation strategy	Plant species	Amendments	Reference
Pilot scale	Polluted dredged marine sediments	Livorno, Italy	Zn, Pb, Ni, Cu, Cd	Phytoextraction Phytostabilization	*Paspalum vaginatum, Spartium junceum, Tamarix gallica*	Compost	Doni et al., 2015
Pilot scale	Vehicle dismantler soil	Pisa, Italy	Cd, Cu, Pb, Zn	Phytoextraction	*Paulownia Tomentosa*	Tartaric acid, glutamic acid, EDTA	Doumett et al., 2008
Field scale	Mine tailings	Dewey–Humboldt, United States	As, Pb, Cu, Cd, Cr, Zn, Pirite	Phytostabilization	*Buchloe dactyloides, Festuca arizonica, Atriplex lentiformis, Cercocarpus montanus, Prosopis juliflora, Acacia greggi*	Compost	Gil–Loaiza et al., 2016
Field scale (real scale)	Soil from a common dump for a range of wastes	Pisa, Italy	Zn, Pb, Cd, Ni, Cu, Cr	Phytoextraction Phytostabilization	*Cytisus scoparius, Populus alba, Paulownia tomentosa*	Compost	Macci et al., 2016
Field scale	Cd-contaminated agricultural field	Mae Sot, Thailand	Cd	Phytoextraction	*Crassocephalum crepidioides, Conyza sumatrensis, Chromolaena odorata, Nicotiana tabacum, Gynura pseudochina*	No	Khaokaew and Landrot, 2015
Field scale	Metallurgical soil	Warsaw, Poland	Zn	Phytoextraction	*Betula pendula*	No	Dmuchowski et al., 2014
Field scale	Uranium mining	Ronneburg, Germany	Hg, As	Phytoextraction	*Shorgum bicolor*	Calcareous topsoil	Phieler et al., 2015

actual development of phytoremediation implies necessarily that the large biomass generated as a by-product of the clean-up process must be used or reprocessed in an integrated and cost-effective approach. Thereby, deriving valuable phyto-products from the biomass generated after the clean-up process is a global technological challenge to bring these phytotechnologies into the gear of a bio-based economy (Tripathi et al., 2016b) and references therein. Some of the products that are being obtained from contaminant enriched biomass include biosurfactants, biocomposites, industrially important solvents, bioplastics, biofortified products, pharmacologically active products, and biofuels (Fig. 2) (Conesa, 2012; Tripathi et al., 2016a,b).

Regarding biofuels, they have been identified as essential components of our future energy supply because they are renewable, efficient, and clean burning (Sainger et al., 2017). According to FAO, bioenergy is energy derived from biofuel, which is fuel produced directly or indirectly from biomass. The arrival of the concept of bioenergy has originated what has been called energy crops, which are plant species that combine both capacities: they are able to achieve energy demands and also have a high phytoextraction potential. This ability allows using land that is currently set aside or polluted for the production of energy crops which is an important consideration since bioenergy production from biomass is continuously under criticism as it requires large tracks of arable lands for plantation (Gomes, 2012). Therefore, the use of energy crops would avoid the conflict of interest between food and biofuel production in the societies (Edrisi and Abhilash, 2016). There are several energy crops in the world but four most promising are *Miscanthus, Ricinus, Jatropha* and *Populus* (Pandey et al., 2016). Other edible vegetable oil crops such as soybean, palm, rape seed, groundnut, sunflower, and flax, and non-edible plant oils such as cotton are also used for biodiesel production. Currently, palm, soybean, rape seed, and sunflower are the major biodiesel

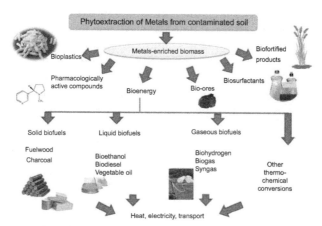

Figure 2. Economically valuable phyto-products that could be obtained from metal-enriched biomass in the future for the development of sustainable phytoremediation.

producing plants accounting 79% of the total world production of biodiesel (Sainger et al., 2017). Hence, biomass energy conversion is an interesting option for biomass management and profits from energy production can significantly reduce remediation costs and even turn phytoextraction into a money-making activity (Vigil et al., 2015).

Although many questions regarding metal-enriched biomass' final disposal have been addressed, some crucial questions about pollution transfer and metal content in the biomass remain unanswered. Is there any chance of occupational exposure to those who have a direct contact with biomass processing and utilization? Will it affect the process during industrial production? (Abhilash and Yunus, 2011). There is urgency for detailed ecotoxicological studies on the fate of accumulated pollutants in biomass during the various stages of its utilization and the potential environmental, occupational, and industrial risk associated with utilization of such harvested biomass from contaminated site.

Many remarkable examples of biobased-economy programs are already taking place in many parts of the world to sustainably produce bioenergy and phyto- and bio-products (Tripathi et al., 2016b). Moreover, the application of these programs for biomass from bioremediation projects could be further improved by suitable biotechnological interventions (Sainger et al., 2017). However, the production of phyto-products from metal-enriched biomass remains to be explored in detail and proper ecotoxicological risk assessment and certification of the phyto-products are necessary before they can be used (Tripathi et al., 2016). Therefore, more research is required for the development of an improved final disposal management system that fits well in a biobased-economy programs.

5. Conclusion

As it was highlighted in this chapter, the strategic use of plants to remediate metals from polluted environments has been extensively studied. From several years ago, a gradual shift in scientific thinking towards a more holistic and wider vision of science resulted in the integration of phytoremediation into the broader concept of "phytotechnologies", which includes interdisciplinary studies of the entire ecosystem processes with the aim of providing solutions using plants. However, there are still some challenges that need to be addressed to effectively apply these remediation techniques to restore contaminated sites. In this sense, we agree that commercial success of phytotechnologies depend on the generation of valuable biomass on contaminated land, rather than conceiving these "green technologies" as merely pure remediation techniques.

References

Abhilash P.C., Yunus M. Can we use biomass produced from phytoremediation? Biomass and Bioenergy 35, 1371–1372, 2011.

Adki V.S., Jadhav J.P., Bapat V.A. *Nopalea cochenillifera*, a potential chromium (VI) hyperaccumulator plant. Environ. Sci. Pollut. Res. 20, 1173–1180, 2013.

Afzal M., Khan Q.M., Sessitsch A. Endophytic bacteria: Prospects and applications for the phytoremediation of organic pollutants. Chemosphere 117, 232–242, 2014.

Ahmadpour P., Ahmadpour F., Sadeghi S.M., Tayefeh F.H., Soleimani, M., Abdu, A.B. Evaluation of four plant species for phytoremediation of copper-contaminated soil. *In*: Hakeem K., Sabir M., Ozturk M., Murmet A. (eds.). Soil Remediation and Plants. Elsevier, pp. 147–205, 2015.

Alfaro-Saldaña E.F., Pérez E., Balch M., Santos-Díaz M.S. Generation of transformed roots of *Scirpusamericanus Pers* and study of their potential to remove Pb^{2+} and Cr^{3+}. Plant Cell Tiss. Organ Cult. 127, 15–24, 2016.

Ali H., Khan E., Sajad M.A. Phytoremediation of heavy metals—concepts and applications. Chemosphere 91, 869–881, 2013.

Alkorta I., Hernández-Allica J., Becerril J.M., Amezaga I., Albizu I., Onaindia M., Garbisu C. Chelate-enhanced phytoremediation of soils polluted with heavy metals. Rev. Environ. Sci. Biotechnol. 3, 55–70, 2004.

Amiard-Triquet C. How to improve toxicity assessment? From single-species tests to mesocosms and field studies. Aquatic Ecotoxicol. 6, 127–151, 2015.

Andreazza R., Bortolon L., Pieniz S., Bento F.M., Camargo F.A. Evaluation of two Brazilian indigenous plants for phytostabilization and phytoremediation of copper-contaminated soils. Braz. J. Biol. 75, 868–877, 2015.

Babu A.G., Kim J.D., Oh B.T. Enhancement of heavy metal phytoremediation by *Alnus firma* with endophytic *Bacillus thuringiensis* GDB-1. J. Hazard. Mater. 250-251, 477–483, 2013.

Barbafieri M., Japenga J., Romkens P., Petruzzelli G., Pedron F. Protocols for applying phytotechnologies in metal-contaminated soils. *In*: Gupta D.K (ed.). Plant-based Remediation Processes. Springer-Verlag, Heidelberg, pp. 19–37, 2013.

Belouchrani A.S., Mameri N., Abdi N., Grib H., Lounici H., Drouiche N. Phytoremediation of soil contaminated with Zn using Canola (*Brassica napus* L.). Ecol. Eng. 95, 43–49, 2016.

Betch J., Duran P., Roca N., Poma W., Sánchez I., Roca-Pérez L., Boluda R., Barceló J., Poschenrieder C. Accumulation of Pb and Zn in *Bidens triplinervia* and Senecio sp. spontaneous species from mine spoils in Peru and their potential use in phytoremediation. J. Geochem. Explor. 123, 109–113, 2012.

Branzini A., Zubillaga M.S. Comparative use of soil organic and inorganic amendments in heavy metals stabilization. App. Environ. Soil Sci. ID 721032, 1–7, 2012.

Buendía-González L., Orozco-Villafuerte J., Cruz-Sosa F., Vernon-Carter E.J. *Prosopis laevigata* a potential chromium (VI) and cadmium (II) hyperaccumulator desert plant. Biores. Technol. 101, 5862–5867, 2010.

Chhotu D.J., Fulekar M.H. Phytoremediation of heavy metals: recent techniques. Afr. J. Biotechnol. 8, 921–928, 2009.

Conesa H.M., Evangelou M.W.H., Robinson B.H., Schulin R. A critical view of current state of phytotechnologies to remediate soils: still a promising tool? The Scientific World Journal Article ID 173829, 1–10, 2012.

Cunningham S., Ow, D.W. Promises and prospects of phytoremediation. Plant Physiol. 110, 715–719, 1996.

Dasgupta S., Satvat P.S., Mahindrakar A.B. Ability of *Cicerarietinum* (L.) for bioremoval of lead and chromium from soil. Int. J. Technol. Eng. Sys. 2, 338–341, 2011.

de Oliveira L.M., Ma L.Q., Santos J.A., Guilherme L.R., Lessl J.T. Effects of arsenate, chromate, and sulfate on arsenic and chromium uptake and translocation by arsenic hyperaccumulator *Pteris vittata* L. Environ. Pollut. 184, 187–192, 2014.

Diwan H., Khan I., Ahmad A., Iqbal M. Induction of phytochelatins and antioxidant defence system in *Brassica juncea* and *Vignaradiata* in response to chromium treatments. Plant Growth Regul. 67, 97–107, 2010.

Dmuchowski W., Gozdowski D., Bragoszewska P., Baczewska A., Suwara I. Phytoremediation of zinc contaminated soils using silver birch (*Betula pendula* Roth). Ecol. Eng. 71, 32–35, 2014.

Doni S., Macci C., Peruzzi E., Iannelli R., Masciandaro G. Heavy metal distribution in a sediment phytoremediation system at pilot scale. Ecol. Eng. 81, 146–157, 2015.

Doumett S., Lamperi L., Checchini L., Azzarello E., Mugnai S., Mancuso S., Petruzzelli G., Del Bubba M. Heavy metal distribution between contaminated soil and *Paulownia tomentosa*, in a pilot-scale assisted phytoremediation study: influence of different complexing agents. Chemosphere 72, 1481–1490, 2008.

Duffus J.H., Nordberg M., Templeton D.M. Glossary of terms used in toxicology. International Union of Pure and Applied Chemistry and Human Health Division (2nd Ed). Pure Appl. Chem. 79, 1153–1344, 2007.

Eapen S., Singh S., D Souza S.F. Phytoremediation of metals and radionuclides. *In*: Singh S.N., Tripathi R.D. (eds.). Environmental Bioremediation Technologies. Springer Science & Business Media, Springer-Verlag, Heidelberg, pp. 1–209, 2007.

Edrisi S.A., Abhilash P.C. Exploring marginal and degraded lands for biomass and bioenergy production: An Indian scenario. Renew. Sustain. Energy Rev. 54, 1537–1551, 2016.

Epelde L., Becerril J., Kowalchuk G., Deng Y., Zhou J., Garbisu C. Impact of metal pollution and *Thlaspi caerulescens* growth on soil microbial communities. App. Environ. Microbiol. 23, 7843–7853, 2010.

Erakhrumen A.A. Phytoremediation: an environmentally sound technology for pollution prevention, control and remediation in developing countries. Educ. Res. Rev. 2, 151–156, 2007.

Garcia-Salgado S., Garcia-Casillas D., Quijano-Nieto M.A., Bonilla-Simon M.M. Arsenic and heavy metal uptake and accumulation in native plant species from soils polluted by mining activities. Water Air Soil Pollut. 223, 559–572, 2012.

Gil-Loaiza J., White S., Root R., Solís-Dominguez F., Hammond C., Chorover J., Maier R. Phytostabilization of mine tailings using compost-assisted direct planting: Translating greenhouse results to the field. Sci. Total Environ. 565, 451–461, 2016.

Gomes H.I. Phytoremediation for bioenergy: challenges and opportunities. Environ. Technol. Rev. 1, 59–66, 2012.

Hadad H.R., Mufarrege M.M., Pinciroli M., Di Luca G.A., Maine M.A. Morphological response of *Typha domingensis* to an industrial effluent containing heavy metals in a constructed wetland. Arch. Environ. Cont. Toxicol. 58, 666–675, 2010.

Hu Y., Nan Z., Su J., Wang N. Heavy metal accumulation by poplar in calcareous soil with various degrees of multi-metal contamination: implications for phytoextraction and phytostabilization. Environ. Sci. Pollut. Res. Int. 20, 7194–7203, 2013.

Ismail A.M., Theodor P.A. The effect of heavy metals Zn and Ni on growth of *in vitro* hairy root cultures of indian mustard *Brassica juncea* L. Int. J. Adv. Biotech. Res. 3, 688–697, 2012.

Jun R., Ling T. Increase of Cd accumulation in five poplar (*Populus* L.) with different supply levels of Cd. Int. J. Phytorem. 14, 101–113, 2012.

Khaokaew S., Landrot G. A field-scale study of cadmium phytoremediation in a contaminated agricultural soil at Mae Sot District, Tak Province, Thailand: (1) Determination of Cd-hyperaccumulating plants. Chemosphere 13, 883–887, 2015.

Koelmel J., Prasad M.N.V., Pershell K. Bibliometric analysis of phytotechnologies for remediation: Global scenario of research and applications. Int. J. Phytorem. 17, 145–153, 2015.

Kohler J., Caravaca F., Azcón R., Díaz G., Roldán A. Selection of plant species-organic amendment combinations to assure plant establishment and soil microbial function recovery in the phytostabilization of a metal-contaminated soil. Water Air Soil Pollut. 225, 1–13, 2014.

Ma Y., Rajkumar M., Rocha I., Oliveira R., Freitas H. Serpentine bacteria influence metal translocation and bioconcentration of *Brassica juncea* and *Ricinus communis* grown in multi-metal polluted soils. Front Plant Sci. 5, 1–13, 2015.

Macci C., Peruzzi E., Doni S., Poggio G., Masciandaro G. The phytoremediation of an organic and inorganic polluted soil: A real scale experience. Int. J. Phytorem. 18, 378–386, 2016.

Maestri E., Marmiroli N. Transgenic plants for phytoremediation. Int. J. Phytorem. 13, 264–279, 2011.

Maine M.A., Suñe N., Hadad H., Sánchez G., Bonetto C. Influence of vegetation on the removal of heavy metals and nutrients in a constructed wetland. J. Environ. Manage. 90, 355–363, 2009.

Manousaki E., Kalogerakis N. Phytoextraction of Pb and Cd by the Mediterranean saltbush (*Atriplexhalimus* L.): metal uptake in relation to salinity. Environ. Sci. Pollut. Res. Int. 16, 844–854, 2009.

Marmiroli N., Marmiroli M., Maestri E. Phytoremediation and phytotechnologies: a review for the present and the future. *In*: Twardowska I., Allen H.E., Haggblom M.H. (eds.). Soil and Water Pollution Monitoring, Protection and Remediation. Springer, Dordretch, pp. 403–416, 2006.

Marrugo-Negrete J., Marrugo-Madrid S., Pinedo-Hernández J., Durango-Hernández J., Díez S. Screening of native plant species for phytoremediation potential at a Hg-contaminated mining site. Sci. Total Environ. 542, 809–816, 2015.

Mkumbo S., Mwegoha W., Renman G. Assessment of the phytoremediation potential for Pb, Zn and Cu of indigenous plants growing in a gold mining area in Tanzania. Int. J. Environ. Sci. 2, 2425–2434, 2012.

Muehe E., Weigold P., Adaktylou I., Planer-Friedrich B., Kraemer U., Kappler A., Behrens S. Rhizosphere microbial community composition affects cadmium and zinc uptake by the metal-hyperaccumulating plant *Arabidopsis halleri*. Appl. Environ. Microbiol. 81, 2173–2181, 2015.

Nwoko C.O. Trends in phytoremediation of toxic elemental and organic pollutants. Afr. J. Biotech. 9, 6010–6016, 2010.

Ono N.N., Tian L. The multiplicity of hairy root cultures: prolific possibilities. Plant Sci. 180, 439–446, 2011.

Orroño D.I., Schindler V., Lavado R.S. Heavy metal availability in *Pelargonium hortorum* rhizosphere: interactions, uptake and plant accumulation. J. Plant Nutrition 35, 1374–1386, 2012.

Pandey V.M., Bajpai O., Singh N. Energy crops in sustainable phytoremediation. Renew. Sustain. Energy Rev. 54, 58–73, 2016.

Panich-Pat T., Upatham S., Pokethitiyook P., Kruatrachue M., Lanza G. Phytoextraction of metal contaminants by *Typha Angustifolia*: interaction of lead and cadmium in soil-water microcosms. J. Environ. Protection 1, 431–437, 2010.

Petruzzelli G., Pedron F., Barbafieri M., Tassi E., Gorini F., Rosellini I. Enhanced bioavailable contaminant stripping: a case study of Hg contaminated soil. Chem. Eng. Transact. 28, 211–216, 2012.

Petruzzelli G., Pedron F., Tassi E., Franchi E., Bagatin R., Agazzi G., Barbafieri M., Rosellini I. The effect of thiosulphate on arsenic bioavailability in a multi contaminated soil. A novel contribution to phytoextraction. Res. J. Environ. Earth Sci. 6, 38–43, 2014.

Phieler R., Merten D., Roth M., Büchel G., Kothe E. Phytoremediation using microbially mediated metal accumulation in *Sorghum bicolor*. Environ. Sci. Pollut. Res. Int. 22, 19408–19416, 2015.

Redondo-Gómez S., Mateos-Naranjo E., Vecino-Bueno I., Feldman S.R. Accumulation and tolerance characteristics of chromium in a cordgrass Cr-hyperaccumulator, *Spartina argentinensis*. J. Hazard. Mater. 185, 862–869, 2011.

Roccotiello E., Serrano H.C., Mariotti M.G., Branquinho C. Nickel phytoremediation potential of the Mediterranean *Alyssoides utriculata* (L.) Medik. Chemosphere 119, 1372–1378, 2015.

Sainger M., Jaiwal A., Sainger P.A., Chaudhary D., Jaiwal R, Jaiwal P.K. Advances in genetic improvement of *Camelina sativa* for biofuel and industrial bio-products. Renew. Sustain. Energy Rev. 68, 623–637, 2017.

Sarandon R., Gaviño Novillo M., Muschong D., Guerrero Borges V. Lacar lake demonstration project for ecohydrology: Improving land use policy at Lacar lake watershed based on an ecohydrological approach (San Martin de los Andes-Neuquén-R. Argentina). Ecohydrol. Hydrobiol. 9, 125–134, 2009.

Sarma, H. Metal hyperaccumulation in plants: a review focusing on phytoremediation technology. J. Env. Sci. Technol. 4, 118–138, 2011.

Sas-Nowosielska A., Kucharski R., Małkowski E., Pogrzeba M., Kuperberg J.M., Kryński K. Phytoextraction crop disposal—an unsolved problem. Environ. Pollut. 128, 373–379, 2004.

Sheoran V., Sheoran A., Poonia P. Role of hyperaccumulators in phytoextraction of metals from contaminated mining sites: a review. Crit. Rev. Environ. Sci. Technol. 41, 168–214, 2011.

Sidhu G.P., Singh H.P., Batish D.R., Kohli R.K. Tolerance and hyperaccumulation of cadmium by a wild, unpalatable herb *Coronopus didymus* (L.) Sm. (*Brassicaceae*). Ecotoxicol. Environ. Saf. 135, 209–215, 2017.

Soudek P., Petrová S., Benesová D., Vanek T. Uranium uptake and stress responses of *in vitro* cultivated hairy root culture of *Armoracia rusticana*. Agrochimica 55, 15–28, 2011.

Straczek A., Wannijn J., Van Hees M., Thijs H., Thiry Y. Tolerance of hairy roots of carrots to U chronic exposure in a standardized *in vitro* device. Environ. Exp. Bot. 65, 82–89, 2009.

Subhashini V., Swamy A.V.V.S. Phytoremediation of metal (Pb, Ni, Zn, Cd and Cr) contaminated soils using *Canna Indica*. Curr. World Environ. 9, 780–784, 2014.

Subroto M.A., Priambodo S., Indrasti N.S. Accumulation of Zinc by hairy hoot cultures of *Solanum nigrum*. Biotechnology 6, 344–348, 2007.

Talano M.A., Wevar Oller A.L., González P., Oliva González S., Agostini E. Effects of arsenate on tobacco hairy root and seedling growth, and its removal. *In vitro* Cell Dev. Biol. Plant 50, 217–252, 2014.

Tingey D.T., Lee E.H., Lewis J.D., Johnson M.G., Rygiewicz P.T. Do mesocosms influence photosynthesis and soil respiration? Environ. Exp. Botany 62, 36–44, 2008.

Torri S.I., Lavado R.S. Plant absorption of trace elements in sludge amended soils and correlation with soil chemical speciation. J. Hazard. Mater. 166, 1469–1465, 2009.

Tripathi V., Edrisi S.A., Abhilash P.C. Towards the coupling of phytoremediation with bioenergy production. Renew. Sustain. Energy Rev. 57, 1386–1389, 2016a.

Tripathi V., Edrisi S.A., O'Donovan A., Gupta V.K., Abhilash P.C. Bioremediation for fueling the biobased economy. Trends Biotechnol. 34, 775–777, 2016b.

Vamerali T., Bandiera M., Mosca G. Field crops for phytoremediation of metal-contaminated land. A review. Environ. Chem. Lett. 8, 1–17, 2010.

Vangronsveld J., Herzig R., Weyens N., Boulet J., Adriaensen K., Ruttens A., Thewys T., Vassilev A., Meers E., Nehnevajova E., van der Lelie D., Mench M. Phytoremediation of contaminated soils and groundwater: lessons from the field. Environ. Sci. Pollut. Res. 16, 765–794, 2009.

Varun M., D'Souza R., Pratas J., Paul M.S. Metal contamination of soils and plants associated with the glass industry in North Central India: prospects of phytoremediation. Environ. Sci. Pollut. Res. Int. 19, 269–281, 2012.

Vasavi A., Usha R., Swamy P. Phytoremediation—An overview review. Jr. Ind. Poll. Control 26, 83–88, 2010.

Velez P.A., Talano M.A., Paisio C.E., Agostini E., González P.S. Synergistic effect of chickpea plants and *Mesorhizobium* as a natural system for chromium phytoremediation. Environ. Technol. doi:10.1080/09593330.2016.1247198, 2016.

Verbruggen, N., Hermans C., Schat, H. Molecular mechanisms of metal hyperaccumulation in plants. New Phytologist 181, 759–776, 2009.

Vigil M., Marey–Pérez M.F., Martinez Huerta G., Álvarez Cabal V. Is phytoremediation without biomass valorization sustainable? Comparative LCA of landfilling vs. anaerobic co-digestion. Sci. Total Environ. 505, 844–850, 2015.

Vinterhalter B., Savic J., Platis J., Raspor M., Ninkovic S., Mitic N., Vinterhalter D. Nickel tolerance and hyperaccumulation in shoot cultures regenerated from hairy root cultures of *Alyssum murale* Waldstet Kit. Plant Cell Tiss. Organ. Cult. 94, 299–303, 2008.

Weyens N., Van der Lelie D., Taghavi S., Vangronsveld J. Phytoremediation: plant-endophyte partnerships take the challenge. Curr. Opin. Biotechnol. 20, 248–254, 2009.

Willscher S., Jablonski L., Fona Z., Rahmi R., Wittig J. Phytoremediation experiments with *Helianthus tuberosus* under different pH and heavy metal soil concentrations. Hydrometallurgy 168, 153–158, 2017.

Wu G., Kang H., Zhang X., Shao H., Chu L., Ruan C. A critical review on the bio-removal of hazardous heavy metals from contaminated soils: issues, progress, eco-environmental concerns and opportunities. J. Hazard. Mater. 174, 1–8, 2010.

Zacarías M., Beltrán M., Torres L.G., González A.A. Feasibility study of perennial/annual plant species to restore soils contaminated with heavy metals. Phys. Chem. Earth. 37, 37–42, 2012.

Zhu T., Fu D., Yang F. Effect of saponin on the phytoextraction of Pb, Cd and Zn from soil using Italian ryegrass. Bull. Environ. Contam. Toxicol. 94, 129–133, 2015.

Zubillaga M.S., Bressan E., Lavado R.S. Effects of phytoremediation and application of organic amendment on the mobility of heavy metals in a polluted soil profile. Int. J. Phytoremediation 14, 12–220, 2012.

    ... ...
    ... ...

... ...
    ... ...

PART III

BIOREMEDIATION OF RELEVANT METAL(LOID)S

CHAPTER 10

Bioremediation of Arsenic Using Bioflocculants and Microorganisms

K.A. Natarajan

1. Introduction

Microbial community inhabiting mining environments, waste disposal sites, and mine waters participate in several metal-microbe redox cycles. In the case of refractory sulfidic ores containing precious metals, the gold particles are finely disseminated in pyrite–arsenopyrite matrices. Biooxidation of pyrite–arsenopyrite in the presence of acidophilic autotrophs such as *At. ferrooxidans* and *L. ferrooxidans* leads to dissolution of arsenopyrite and pyrite liberating entrapped gold particles for subsequent cyanidation recovery. Microorganisms such as *At. ferrooxidans, At. thiooxidans, L. ferrooxidans*, Sulfate Reducing Bacteria, *Bacillus* spp. as well as arsenic-tolerant *Thiomonas* spp. inhabit such sulfide ore deposits, mine waters, and processed mill tailings. Acid mine drainage emanating from such mines, mined over burden and tailing dams containing pyrite and arsenopyrite is a source of ground water arsenic contamination brought out by indigenous acidophilic microorganisms. In this paper, microbiological aspects of arsenic dissolution from arsenopyrite in the presence of acidophilic microorganisms are brought out with special reference to refractory sulfidic ore mineralization, abandoned mines, and tailing dams.

Detoxification and remediation of arsenic water pollution are discussed in terms of:

- Biooxidation of arsenite to arsenate
- Ferric-mediated oxidation of arsenite and precipitation of ferric arsenates in presence of *At. ferrooxidans*.

Department of Materials Engineering, Indian Institute of Science, Bangalore–560 012, India.
E-mail: kan@materials.iisc.ernet.in

- Adsorption of arsenic species onto metabolic precipitates generated during growth of *At. ferrooxidans*.
- Precipitation of arsenic as sulfides in the presence of sulfate reducing bacteria.
- Arsenic removal using biosorbents such as bioflocculants and sea shell composites.

2. Role of indigenous microorganisms in arsenic dissolution

Indigenous microbes present in mining environments were isolated, subcultured and identified as detailed in Table 1:

Table 1. Isolation of microorganisms from sulfide ores, mine water, and tailing dams.

Source	Microorganisms isolated
• Pyrite-arsenopyrite mineralization containing gold, silver, copper, lead, and zinc (ore + mine water samples)	*At. ferrooxidans* *At. thiooxidans* *L. ferrooxidans* *Thiomonas* sp. *Bacillus* sp.
• Processed sulfidic gold ore tailings	*Acidiphilium* sp. *Desulfovibrio desulfuricans* *Desulfotomaculum* sp.

Initial tests were focussed on establishing the role of *At. ferrooxidans* in the biooxidation of pyrite and arsenopyrite present in the ore and tailing samples to generate acid and dissolve arsenic and iron (Natarajan, 2008; Natarajan and Ambika, 2008).

Decrease in pH with time in the presence and absence of *At. ferrooxidans* from the sulfide ore and tailing samples was established as shown in Fig. 1. Higher acid formation only in presence of acidophilic iron-sulfur-oxidizing organisms becomes evident. Arsenic dissolution from the above samples was also monitored similarly. Arsenic dissolution from arsenopyrite present in the ore and tailing samples was accelerated in the presence of *At. ferrooxidans* as shown in Fig. 2.

Arsenopyrite, the most abundant arsenic-containing mineral is the potential source of arsenic contamination in the environment. Microbial oxidation of arsenopyrite in the presence of oxygen and water results in the production of arseneous (H_3AsO_3) and arsenic (H_3AsO_4) acids.

$$2FeAsS + 3H_2O + 5.5O_2 \leftrightarrows 2H_3AsO_3 + 2FeSO_4 \qquad (1)$$

$$2FeAsS + 3H_2O + 6.5\,O_2 \leftrightarrows 2H_3As\,O_4 + 2FeSO_4 \qquad (2)$$

$$2FeAsS + H_2SO_4 + 0.5\,O_2 \leftrightarrows 2Fe_2\,(SO_4)_3 + H_2O \qquad (3)$$

Indirect mechanism of microbial oxidation of arsenopyrite also occurs as

$$FeAs\,S + 5\,Fe^{3+} + 3H_2O \leftrightarrows 6\,Fe^{2+} + H_3AsO_3 + 3H^+ + S^\circ \qquad (4)$$

Under strong oxidizing environments, sulfur will be further oxidized to sulfate and As(III) to As(V).

$$FeAs\,S + 8H_2O + 13\,Fe^{3+} \leftrightarrows H_3AsO_4 + 14\,Fe^{2+} + 13\,H^+ + SO_4^{2-} \qquad (5)$$

Acid water containing dissolved arsenite and arsenate species along with ferrous, ferric and sulfate ions are thus formed due to microbial arsenopyrite oxidation.

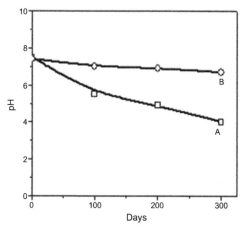

Figure 1. pH decrease with time in the presence (A) and absence (B) of *At. ferrooxidans* (Natarajan, 2008; Natarajan and Ambika, 2008).

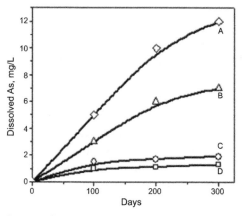

Figure 2. Arsenic dissolution with time in the presence (A,B) and absence (C,D) of *At. ferrooxidans* from different sulfide ores (pyrite, arsenopyrite) (Natarajan, 2008; Natarajan and Ambika, 2008).

Arsenic minerals can also be oxidized by neutrophilic organisms. For example, arsenite can be oxidized by *Thiomonas* sp. and various *Bacillus* sp. Microbes involved in arsenic mobilization also need to be tolerant (resistant) to higher concentrations of arsenite and arsenate.

Biogeochemical cycles in mining environments have been explained through conceptual models (Drewniak and Sklodowska, 2013). Primary arsenic mineral dissolution is mediated by acidophilic microorganisms using arsenic, iron, and sulfur as energy substrates and also by arsenic-tolerant organisms. Secondary minerals formed by As(III) and As(V) adsorption on iron oxides can be solubilized by reductive dissolution. Depending on indigenous microbial communities, transformation between arsenite-arsenate species can occur. Presence of nobler pyrite along with electrochemically and chemically active arsenopyrite can also promote accelerated galvanic dissolution of arsenic. Oxidative dissolution of arsenopyrite by mesophilic and moderately thermophilic acidophiles has been reported (Tuovinen et al., 1994). Arsenic accumulation from acid mine waters by ferruginous bacterial accretions becomes possible (Leblanc et al., 1996).

3. Microbially-induced flotation separation of arsenopyrite from pyrite

Since the presence of arsenopyrite in ore deposits, mine wastes and tailing storages are the major causes for arsenic contamination, its removal from the source itself would be the best preventive strategy. *At. ferrooxidans* isolated from the mine waters can be used for the purpose to bring about selective dissolution or flotation separation of arsenopyrite (Chandraprabha, 2007; Chandraprabha et al., 2004). In the presence of pyrite, galvanic dissolution of anodic (more active) arsenopyrite will be accelerated and can be selectively removed. Yet another approach is through selective flotation separation using *At. ferrooxidans*. *At. ferrooxidans* exhibit different interaction behavior with arsenopyrite and pyrite facilitating their flotation separation.

- Interaction with cells of *At. ferrooxidans* brought about more significant surface chemical changes such as shift in zeta potentials and isoelectric point on pyrite than on arsenopyrite.

- Bacterial cells exhibited higher surface affinity towards pyrite than arsenopyrite. Adsorption density of *At. ferrooxidans* onto pyrite was found to be 10 times higher than that of arsenopyrite. Profuse attachment of *At. ferrooxidans* was observed on pyrite surface, while only sparse cell adhesion occurred on arsenopyrite.

- Bacterial interaction rendered pyrite surfaces more hydrophilic (promoting settling and depression) while arsenopyrite surfaces turned increasingly hydrophobic (promoting dispersion and flotation).

Differential flotation tests, using pyrite-arsenopyrite mixtures, after being interacted with *At. ferrooxidans* and further followed by conditioning with copper sulfate and xanthate collector resulted in more than 95% recovery of arsenopyrite in the concentrate, while pyrite was effectively depressed. Similarly, prior interaction with *At. ferrooxidans* promoted rapid and significant dispersion of arsenopyrite particles while pyrite particles were settled in the aqueous solution. It then becomes possible to separate arsenopyrite from pyrite either through selective flocculation or selective flotation after bacterial interaction. Such separation methods can be used for beneficiation of pyrite-arsenopyrite containing ores or their processed tailings.

4. Arsenic bioremediation in presence of *At. ferrooxidans*

Arsenite and arsenate concentrations in the solution, during the growth of *At. ferrooxidans*, in 9K medium were measured along with ferric-ferrous ion levels and cell population as shown in Fig. 3. In the presence of 2.5 g/L inital concentration of arsenate, arsenic level decreased during exponential bacterial growth to about 1.0 g/L and ultimately to 0.5 g/L after 40 hours. No arsenite was detected in the solution during bacterial growth. In presence of bacterially oxidized ferric ions, arsenate will be precipitated as ferric arsenate over wide pH levels.

Decrease in arsenate from the 9K medium during growth of *At. ferrooxidans* can thus be explained. In the presence of similar concentrations of arsenite ions in the 9K medium also, time-wise decrease in its concentration with bacterial growth could be observed. For example, arsenic concentration decreased to 1.0 g/L from an initial value of 2.5 g/L during the log phase of

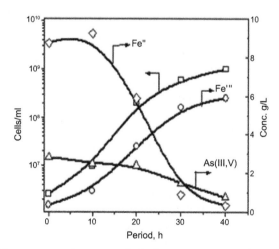

Figure 3. Arsenite, arsenate, ferrous, and ferric concentrations during the growth of *At. ferrooxidans* (Chandraprabha, 2007; Chandraprabha and Natarajan, 2011).

bacterial growth. Arsenate concentrations could not be detected in solution (Chandraprabha, 2007; Chandraprabha and Natarajan, 2009, 2011). Any decrease in added arsenite concentrations in the presence of cell-free acidic metabolic products was also tested. No significant decrease in arsenite concentration in the solution could be observed indicating that ferric ions present in the metabolite solution were unable to precipitate arsenite. Extracellular polymeric substances secreted during bacterial growth could not either oxidize or precipitate arsenite ions. However, arsenite could be effectively removed during growth of *At. ferrooxidans* as observed earlier. The presence of active growing cells may be essential for arsenic removal from solution. The role 9K medium in the absence of bacterial cells on arsenite precipitation was also examined and no significant decrease in arsenite level was observed (Chandraprabha, 2007; Chandraprabha and Natarajan, 2011). The above experimental observations indicate that the presence of ferric ions cannot directly oxidize and precipitate arsenite. The observed decrease in arsenite in solution during the growth of *At. ferrooxidans* could well be due to oxidation of arsenite to arsenate and subsequent precipitation of ferric arsenate compounds. Both arsenite and arsenate ions are co-precipitated with ferric ions formed during growth of *At. ferrooxidans.*

Dissolved arsenic in acid drainage waters was seen to be precipitated in the presence of *At. ferrooxidans*. *At. ferrooxidans* did not oxidize arsenite to arsenate directly or indirectly. One strain precipitated arsenic as arsenite (not arsenate) with ferric ion. Arsenite was found to be associated with schwertmannite ($Fe_8O_8 (OH)_6 SO_4$) and not adsorbed on jarosite. On the contrary, arsenate is known to be efficiently precipitated with ferric iron and sulfate as ordered schwertmannite depending on As:S ratio. Co-precipitation of arsenite with schwertmannite could also be a potential mechanism. Arsenite removal by co-precipitation with ferric iron could well be the common property of *At. ferrooxidans* (Duquesne et al., 2003). Inhibition of ferrous ion oxidation by *At. ferrooxidans* in presence of As(III) and As(V) should also to be considered. As(III) is more inhibitory than As(V). Through serial subculturing in the presence of increasing concentrations of As(III) and As(V), arsenic tolerant strains of *At. ferrooxidans* can be developed.

Growth behavior of sulfur-grown, arsenite-adapted strains *At. ferrooxidans* in the presence of 2.5 g/L of arsenite was studied. However, no significant decrease in arsenite was observed unlike in the case with ferrous–iron grown cells. The cells were unable to oxidize and precipitate arsenite directly and the presence of ferrous-ferric iron becomes essential for the purpose (Chandraprabha and Natarajan, 2011). EPS of *At. ferrooxidans* has been shown to contain firmly bound ferric ions (Gherke et al., 2001). Ferric ions were detected in the EPS of ferrous-grown *At. ferrooxidans* and not sulfur-grown cells. Arsenite- and arsenate-adapted cells of *At. ferrooxidans* contain arsenic in the EPS. Binding of arsenic to the cell EPS influence the zeta potential of cells. Presence of ferric ions is essential for binding of arsenic species to the EPS. EPS extracted from arsenic-adapted and unadapted cells of *At. ferrooxidans* was tested for its ability to bind arsenic. Extracted EPS could not bind to arsenite ions, while arsenate was found to be bound to EPS. Strong affinity between ferric ions and arsenate is once again ascertained by the above observations.

Experimental results further suggested that arsenic removal from growth solutions coincided with the formation of ferric ions through bacterial oxidation. At acidic pH, arsenite was shown to be co-precipitated with ferric ions (Kirk, 1993; Wilki and Hering, 1996).

Oxidation of ferrous ions in acid mine drainage results in precipitation of ferric compounds incorporating arsenic. Formation of nanocrystalline tooeliete, $Fe_6(AsO_3)_4(SO_4)(OH)_4.42H_2O$ was observed together with amorphous mixed As(III) As(V)-Fe(III) oxyhydroxide compounds. Metabolic activity *At. ferrooxidans* and *Thiomonas* sp. in the acidic drainage could promote the above precipitation reactions (Morin et al., 2003). With another strain of *At. ferrooxidans*, formation of schwertmannite in mine water was observed (Duquesne et al., 2003). In our studies, growth of *At. ferrooxidans* in the presence of arsenate ions was found to promote scorodite ($FeAsO_4.2H_2O$) formation. Both arsenite and arsenate ions could get co-precipitated with biogenic ferric ions at acidic pH, resulting in the formation of scorodite and amorphous ferric arsenate, or tooeliete or schwertmannite as the case may be. Arsenic remediation from contaminated acid drainage could be brought about through the above co-precipitation process promoted by *At. ferrooxidans*.

5. Role of metabolic precipitates and *Thiomonas* sp.

Oxidation of As(III) to As (V) in the presence of *Thiomonas* sp. (isolated from mine sites) is illustrated in Fig. 4. Arsenite concentrations decreased from 30 mg/L to less than 5 mg/L after about 190 hours.

Proportionately, arsenate concentration increased in the solution. Since As(III) is more toxic and inhibitory compared to As(V), the above bacterial oxidation can in fact bring about arsenic detoxification. This native organism indigenous to acid drainage waters can play an important role in the oxidation and immobilization of arsenic (Natarajan and Ambika, 2008). In the synergistic presence of *At. ferrooxidans* in acid mine waters, arsenic can be precipitated in the presence of biogenic As(V) and Fe(III) (Duquesne et al., 2007). Arsenite oxidase present in *Thiomonas* is a molybdopterin and can be very effective in the cells grown in presence of As(III). Biofilms development in the synergistic presence of *Thiomonas* sp. and *At. ferrooxidans* in arsenic-containing mine waters can play a key role in natural arsenic bioremediation (Marchal et al., 2007). As(III) exposure impacts *Thiomonas* biofilm maturation through synthesis of extracellular matrix. *Thiomonas* can thus survive in extreme environments overcoming As(III) stress. *Thiomonas arsenivorans* is capable of oxidizing up to 100 mg/L of As(III) and using fixed–bed bioreactors,

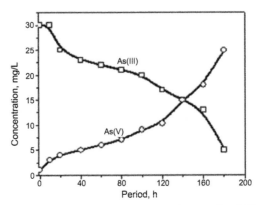

Figure 4. As(III), As(V) concentrations with time during the growth of *Thiomonas* sp. (Natarajan and Ambika, 2008).

efficient biological As(III) oxidation of drinking water could be performed (Guezennec et al., 2010). A bacterial community consisting of As(III) oxidising *Thiomonas*-like organisms and ferrous iron-oxidising *At. ferrooxidans* or *L. ferrooxidans* could be efficiently harvested to develop cost-effective arsenic-remediation technologies. Precipitates formed in the metabolites due to the growth of *At. ferrooxidans* in a 9K medium contain jarosites and various iron oxyhydroxides. Adsorption of arsenite onto the above precipitates as a function of time and pH is illustrated in Fig. 5. Significant arsenite removal through interaction with metabolic precipitates of *At. ferrooxidans* could be seen (Natarajan and Ambika, 2008; Chandraprabha and Natarajan, 2009). Arsenite adsorption (removal) at different pH values onto the biogenic precipitates is shown in Table 2.

SEM micrographs illustrating morphological features of metabolic precipitates generated by the growth of *At. ferrooxidans* before and after exposing to As(III) are shown in Fig. 6. Clustered, dendritic and heterogeneous precipitates could be seen. Large exposed area of the precipitate materials efficiently adsorbs arsenic. EDX analysis indicated profound peaks due to arsenic adsorption on the precipitate surfaces.

6. Arsenic removal using sulfate reducing bacteria

Desulfovibrio desulfuricans and *Desulfotomaculum* sp. isolated from mine waters were also used to remove dissolved As(III) and As(V) from solutions at different pH levels.

$$SO_4^{--} + 2CH_2O \leftrightarrows H_2S + 2HCO_3 \tag{6}$$

$$M^{++} + H_2S = MS + 2H^+ \text{ (M is As(III) or As(V))} \tag{7}$$

Bicarbonate alkalinity would neutralize acidic water.

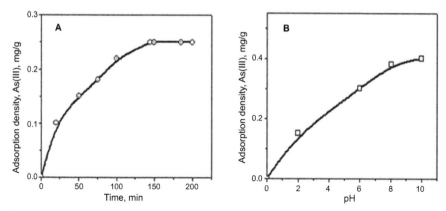

Figure 5. (A) Adsorption of As(III) on metabolic precipitates formed during growth of *At. ferrooxidans*. (B) Effect of pH on As(III) adsorption (Natarajan and Ambika, 2008; Chandraprabha and Natarajan, 2009).

Table 2. Arsenite adsorption on metabolic precipitates of *At. ferrooxidans* at different pH values (Natarajan and Ambika, 2008; Chandraprabha and Natarajan, 2009).

pH	Percent arsenite adsorbed
2–2.5	50
5	61
7–8	65
9–10	72–75

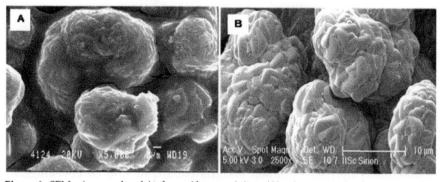

Figure 6. SEM micrographs of *At. ferrooxidans* precipitate (A) prior to arsenic adsorption and (B) after adsorption of arsenic.

Efficient precipitation of arsenic sulfides was observed as illustrated in Fig. 7. Co-precipitation of arsenic sulfides with iron sulfides in the presence of sulfate reducing bacteria would promote efficient arsenic removal. Sorption of dissolved arsenic by the sulfide precipitates formed by biogenic sulfate reduction would be an added advantage. Fresh precipitates were found to

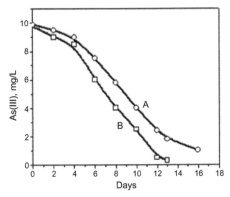

Figure 7. Arsenic precipitation in presence of (A) *Desulfovibrio desulfuricans* and (B) *Desulfotomaculum* sp.

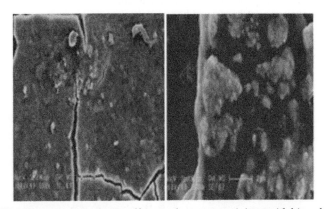

Figure 8. SEM photographs of the *Desulfotomaculum* spp. precipitate with biosorbed arsenic.

be highly surface-active and could remove more than 80% arsenite from aqueous solutions within a week. As(III) removal as a function of time using the above SRB species is represented in Fig. 7. Scanning electron micrographs of arsenic-biosorbed sulfide precipitates are shown in Fig. 8.

7. Use of bioflocculants and sea shells for arsenic removal

Bioflocculants derived from soil bacteria could prove to be cost-effective and environment-friendly biosorbents for arsenic species. Pure strains of *Bacillus licheniformis* and *Bacillus megaterium* isolated from mine soils were subcultured in the laboratory as per procedures described earlier (Karthiga Devi and Natarajan, 2015a,b,c; Natarajan, 2017). Standard procedures were followed to precipitate and extract pure bioflocculants from the above cultures. Bioflocculant yields and amounts of exopolysaccharides and proteins present in the different bioflocculants are given in Table 3.

Table 3. Bioflocculant yields with exopolysaccharides and protein contents (Karthiga Devi and Natarajan, 2015a,b,c).

Microorganism	Yield of bioflocculant (g/L)	Exopolysaccharides (µg/ml)	Exoproteins (µg/ml)
Bacillus megaterium	8	140–150	80–90
Bacillus licheniformis	18	170–180	45–50

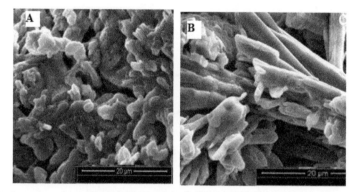

Figure 9. Scanning electron micrographs illustrating surface morphology of bioflocculants generated from (A) *B. licheniformis* (L) (B) *B. megaterium* (M) (Karthiga Devi and Natarajan, 2015a,b,c).

Surface morphologies of the extracted bioflocculants are evident from scanning electron micrographs shown in Fig. 9. Flaky, fibrous and crystalline acicular nature of the bioflocculants is evident. Due to surface heterogeneity and large surface area, they are capable of high adsorption of metal ions. FTIR and DSC spectral analysis showed the presence of carboxyl, hydroxyl and amino-functional groups in the bioflocculants having high thermal stability. As(III) adsorption as a function of bioflocculant dosage, pH and initial arsenic concentration on the two types of bioflocculants M and L (M = bioflocculant from *B. megaterium*, L = bioflocculant from *B. licheniformis*) is illustrated Fig. 10. Bioflocculant M was superior to L in arsenic removal. More than about 85% of arsenic could be removed at neutral pH levels. Both at high acidic and high alkaline pH levels, arsenic adsorption was seen to decrease. There was an optimum dosage for the bioflocculant for maximum arsenic adsorption. Higher initial arsenic concentrations decreased removal rates (Karthiga Devi and Natarajan, 2015b,c; Natarajan, 2017). Bioflocculants as different from chemical derivatives are beneficial for detoxification of contaminated waters since they are cheaper, biodegradable, and free from secondary pollution risk.

Similarly, the use of naturally available sea shells could also prove to be efficient and environment-friendly biosorbents for arsenic (Tsiamis, 2007;

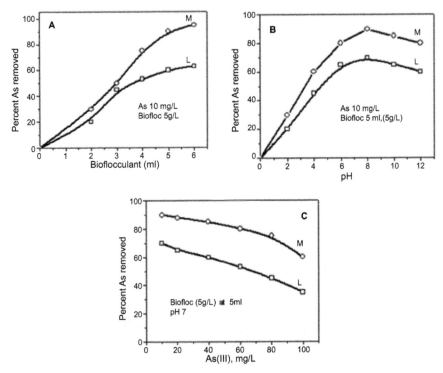

Figure 10. Arsenic (III) removal using bioflocculants L, M. Effect of (A) bioflocculant dosage, (B) pH, and (C) As(III) concentration (Karthiga Devi and Natarajan, 2015a,b,c).

Chowdhury and Saha, 2010). Sea shells were collected from the beach areas of Tuticorin and Kanyakumari in Tamil nadu, India and cleaned well with fresh water. The shells were dried in an oven and subsequently in open air at room temperature and powdered using ball mill. The shell powder was graded as per particle size and used in adsorption tests. The sea shell samples were characterized by chemical analysis, X-ray diffraction, FTIR spectroscopy and scanning electron microscopy.

A stock solution of sodium arsenite, As (III) and sodium arsenate, As (V) of 1000 mg L⁻¹ was prepared in double distilled water and working solutions were prepared by appropriate dilution. The pH of the solution was adjusted before adsorption tests. Batch tests were carried out in 250 mL Erlenmeyer flasks with 100 mL of working volume, with a concentration of 50 mgL⁻¹. A weighed amount of adsorbent was added to the solution. The flasks were agitated at a constant speed of 150 rpm for 2 h in an orbital shaker. The influence of particle size and adsorbent dose were evaluated. Samples were collected from the flasks at predetermined time intervals for analyzing arsenic concentration in the solution.

Table 4. Chemical composition of sea shells.

Sea Shell Samples	Percent composition				
	SiO_2	Fe_2O_3	Al_2O_3	CaO	MgO
Tuticorin	0.6	0.1	0.2	53.2	1.4
Kanyakumari	0.3	0.1	0.10	53.3	1.8

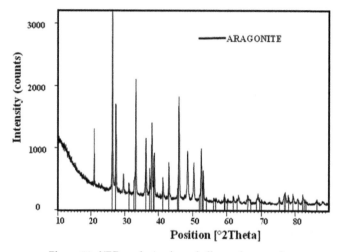

Figure 11. XRD analysis of sea shell powder samples.

Chemical analysis showed that both sea shells are rich in calcium oxide. Typical chemical compositors of collected sea shells are shown in Table 4. The main constituent of sea shells is $CaCO_3$ as can be seen from X-Ray analysis in Fig. 11. Planes corresponding to aragonite could be observed and indexed. FTIR spectral analysis indicated the presence of—OH and CH_2—stretching vibrations. Strong carbonate group was observed. Phosphate group corresponding to O–P–O could be observed in some samples. Negatively charged functional groups such as $-CH_2$, –OH, $-PO_4$ and $-CO_3$ exist on the surface of sea shell powders.

Scanning electron micrographs revealed surface morphology of sea shell powder samples as porous and irregular surface structure. Heterogeneous cavities and porosities provided very high surface area for adsorption.

As (III) removal using different sea shell samples is illustrated in Tables 5–6.

Both As(II) and As(V) could be effectively removed from aqueous solutions using sea shell biosorbents. It becomes possible to design series of column reactors where sea shell powder composites can be packed and arsenic contaminated water is percolated and recycled to maximize toxic metal removal.

Table 5. As(III) removal after interaction with different sea shells.

Sea	Percent Removal of As(III)	
shell powder (g/L)		
	Tuticorin	Kanyakumari
1	30	20
2	45	30
3	85	70

Table 6. As(V) removal after interaction with different sea shells.

Sea	Percent Removal of As(V)	
shell powder (g/L)		
	Tuticorin	Kanyakumari
1	35	25
2	53	38
3	76	62

8. Conclusion

Indigenous microorganisms such as *At. ferrooxidans* and *Thiomonas* sp. take part in arsenic mineral dissolution using arsenic, iron and, sulfur as energy substrates. Detoxification and removal of dissolved arsenic species can be brought about through various strategies such as removal of arsenopyrite from acid drainage causing mineral wastes and tailing dams, co-precipitation of arsenic and iron compounds promoted by microorganisms such as *At. ferrooxidans* and *Thiomonas* spp., adsorption onto metabolic precipitates formed during growth of *At. ferrooxidans,* and sorption onto bioflocculants and sea shell composites. Sulfate reducing bacteria could also be used to precipitate dissolved arsenic as sulfides.

Acknowledgement

The author thanks the National Academy of Sciences, India (NASI), Allahabad for award of NASI Honorary Scientist contingency grant.

References

Chandraprabha M.N. Ph. D thesis, Surface studies on sulphide minerals and *Acidithiobacillus* bacteria with respect to biobeneficiation and bioremediation. Indian Institute of Science, Bangalore, 2007.

Chandraprabha M.N., Natarajan K.A. Mechanism of arsenic tolerance and bioremoval of arsenic by *Acidithiobacillus ferrooxidans*. J. Biochem. Technol. 3, 257–265, 2011.

Chandraprabha M.N., Natarajan K.A. Microbially-induced separation of arsenopyrite and bioremediation of arsenic. Miner. Metall. Proc. 26, 217–221, 2009.

Chandraprabha M.N., Natarajan K.A., Somasundaran P. Selective separation of arsenopyrite from pyrite by biomodulation in the presence of *Acidithiobacillus ferrooxidans*. J. Coll. Interf. Sci. 276, 323–332, 2004.

Chowdhury S., Saha P. Sea shell powder as a new adsorbent to remove basic green 4 (Malachite Green) from aqueous solutions: equilibrium, kinetic and thermodynamic studies. Chem. Eng. J. 164, 168–177, 2010.

Drewniak L., Sklodowska A. Arsenic-transforming microbes and their role in biomining processes. Environ. Sci. Pollut. Res. Int. 20, 7728–7739, 2013.

Duquesne K., Lebrun L., Casiot C., Bruneel O., Personné J.C., Leblanc M., Poulichet F.E., Morin G., Bonnefoy V. Immobilization of arsenite and ferric ion by *At. ferrooxidans* and its relevance to acid mine drainage. Appl. Environ. Microbiol. 69, 6165–6173, 2003.

Duquesne K., Lieutaud A., Ratouchniak J., Yarzabal A., Bonnefoy V. Mechanisms of arsenite elimination by *Thiomonas* sp. isolated from Carnoules acid mine drainage. Eur. J. Soil Biol. 43, 351–355, 2007.

Gherke T., Hallmann R., Kinzler K., Sand W. The EPS of *Acidithiobacillus ferrooxidans*—a model for structure—function relationships of attached bacteria and their physiology. Water Sci. Technol. 37, 159–167, 2001.

Guezennec A.G., Michel C., Dictor M.C., Deluchat V., Dagot C., Klin J.K., Simon S. Biological As(III) oxidation of drinking water supply using *Thiomonas arsenivorans* in fixed bed bioreactor. The Water Research Conf. Lisbon, 2010.

Karthiga Devi K., Natarajan K.A. Production and characterization of bioflocculants for mineral processing applications. Int. J. Miner. Process 137, 15–25, 2015a.

Karthiga Devi K., Natarajan K.A. Isolation and characterization of a bioflocculant from *Bacillus megaterium* for turbidity and arsenic removal. Miner. Metall. Proc. 32, 222–229, 2015b.

Karthiga Devi K., Natarajan K. Isolation and characterization of toxic metal removing bacterial bioflocculants. Adv. Mat. Res. 1130, 585–588, 2015c.

Kirk T.L, Master Thesis, Department of Civil Engineering, New Mexico State University, Las Cruces, NM. 1993.

Leblanc M., Achard B., Othman D., Luck J.M., Bertrand-Sarfati J., Personne J.C. Accumulation of arsenic from acidic mine waters by ferruginous bacterial accretions (stromatolites). Appl. Geochem. 11, 541–554, 1996.

Marchal M., Briandet R., Halter D., Koechler S., DaBow M.S. Mechanisms of arsenite elimination by *Thiomonas* sp. isolated from Carnoules acid mine drainage. Eur. J. Soil Biol. 43, 351–355, 2007.

Morin G., Juillot F., Casiot C., Bruneel O., Personne J.C., Poulichet F.E., Leblanc M., Ildefonse P., Calas G. Bacterial formation of tooeleite and mixed As(III)/(V)-Fe(III) gels in the Carnoulès acid mine drainage of France—A XANES, XRD and SEM study. Env. Sci. Tech. 37, 1705–1712, 2003.

Natarajan K.A. Use of bioflocculants for mining environmental control. Trans. Indian Inst. Metals 70, 519–525, 2017.

Natarajan K.A., Ambika S.N. Biological aspects of arsenic speciation and remediation. *In*: Young C.A., Taylor P.R., Anderson C.G., Choi Y. (eds.). Hydrometallurgy. SME, Littleton, Colorado, 108–116, 2008.

Natarajan K.A. Microbial aspects of acid mine drainage and its bioremediation. Trans. Nonferrous. Met. Soc. China 18, 1355–1360, 2008.

Tsiamis D. Removal of arsenic from water using ground clam shells. J. USSIWP, 80–93, 2007.

Tuovinen O.H., Bhatti T.M., Bigham J.M., Hallberg K.B., Garcia JR.o., Lindström B.E. Oxidative dissolution of arsenopyrite by mesophilic and moderately thermophilic acidophiles. Appl. Environ. Microbiol. 60, 3268–3274, 1994.

Wilkie J.A., Hering J.G. Adsorption of arsenic onto hydrous ferric oxide: effects of adsorbate/adsorbent ratios and cooccurring solutes. Colloids and Surfaces A: Phys. Eng. Aspects 107, 97–124, 1996.

CHAPTER 11

Bioremediation
A Powerful Technique for Cadmium Removal from the Environment

Abhishek Mukherjee

1. Introduction

Cd is one of the major heavy metal contaminants of the environment. The primary sources of Cd are the anthropogenic activities and hard rock mining (Wahsha et al., 2012). Cd is used in essential daily life requirements such as rechargeable batteries, electronic equipments, bearing alloys, pigments for ceramic glazes, paints, and plastics (Adamis et al., 2003). Phosphate fertilizers are also a major source of Cd for soil bodies (Perez and Anderson, 2009). Industrial effluents introduce large amount of Cd to the water bodies (Jarup and Akesson, 2009). The farmers in Japan were diagnosed with Itai–itai disease that resulted from rice grains containing elevated levels of Cd. This remarkable finding led the researchers to characterize the toxic potential of Cd on life forms. Cd has been accepted as a category 1 (human) carcinogen by the International Agency for Research on Cancer (Hossain and Huq, 2002).

Cd can easily enter the plant root and is subsequently translocated to the plant shoot, thus finally introduced into the food chain (Zhou and Qiu, 2005; Chakraborty et al., 2014). As a non-essential metal, Cd interferes with the cellular biochemical and physiological processes that finally results in cell mortality (Wu et al., 2014; Ahmad et al., 2015).

Traditional remediation techniques include chemical precipitation, filtration, electrochemical treatments, reverse osmosis, ion exchange, and adsorption which require high cost and are inadequate for the removal of

Special Centre for Molecular Medicine, Jawaharlal Nehru University, New Delhi, India.
E-mail: abhi_biochem81@yahoo.in

heavy metals from the environment (Mukherjee et al., 2010). Also, processes applied to remediate high metal concentrations have been found to be ineffective for low metal contaminated sites (Lodeiro et al., 2005). For the past two decades, researchers are primarily focussing on sequestering toxic metals and other environmental contaminants by using plants and microorganisms. The process, known as bioremediation, has shown extraordinary efficacy in removing heavy metals including Cd to detoxify the contaminated site. Although most of the experiments have been conducted in the laboratory environment, practical implications are being reported at an increasing pace to validate the suitability of this eco-friendly method for environmental cleaning.

2. Cd toxicity to life forms

The environmental Cd concentration is steadily increasing due to the non-destructive nature of Cd. Atmospheric deposition and the use of phosphate fertilizers are the primary sources of Cd in agricultural soil. Rice, wheat grains, and potato tubers are some of the most important food sources of Cd (Clemens et al., 2013). Also cigarette smoking is another contributor of Cd poisoning in humans as tobacco plant's leaves have extraordinary capacity to accumulate Cd (Lugon-Moulin et al., 2004). Dietary Cd intake depends upon the food items and varied food habits of young and aged individuals. According to the European Food Safety Authority, vegetarians take up more Cd than the non-vegetarians due to the presence of more grains and vegetables in the foodstuffs.

Cd is generally taken up by plant roots and distributed to different tissues thereby interfering with the plant's growth and development (Pagani et al., 2012). Surprisingly, being a non-redox metal, Cd induces the generation of reactive oxygen species (ROS) in living cell by interfering with enzyme activities responsible for maintaining cellular redox homeostasis (Schutzendubel et al., 2001; Yan et al., 2013). The interaction of Cd with antioxidative defence system has been well documented (Vitoria et al., 2001). The generated ROS leads to the peroxidation of membrane lipids to disturb the cellular integrity (Wu et al., 2014). Also ROS causes protein oxidation, DNA damage, chlorophyll destruction, and disturbs the carbon assimilation mechanism, thereby resulting in plant cell's death (Singh and Prasad, 2014; Ahmad et al., 2015). The unusually high affinity of Cd with the sulfhydryl groups is primarily responsible for abnormal enzyme activities in living cells (Wada et al., 2014).

Cells have evolved antioxidative defence mechanism to counterbalance heavy metal toxicity. The efflux of toxic metal ions is another part of combat mechanism exhibited by living cells (Mukherjee et al., 2010; Chakraborty et al., 2014). However, Cd is retained in the human kidney and its biological half-life is almost 10–30 years. Hence, gradual increase of Cd concentration is observed in proportionate amount with age in humans (Clemens et al., 2013). Long-term exposure to Cd leads to renal failure, osteoporosis and cancer (Nawrot et al., 2010; Satarug et al., 2010). Diabetic individuals are highly affected due to renal Cd retention (Satarug et al., 2010). Cd uptake has been shown to be highly correlated with physiological iron status. Divalent metal transporter 1 expressed on the epithelial cell surface can efficiently lead to the influx of Cd during low iron condition (Kim et al., 2007). As anaemia is highly prevalent among young women, they are believed to be more prone to physiological Cd accumulation and toxicity (Vahter et al., 2007).

3. Chemical methods of cadmium remediation: their limitations

Conventional methods for the remediation of Cd-contaminated soil have been classified into two major types. *In-situ* bioremediation doesn't need the removal of soil from the contaminated sites. Once the contaminated part is identified, various treatment procedures are followed to remove Cd from the soil. Lime and acid treatment is an easy procedure to change the pH of the soil to alter the mobility of various Cd species present within. The treatment of soil with Cd-chelating agents is also useful in decreasing Cd content of the contaminated soil. The change in soil pH sometimes destroys its natural properties and removes soil microbiota to a significant extent. Chelation techniques non-specifically remove essential metals to result in soil infertility. In *ex-situ* bioremediation, the areas with greatest Cd contents are identified and removed. The soil is further processed for Cd removal. Digging up the contaminated site generates dust particles and increases the risk of exposure. Also, this method is cost prohibited for large areas and suffers from the drawback of large scale burial of waste materials.

Cd is accumulated in the water bodies by mining and industrial effluents. Various techniques such as precipitation and cementation have been developed for removing Cd from those effluents. Cd can be precipitated as insoluble carbonates, sulphides, or hydroxides (Karthikeyan et al., 1996). Precipitation of Cd as hydroxides by increasing the pH of wastewater is the most common technique due to its easy operation and low cost. Removal of Cd by cementation with magnesium or zinc powder has been well documented (Ku et al., 2002; Younesi et al., 2006). Liquid membrane process for separating Cd from effluents suffers due to the instability of membranes in highly acidic or Saline conditions. The use of ion-exchange and solvent extraction methods is not suitable due to high operational cost.

4. Bioremediation: its necessity

Bioremediation technique uses biological species such as plants and microorganisms to convert environmental contaminants to less toxic forms or to take them out from the affected sites. Plants and microorganisms automatically find their roles in bioremediation as they have to survive within toxic environmental conditions. The choice of bioremediators depends on the situation and extent of toxicity of the affected sites and this situation limits the choice of organisms. The isolation and characterization of bioremediators are important for their suitability of use at the affected sites. However, their practical implication in contaminated soil and water is still lacking and needs to be extensively studied to assess their effectiveness. *In-situ* bioremediation offers direct treatment strategies at the contaminated sites (soil or water). *Ex situ* bioremediation requires removal of the contaminated moieties prior to the initiation of treatment. *In-situ* techniques are cost-effective; they generate less dust and debris thus minimizing the release of contaminants.

Bioremediation is advantageous over traditional methods for Cd removal as it tends to detoxify Cd of the contaminated moieties. The cost of bioremediation technologies can be considerably lower than that of conventional treatment methods. Although bioremediation requires typical longer time in comparison to the traditional techniques, it doesn't hamper the quality of treated soil or water. Therefore, the process may especially be suitable for agricultural soil and aquatic bodies with habitats.

4.1 Phytoremediation of Cd

Phytoremediation is an *in situ* approach that uses plants to clean the contaminated sites. This green technique offers low cost alternative over traditional remediation methodologies for heavy metal removal (Ali et al., 2013). Different types of phytoremediation include phytoextraction, phytovolatilization, phytofiltration, phytodegradation, and phytostabilization. Phytoremediation is mainly used to remove Cd from contaminated soil. The high mobility of soil Cd due to its weak interaction with soil colloids makes it easily available for being taken up by plant roots (Alloway, 1995). Phytochelatins and other organic acids in plants form complexes with Cd that are translocated to different tissues via xylem (Salt et al., 1995). Cd along with other heavy metals interferes with the nutrient uptake and damages plant health. Therefore, plants take up heavy metals and sequester them by binding with certain biomolecules and store them in vacuoles and tonoplasts. In spite of this natural ability of plants to take up Cd, most of them cannot be used as phytoremediators because of their limited biomass production capability at the contaminated site. Ideally, plants used for Cd remediation are considered to be based on high growth potential, deep rooting, and easy propagation. Certain plants, termed as hyperaccumulators, offer significant tolerance potential and accumulate Cd to unusually high concentrations. Hyperaccumulators exhibit transporters at the root plasma membrane to take up Cd even from low concentration areas and effectively quench its toxic effects by storing high amount of the metal within vacuoles (Salt and Kramer, 2000). These plants employ a number of defence mechanisms such as chelation of Cd with metallothioneins and phytochelatins; upregulation of the antioxidant enzyme activities to counterbalance ROS; and synthesis of glutathione and proline for quenching the toxic effects of Cd (Sharma and Dietz, 2009). Commonly, Cd detoxification is achieved by binding of Cd to the cell wall, complexation with phytochelatins, and subsequent compartmentalization in the vacuole and complexation with metallothioneins and reduced glutathione in cell cytosol.

In recent years, Cd phytoextraction has gained considerable research attention. Hyperaccumulators for Cd are not frequently found in the nature. *Thlaspi caerulescens* and *Arabidopsis halleri* have been identified as Cd hyperaccumulator (Baker et al., 2000; Küpper et al., 2000). The ability of *T. caerulescens* to hyperaccumulate Cd is known for a long time. It was reported that Cd accumulation for *T. caerulescens* leaves was up to 1600 mg Cd kg^{-1} DW without detectable decrease of its dry biomass upto 50 mg extractable Cd kg^{-1} soil (Robinson et al., 1998). A French population of *T. caerulescens* was able to accumulate Cd in the shoots over 10000 mg kg^{-1} without significant changes in biomass production (Lombi et al., 2000). The potential of *T. caerulescens* for Cd extraction, in the field trial was calculated as 2 kg.ha^{-1}yr^{-1} under optimum growth conditions (Saxena et al., 1999).

The use of ornamental plants for Cd removal has dual advantages of beautification and remediation. The Cd tolerance and accumulation potential was tested in *Impatiens balsamina*, *Calendula officinalis* and *Althea rosea*. Results indicated that for *Calendula officinalis*, Cd concentration in shoots was lower than in roots indicating its limited potential for Cd translocation from roots to shoots. Therefore, this plant may be suitable for phytostabilization of Cd contaminated soils. *Althea rosea* exhibited high Cd tolerance and its significant translocation from roots to other tissues (Liu et al., 2008).

Mosses have potential role in removing Cd from soil. The properties such as less developed cuticle, more proteins, and less fibres in the cell wall allow easy exchange of cations to boost the bryophytes with great capacity to remove metal ions from the surroundings (Boudet et al., 2011). Exposure to Cd induces overexpression of reduced glutathione that chelates and transports Cd to vacuoles for storage and immobilization (Bleuel et al., 2005). Esposito et al. reported that the induction of heat-shock protein response may be responsible for high resistance of *Leptodictyum riparium* against Cd. *Rhytidiadelphus squarrosus* biomass was found to be successful in removing Cd from single and binary solutions (Pipiska et al., 2013). *Spagnum peat* was successfully used to remove Cd from aqueous solutions (Balan et al., 2010).

Many fern sp. have been identified that are able to absorb Cd and other toxic metals (Ma et al., 2001; Srivastava et al., 2010). *Salvinia minima* and *S. herzogii* have been considered as hyperaccumulators of Cd as they showed high bioconcentration factors for taking up Cd from solutions (Olguin et al., 2002). *Azolla caroliniana* was reported to accumulate high amount of Cd from liquid media (Stepniewska et al., 2005). Other *Azolla* sp. was also reported to take up Cd, the extent of which decreased with an increase in Cd dosage (Arora et al., 2004; Calabrese and Blain, 2009).

Woody biomass plants such as *Salix* and *Populus* are also suitable for Cd removal due to their capability of high biomass production. *Salix* sp., although not hyperaccumulators, could take up large amount of Cd with its effective translocation and immobilization in their shoots (Landberg and Greger, 1996; Rulford et al., 2002). Short-rotation coppice of *Salix* was capable to accumulate very high levels of Cd (Greger and Landberg, 1999; Rulford et al., 2002).

Several non-woody plants such as *Brassica juncea, Zea mays, Nicotiana tobacum*, etc. have been taken into consideration for their potential to remove Cd from soil (Blaylock et al., 1997; Kayser et al., 2000; Gupta et al., 2001). Ecotypes of *Silena vulgaris* are found to be Cd accumulator (Ernst et al., 2000).

Aquatic macrophytes such as *Elodea canadensis* and *Lemna minor* are fast-growing, heavy metal-tolerant species that are able to accumulate Cd (Fritioff and Greger, 2007; Hou et al., 2007) and hence suitable for Cd removal from industrial effluents. *Pistia stratoites* is for Cd removal from surface waters (Das et al., 2013). Experiments with water hyacinth showed that it could accumulate high amount of Cd in leaves and stems (Stratford et al., 1984). High bioaccumulation values for Cd in *Elodea nuttallii* were also observed (Nakada et al., 1979).

4.2 Cd removal by microorganisms

During the last two decades, there have been growing interests in removing heavy metals by the use of microorganisms (Perez-Rama et al., 2002; Mukherjee et al., 2010). Numerous reports indicate the efficacy of biomass of dead microorganisms in adsorbing high amount of Cd to cell walls. Therefore, it is likely that dead biomass would be suitable for bioremediation as there is no need for nutrient supply and no toxicity issues to the biosorbents. However, this process is dependent on chemical parameters such as composition and pH of the contaminated site as well as the structure and morphology of the cell surface used as biosorbents (Sag and Kutsal, 1996; Zouboulis et al., 2004). The use of live and growing microorganisms seems to be more promising as Cd may be removed by adsorption as well as intracellular accumulation by the growing cells (Guo et al., 2010; Rojas et al., 2011). In addition, microbial reproduction can supply cells for continuous operation of Cd removal thus leading to simple controlling and cost-effective manipulation (Wang and Hu 2008). The use of live biomass for bioremediation suffers due to the toxicities of heavy metals. Therefore, screening Cd-resistant strains and their use can provide high removal efficiency under varied operational conditions (Zhou et al., 2013). Preliminery studies have shown that Cd could be removed by live microorganisms such as entophytic bacterium, microalgae, and fungi (Guo et al., 2010; Haq et al., 1999; Roane et al., 2001; Chakraborty et al., 2014).

Bacteria have proven efficacy of Cd removal from contaminated bodies. *Pseudomonas aeruginosa* and *Rhodobacter sphaeroides* have been found to remove large amount of Cd in fixed temperature conditions (Wang et al., 1997; Bai et al.,

2008). A deep-sea strain of *Pseudoalteromonas* sp. efficiently removed Cd by means of biosorption as well as intracellular uptake from liquid media (Zhou et al., 2013). In a very recent work, Zu et al. (2017) developed Cd-resistant mutants of *Enterobacter cloacae* TU to test their biostabilization mechanism and concluded that live biomass of the mutants were better remediator of Cd. Also pot experiments showed that these strains increased the Cd removal capacity of cultivated tobacco plants and the removed Cd was biostabilized within plants' shoots and leaves. Bacterial strains isolated from Cd-contaminated environment were reported to have intrinsic mechanisms for combating Cd-induced toxicity and significant potential for Cd removal (Abbas et al., 2014). Ahmad et al. (2014) investigated the effects of Cd-tolerant bacteria on Cd-stress tolerance in cereals. Results showed that *Klebsiella* sp. was the most effective in lowering Cd uptake in maize and wheat. In another study, A Cd-tolerant *Bacillus* sp. with urease production capacity was isolated from mining soil. The strain exhibited high Cd removal efficiency by converting soluble Cd into insoluble carbonate crystals, the formation of which was affected by initial Cd concentrations, pH, and contact time (Zhao et al., 2017).

Generally, bacterial isolates from polluted areas show high capacity for Cd removal. *Burkholderia cenocepaci* can tolerate higher Cd concentrations by pumping out Cd^{2+} ions (Siudek et al., 2011). *Thalassiosira wessflogii* can use Cd as a nutrient and is likely to detoxify significant amount of Cd from contaminated sites (Lee et al., 1995; Siudek et al., 2011). *Klebsiella pneumoniae* CBL-1 was repoted to remove Cd at a concentration of 1500 mg/ml (Shamim and Rehman, 2012). Chovanová et al. (2004) revealed the significant potential of Cd-resistant bacterial isolates; *Comamonas testosteroni, Klebsiella planticola, Alcaligenes xylosoxidans, Pseudomonas fluorescens, Pseudomonas putida,* and *Serratia liquefaciens,* isolated from Cd-contaminated sewage sludge, for the removal of Cd. *Enterobacter agglomerans* SM 38 and *B. subtilis* WD 90 were also able to remove high amount of Cd (Kaewchai and Prasertsan, 2002). A recent study showed that *Penicillium chrysogenum* XJ-1 endured high levels of Cd contamination via biosorption, metal sequestration, and antioxidant defense systems (Xu et al., 2015). The strain could colonize Cd-polluted soils and reduce bioavailable soil-Cd fractions via high-affinity biosorption on fungal biomass. The use of such strains with other soil amendments may be helpful for inducing the growth of plants.

Fungal cell wall contains chitins, glucans, mannans and proteins along with other polysaccharides, lipids, and pigments (melanin) which facilitate binding of Cd ions onto the mycelial surface (Latge, 2010; Wang and Chen, 2009). In recent years, filamentous fungi are gaining prime importance as candidates for Cd bioremediation due to their capability of growing in drastic environments and taking up Cd by bioadsorption as well as intracellular uptake (Chakraborty et al., 2014).

Aspergillus sp. have proven efficacy in removing Cd and other heavy metals (Aung and Ting, 2005; Santhiya and Ting, 2006). *A. niger* biomass pretreated by boiling in NaOH solution exhibited high capacity of Cd removal (Kapoor et al., 1999). *A. niger* was successfully used to remove Cd from oil field water (Barros Jnior et al., 2003). *A. clavatus* could immobilize high amount of Cd from aqueous solution (Cernansky et al., 2008). *A. foetidus* strain isolated from wastewater treatment plant was found to be highly resistant to Cd (Chakraborty et al., 2014). The strain could retain Cd onto its biomass by converting soluble Cd to insoluble Cd-oxalate crystals.

Cadmium tolerance and bioremediation capacity of seven isolates including *Aspergillus versicolor*, *A. fumigatus*, *Paecilomyces* sp., *Paecilomyces* sp., *Terichoderma* sp., *Microsporum* sp., and *Cladosporium* sp. were tested (Soleimani et al., 2015). Their unusually extremely minimum inhibitory concentration values (1,000–4,000 mg.L^{-1}) indicated that the isolated strains had the capability to survive in Cd-polluted environments. Among them, *Aspergillus versicolor* showed the highest tolerance index. Hashem et al. (2016) showed that the inoculation of arbuscular mycorrhizal fungi improved the Cd tolerance potential of *Cassia italica* Mill.

Algal sp. are suitable for Cd removal mainly due to the cell wall structure containing functional groups such as amino, hydroxyl, carboxyl and sulphate which can act as binding sites for metals (Beveridge and Murray, 1980). Experiments with micro-algae showed their high Cd biosorption capability in laboratory conditions (Harris and Ramelow, 1990; Fehrmann and Pohl, 1993). Brown algae *Fucus vesiculosus* was proved to be the better choice for Cd removal in comparison to red and green algae (Mata et al., 2008). The green algae *Chlorella emersonii*, *Sargassum muticum*, *Ascophyllum sargassum*, and red algae *Ceramium virgatum* are some of the candidates used for removing Cd from wastewater (Arkipo et al., 2004; Loderro et al., 2004; Volesky and Holan, 1995; Hamdy, 2000).

Apart from the above mentioned strains, several other microorganisms are being isolated and tested for their Cd removal efficacy in order to find suitable candidates for Cd removal from soil and aquatic environments.

5. Future perspective

Bioremediation is a very safe and green technology used for Cd removal without causing any harm to the soil and aquatic habitats as well as the surrounding environment. Recent trends rely on developing genetically modified strains with increased tolerance and bioaccumulation potential. Incorporation of metallothionein genes results in highly Cd tolerant species which may be suitable for use at the contaminated sites. Recent research focuses on increasing the antioxidative defence capacity of microorganisms by biotechnological modification to combat high concentration of Cd generally present in the contaminated sites. Agronomy based improvement techniques rely on the use of a small amount of chemicals or growth factors to increase Cd uptake by plants and microorganisms. Consortia of microorganism are applied along with plants to increase their phytoremediation potential.

However, bioremediation strategies suffer from a drawback. Most experiments are carried out in laboratory conditions thus posing a question mark regarding the capability of the strains for practical use. As this natural method is time consuming, bioremediation may take another decade for its fruitful field application. Once experimental approaches lead to the optimization of plants and microbes for site-specific use, bioremediation technique will offer the best alternative as an efficient and cost-effective way to treat Cd contaminated ground water and soil.

References

Abbas S.Z., Rafatullah M., Ismail N., Lalung J. Isolation, identification and characterization of cadmium resistant *Pseudomonas* sp. M3 form industrial wastewater. J. Waste Manage. 2014, 1, 2014.

Adamis P.D.B., Panek A.D., Leite S.G.F., Eleutherio E.C.A. Factors involved with cadmium absorption by a wild-type strain of *Saccharomyces cerevisiae*. Braz. J. Microbiol. 34, 55–60, 2003.

Ahmad A., Hadi F., Ali N. Effective phytoextraction of cadmium (Cd) with increasing concentration of total phenolics and free proline in *Cannabis sativa* (L.) plant under various treatments of fertilizers, plant growth regulators and sodium salt. Int. J. Phytoremed. 17, 56–65, 2015.

Ahmad I., Akhtar M.J., Zahir Z.A., Naveed M., Mitter B., Sessitsch A. Cadmium-tolerant bacteria induce metal stress tolerance in cereals. Environ. Sci. Pollut. Res. 21, 11054–11065, 2014.

Ali H., Khan E., Sajad M.A. Phytoremediation of heavy metals-concepts and applications [Review]. Chemosphere 91, 869–881, 2013.

Alloway B.J. (ed.) Cadmium, Heavy Metals in Soil. Blackie and Academic Professional, New York, pp. 121–151, 1995.

Arkipo G.E., Kja M.E., Ogbonnaya L.O. Cd uptake by the green alga *Chlorella emersonii*. Global J. Pure Appl. Sci. 10, 257–262, 2004.

Arora A., Sood A., Singh P.K. Hyperaccumulation of cadmium and nickel by *Azolla* species. Ind. J. Plant Physiol. 9, 302–304, 2004.

Aung K.M., Ting, Y.P. Bioleaching of spent fluid catalytic cracking catalyst using *Aspergillus niger*. J. Biotechnol. 116, 159–170, 2005.

Bai H.J., Zhang Z.M., Yang G.E., Li B.Z. Bioremediation of cadmium by growing *Rhodobacter sphaeroides*: kinetic characteristic and mechanism studies. Bioresour. Technol. 99, 7716–7722, 2008.

Baker A.J.M., McGrath S.P., Reeves R.D., Smith J.A.C. Metal hyperaccumulator plants: a review of the ecology and physiology of a biochemical resource for phytoremediation of metal-polluted soils. *In*: Terry N., Banuelos G. (eds.). Phytoremediation of Contaminated Soil and Water. Lewis Publishers, Boca Raton, Florida, pp. 85–107, 2000.

Balan C., Bilba D., Macoveanu M. Removal of cadmium from aqueous solutions by Sphagnum moss peat equilibrium study. Environ. Eng. Manag. J. 1, 17–23, 2010.

Barros Jnior L.M., Macedo G.R., Duarte M.M.L., Silva E.P., Lobato A.K.C.L. Biosorption of cadmium using the fungus *Aspergillus niger*. Braz. J. Chem. Eng. 20, 229–239, 2003.

Beveridge T.J., Murray R.G.E. Sites of metal deposition in the cell wall of *Bacillus subtilis*. J. Biotechnol. 141, 876–887, 1980.

Blaylock M.J., Salt D.E., Dushenkov S., Zakharova O., Gussman C., Kapulnik Y., Ensley B.D., Raskin I. Enhanced accumulation of Pb in Indian mustard by soil applied chelating agents. Environ. Sci. Technol. 31, 860–865, 1997.

Bleuel C., Wesenberg D., Sutter K., Miersch J., Braha B., Barlocher F., Krauss G. The use of the aquatic moss *Fontinalis antipyretica* L. ex Hedw. as a bioindicator for heavy metals Cd²⁺

accumulation capacities and biochemical stress response of two *Fontinalis* species. Sci. Total Environ. 345, 13–21, 2005.

Boudet L.C., Escalante A., von Haeften G., Moreno V., Gerpe M. Assessment of heavy metal accumulation in two aquatic macrophytes: a field study. J. Braz. Soc. Ecotoxicol. 6, 57–64, 2011.

Calabrese E.J., Blain R.B. Hormesis and Plant Biology. Environ. Pollut. 157, 42–48, 2009.

Cernansky S., Urik M., Sevc J., Littera P., Hiller E. Biosorption of arsenic and cadmium from aqueous solutions. Afr. J. Biotechnol. 6, 1932–1934, 2008.

Chakraborty S., Mukherjee A., Khuda-Bukhsh A.R., Das T.K. Cadmium-induced oxidative stress tolerance in cadmium resistant *Aspergillus foetidus*: its possible role in cadmium bioremediation. Ecotoxicol. Environ. Saf. 106, 46–53, 2014.

Chovanova K., Sladekova D., Kmet V., Proksova M., Harichova J., Puskarova A., Polek B., Ferianc P. Identification and characterization of eight cadmium resistant bacterial isolates from a cadmium-contaminated sewage sludge. Biologia Bratislava. 59, 817–827, 2004.

Clemens S., Aarts M.G.M., Thomine S., Verbruggen N. Plant science: the key to preventing slow cadmium poisoning. Trends Plant Sci. 18, 92–99, 2013.

Das S., Goswami S., Talukdar A.D. A study on cadmium phytoremediation potential of water lettuce, *P. stroites* L. Bull. Environ. Contam. Toxicol. 92, 169–174, 2013.

EFSA (European Food Safety Authority). Cadmium dietary exposure in the European population. EFSA J. 10, 2551–2687, 2012.

Ernst W.H.O., Nilisse H.J.M., Ten Brookum W.M. Combination toxicology of metal-enriched soils: physiological responses of and Cd resistant ecotypes of *Silene vulgaris* on polymetallic soils. Environ. Exp. Bot. 43, 55–71, 2000.

Fehrmann C., Pohl P. Cadmium adsorption by the non-living biomass of micro-algae grown in axenic mass culture. J. Appl. Phycol. 5, 555–562, 1993.

Fritioff A., Greger M. Fate of cadmium in Elodea canadensis. Chemosphere 67, 365–375, 2007.

Greger M., Landberg T. Use of willow in phytoextraction. Inter. J. Phytorem. 1, 115–123, 1999.

Guo H., Luo S., Chen L., Xiao X., Xi Q., Wei W., Zeng G., Liu C., Wan Y., Chen J. Bioremediation of heavy metals by growing hyperaccumulaor endophytic bacterium *Bacillus* sp. L14. Bioresour. Technol. 101, 8599–8605, 2010.

Gupta S.K., Wenger K., Gulz P. *In situ* restoration of soil quality by regulating bioavailable metal concentration in soil: chance or utopia. Extended abstracts of the Proceedings of the 4th workshop of COST action 837, Madrid, 2001.

Hamdy A.A. Biosorption of heavy metals by marine algae. Curr. Microbiol. 41, 232–238, 2000.

Haq R., Zaidi S.K., Shakoori A. Cadmium resistant *Enterobacter cloacae* and *Klebsiella* sp. isolated from industrial effluents and their possible role in cadmium detoxification. World J. Microb. Biot. 15, 283–290, 1999.

Harris P.O., Ramelow G.J. Binding of metal ions by particulate biomass derived from *Chlorella vul-garis* and *Scendesmus quadricauda*. Environ. Sci. Technol. 24, 220–228, 1990.

Hashem A., Abd-Allah E.F., Alqarawi A.A., Egamberdieva D. Bioremediation of adverse impact of cadmium toxicity on *Cassia italica* Mill by arbuscular mycorrhizal fungi. Soudi J. Biol. Sci. 23, 39–47, 2016.

Hossain Z., Huq F. Studies on the interaction between Cd^{2+} ions and nucleobases and nucleotides. J. Inorg. Biochem. 90, 97–105, 2002.

Hou W., Chen X., Song G., Wang Q., Chi Chang C. Effects of copper and cadmium on heavy metal polluted waterbody restoration by duckweed (*Lemna minor*). Plant Physiol. Biochem. 45, 62–69, 2007.

Jarup L., Akesson A. Current status of cadmium as an environmental health problem. Toxicol. Appl. Pharmacol. 238, 201–208, 2009.

Kaewchai S., Prasertsan P. Biosorption of heavy metal by thermotolerant polymer producing bacterial cells and bioflocculant. Songklanakarin J. Sci. Technol. 24, 421–430, 2002.

Kapoor A., Viraraghavan T., Cullimore, D.R. Removal of heavy metals using the fungus *Aspergillus niger*. Bioresour. Technol. 70, 95–104, 1999.

Karthikeyan K.G., Elliott H.A., Cannon F.S. Enhanced metal removal from wastewater by coagulant addition. Proc. 50th Purdue Industrial Waste Conf., 50, 259–267, 1996.

Kayser A., Wenger K., Keller A., Attinger W., Felix H., Gupta S.K., Schulin R. Enhancement of phytoextraction of Zn, Cd, and Cu from calcareous soil: the use of NTA and sulfur amendments. Environ. Sci. Technol. 34, 1778–1783, 2000.

Kim D.W., Kim K.Y., Choi B.S., Youn P., Ryu D.Y., Klaassen C.D., Park J.D. Regulation of metal transporters by dietary iron, and the relationship between body iron levels and cadmium uptake. Arch. Toxicol. 81, 327–334, 2007.

Ku Y., Wu M.-H., Shen Y.-S. A study on the cadmium removal from aqueous solutions by zinc cementation. Sep. Sci. Technol. 37, 571–590, 2002.

Küpper H., Lombi E., Zhao F.J., McGrath S.P. Cellular compartmentaion of cadmium and zinc in relation to other elements in the hyperaccumulator *Arabidopsis halleri*. Planta 212, 75–84, 2000.

Landberg T., Greger M. Differences in uptake and tolerance to heavy metals in *Salix* from unpolluted and polluted areas. Appl. Geochem. 11, 175–180, 1996.

Latge J.P. Tasting the fungal cell wall. Cell. Microbiol. 12, 863–872, 2010.

Lee J.G., Roberts S.B., Morel F.M.M. Cadmium: A nutrient for the marine diatom *Thalassiosira weissflogii*. Limnol. Oceanogr. 40, 1056–1063, 1995.

Liu J., Zhoub Q., Suna T., Ma L.Q., Wang S. Growth responses of three ornamental plants to Cd and Cd-Pb stress and their metal accumulation characteristics. J. Hazard. Mater. 151, 261–267, 2008.

Lodeiro P., Cordero B., Barriada J.L., Herrero R., Vincente M.E.S. Biosorption of cadmium by brown marine microalgae. Bioresour. Technol. 96, 1796–1803, 2005.

Lodeiro P., Cordero B., Grille Z., Herrero R., Vicente M.E.S. Physicochemical studies of Cd(II). Biosorption by the invasive algae in Europe. *Sargassum muticum*. Biotechnol. Bioeng. 88, 237–247, 2004.

Lombi E., Zhao F., Dunham S., McGrath S. Cadmium accumulation in populations of *Thlaspi caerulescens* and *Thlaspi goesingense*. New Phytol. 145, 11–20, 2000.

Lugon-Moulin N. Zhang M., Gadani F., Rossi L., Koller D., Krauss M., Wagner G.J. Critical review of the science and options for reducing cadmium in tobacco (*Nicotiana tabacum* L.) and other plants. Adv. Agron. 83, 111–180, 2004.

Ma L.Q., Komar K.M., Tu C., Zhang W., Cai Y., Kennelley E.D. A fern that hyperaccumulates arsenic. Nature 409, 579, 2001.

Mata Y.N., Blazquez M.L., Ballester A., Gonzalez F., Munoz J.A. Characterization of biosortion of cadmium, lead and copper with the brown alga *Fucus Vesiculosus*. J. Hazard. Mater. 158, 316–323, 2008.

Mukherjee A., Das D., Mandal S.K., Biswas R., Das T.K., Boujedaini N., Khuda-Bukhsh A.R. Tolerance to arsenate-induced stress in *Aspergillus niger*, a possible candidate for bioremediation. Ecotoxicol. Environ. Saf. 73, 172–182, 2010.

Nakada M., Fukaya K., Takeshita S., Wada Y. The accumulation of heavy metals in the submerged plant (*Elodea nuttallii*). Bull. Environ. Contam. Toxicol. 22, 21–27, 1979.

Nawrot T.S., Staessen J.A., Roels H.A., Munters E., Cuypers A., Richart T., Ruttens A., Smeets K., Clijsters H., Vangronsveld J. Cadmium exposure in the population: from health risks to strategies of prevention. Biometals 23, 769–782, 2010.

Olguin J., Hernandez E., Ramos I. The effect of both different light conditions and the pH value on the capacity of *Salvinia minima* BAKER for removing cadmium, lead and chromium. Acta Biotechnol. 22, 121–131, 2002.

Pagani M.A., Tomas M., Carrillo J., Bofill R., Capdevila M., Atrian S., Andreo C.S. The response of the different soybean metallothionein isoforms to cadmium intoxication. J. Inorganic Biochem. 117, 306–315, 2012.

Perez-Rama M., Abalde Alonso J., Herrero Lopez C., Torres Vaamonde E. Cadmium removal by living cells of the marine microalga *Tetraselmis suecica*. Bioresour. Technol. 84, 265–270, 2002.

Perez A.L., Anderson K.A. DGT estimates cadmium accumulation in wheat and potato from phosphate fertilizer applications. Sci. Total. Environ. 407, 5096–5103, 2009.

Pipíška M., Horník M., Remenárová L., Augustín J., Lesný J. Biosorption of cadmium, cobalt and zinc by moss *Rhytidiadelphus squarrosus* in the single and using *Pteris* sp. Ecotoxicol. Environ. Saf. 98, 236–243, 2013.

Roane T., Josephson K., Pepper I. Dual-bioaugmentation strategy to enhance remediation of cocontaminated soil. Appl. Environ. Microbiol. 67, 3208–3215, 2001.

Robinson B.H., Meblanc L., Petit D., Brooks R.R., Kirkman J.H., Gregg P.E.H. The potential of *Thlaspi caerulescens* for phytoremediation of contaminated soils. Plant Soil 203, 47–56, 1998.

Rojas L.A., Yanez C., González M., Lobos S., Smalla K., Seeger M. Characterization of the metabolically modified heavy metal-resistant *Cupriavidus metallidurans* strain MSR33 generated for mercury bioremediation. PLoS ONE 6(3), e17555, 2011.

Rulford I.D., Riddell-Black D., Stewart C. Heavy metal uptake by willow clones from sewage sludge-treated soil: the potential for phytoremediation. Int. J. Phytorem. 4, 59–72, 2002.

Sag Y., Kutsal T. The selective biosorption of chromium (VI) and copper (II) ions from binary metal mixtures by *R. arrhizus*. Process Biochem. 31, 561–572, 1996.

Salt D.E., Prince R.C., Pickering I.J., Raskin I. Mechanism of cadmium mobility and accumulation in Indian mustard. Plant Physiol. 109, 1427–1433, 1995.

Salt O.D.E., Kramer U. Mechanisms of metal hyperaccumulation in plants. *In*: Raskin I., Ensley B.D. (eds.). Phytoremediation of Toxic Metals-Using Plants to Clean up the Environment. Wiley, New York, pp. 231–246, 2000.

Santhiya D., Ting Y.P. Use of adapted *Aspergillus niger* in the bioleaching of spent refinery processing catalyst. J. Biotechnol. 121, 62–74, 2006.

Satarug S., Garret S.H., Sens M.A., Sens D.A. Cadmium, environmental exposure, and health outcomes. Environ. Health Perspect. 118, 182–190, 2010.

Saxena P., KrishnaRaj S., Dan T., Perras M., Vettakkorumakankav N. Phytoremediation of heavy metals contaminated and polluted soils. *In*: Prasad M.N.V., Hagemaer J. (eds.). Heavy Metal Stress in Plants: From Molecules to Ecosystems. Springer-Verlag Berlin Heidelberg, pp. 305–329, 1999.

Schutzendubel A., Schwanz P., Teichmann T., Gross K., Langenfeld-Heyser R., Godbold D.L., Polle A. Cadmium-induced changes in antioxidative systems, hydrogen peroxide content, and differentiation in scots pine roots. Plant Physiol. 127, 887–898, 2001.

Shamim S., Rehman A. Camium resistance and accumulation potential of *Klebsiella pneumoniae* Strain CBL-1 isolated from industrial waste water. Pakistan J. Zool. 44, 203–208, 2012.

Sharma S.S., Dietz K.J. The relationship between metal toxicity and cellular redox imbalance. Trends Plant Sci. 14, 43–50, 2009.

Singh A., Prasad S.M. Effect of agro-industrial waste amendment on Cd uptake in *Amaranthus caudatus* grown under contaminated soil: an oxidative biomarker response. Ecotoxicol. Environ. Saf. 100, 105–113, 2014.

Siudek P., Falkowska L., Urba A. Temporal variability of particulate mercury in the air over the urbanized zone of the southern Baltic. Atm. Poll. Res. 2, 484–491, 2011.

Soleimani N., Fazli M.M., Mehrasbi M., Darabian S., Mohammadi J., Ramazani A. Highly cadmium tolerant fungi: their tolerance and removal potential. J. Environ. Health Sci. Eng. 13, 19–28, 2015.

Srivastava S., Mishra S., Dwivedi S., Tripathi R. Roof thiol metabolism in arsenic detoxification in *Hydrilla verticillata* (L.f.) Royle. Water Air Soil. Pollut. 212, 155–165, 2010.

Stêpniewska Z., Bennicelli R.P., Balakhnina T.I., Szajnocha K., Banach A., Woliñska A. Potential of Azolla caroliniana for the removal of Pb and Cd from wastewaters. Int. Agrophys. 19, 251–255, 2005.

Stratford H.K., William T.H., Leon A. Effects of heavy metals on water hyacinths (*Eichhornia crassipes*). Aquat. Toxicol. 5, 117–128, 1984.

Vahter M., Akesson A., Liden C., Ceccatelli S., Berglund M. Gender differences in the disposition and toxicity of metals. Environ. Res. 104, 85–95, 2007.

Vitoria A.P., Lea P.J., Azevedo R.A. Antioxidant enzymes responses to cadmium in radish tissues. Phytochemistry 57, 701–710, 2001.

Volesky B., Holan Z.R. Biosorption of heavy metals. Biotechnol. Prog. 11, 235–250, 1995.

Wada K.C., Mizuuchi K., Koshio A., Kaneko K., Mitsui T., Takeno K. Stress enhances the gene expression and enzyme activity of phenylalanine ammonia-lyase and the endogenous content of salicylic acid to induce flowering in pharbitis. J. Plant Physiol. 171, 895–902, 2014.

Wahsha M., Bini C., Argese E., Minello F., Fontana S., Wahsheh H. Heavy metals accumulation in willows growing on Spolic Technosols from the abandoned Imperina Valley mine in Italy. J. Geochem. Explor. 123, 19–24, 2012.

Wang B.E., Hu Y.Y. Bioaccumulation versus adsorption of reactive dye by immobilized growing *Aspergillus fumigates* beads. J. Hazard. Mater. 157, 1–7, 2008.

Wang C.L., Michels P.C., Dawson S.C., Kitisakkul S., Baross J.A., Keasling J.D., Clark D.S. Cadmium removal by a new strain of *Pseudomonas aeruginosa* in aerobic culture. Appl. Environ. Microbiol. 63, 4075–4078, 1997.

Wang J., Chen C. Biosorbents for heavy metals removal and their future. Biotechnol. Adv. 27, 195–226, 2009.

Wu Q.S., Zou Y.N., Abd Allah E.F. Mycorrhizal association and ROS in plants. *In*: Ahmad P. (ed.). Oxidative Damage to Plants, Elsevier, pp. 453–475, 2014.

Xu X., Xia L., Zhu W., Zhang Z., Huang Q., Chen W. Role of *Penicillium crysogenum* XJ-1 in the detoxification and bioremediation of cadmium. Front. Microbiol. 6, 1422, 2015.

Yan L., Li X., He M., Zeng F. Behavior of native species *Arrhenatherum Elatius* (Poaceae) and *Sonchus Transcaspicus* (Asteraceae) exposed to a heavy metal-polluted field: plant metal concentration, phytotoxicity, and detoxification responses. Int. J. Phytorem. 15, 924–937, 2013.

Younesi S.R., Alimadadi H., Alamdari E.K., Marashi S.P.H. Kinetic mechanisms of cementation of cadmium ions by zinc powder from sulphate solutions. Hydrometallurgy 84, 155–164, 2006.

Zhao Y., Yao J., Yuan Z., Wang T., Zhang Y., Wang F. Bioremediation of Cd by strain GZ-22 isolated from mine soil based on biosorption and microbially induced carbonate precipitation. Environ. Sci. Pollut. Res. 24, 372–380, 2017.

Zhou W., Zhang H., Ma Y., Zhou J., Zhang Y. Bio-removal of cadmium by growing deep-sea bacterium *Pseudoalteromonas* sp. SCSE709-6. Extremophiles 17, 723–731, 2013.

Zhou W.B., Qiu, B.S. Effects of cadmium hyperaccumulation on physiological characteristics of *Sedum alfredii* Hans (Crassulaceae). Plant Sci. 169, 737–745, 2005.

Zouboulis A.I., Loukidou M.X., Matis K.A. Biosorption of toxic metals from aqueous solutions by bacteria strains isolated from metal-polluted soils. Process Biochem. 39, 909–916, 2004.

Zu C., He S., Liu Y., Zhang W., Lu D. Bioadsorption and biostabilization of cadmium by *Enterobacter cloacae* TU. Chemosphere 173, 622–629, 2017.

CHAPTER 12

Process Oriented Characterization in Bioleaching Co-Cu Minerals

Guy Nkulu[1] and *Stoyan Gaydardzhiev*[2,*]

1. Introduction

The copper-cobalt mineralization of the Katanga basin belonging to the Central African Copper belt and situated between the Democratic Republic of Congo (DRC) and Zambia is famous with its mineral reserves. Apart from the enormous copper deposits, more than 40% of the world cobalt production originates from this region (Laurence, 2005; Yager, 2014). Historically, the polymetallic sulphide ores and concentrates from this region have been treated through traditional pyrometallurgical and acid leaching routes. Nevertheless, environmental constraints imposed by the recent mining legislature coupled with the rising costs of the established metal extraction methods have led to the abandoning of several potentially exploitable deposits. This is the case of Kamoya deposit, characterized by stratified type ore bodies with uniformly disseminated carrollite mineralization. During the times of operation, the ore has been processed through sulphating roasting followed by hydrometallurgy for recovery of copper, cobalt, and nickel. During 1998, however, the operations in the mine had been seized due to both technological issues (e.g., Ni elimination

[1] GECAMINES Sarl, Lubumbashi, DR Congo.
  E-mail: guynkulungoie@yahoo.fr
[2] GEMME–ARGENCO, University of Liege, Allée de la Découverte 9, Sart Tilman, 4000 Liege, Belgium.
[*] Corresponding author: s.gaydardzhiev@ulg.ac.be

from the PLS) and financial downturns. Later on, exploratory studies were launched to find alternative and economically viable ways to extract remaining metals (cobalt mainly) from the ore and from the surrounding tailings (Kitobo, 2009). These studies have been realized in majority on lab scale with very a few of them being further up-scaled. Given the fact that carrollite is widely present elsewhere in the polymetallic deposits of the Katanga metallogenic zone, there is a strong incentive to render biometallurgy, a commercially justifiable option for cobalt extraction. In order to meet this need, the dissolution mechanisms of the carrollite have to be identified better. While bioleaching reactions for copper minerals are broadly-well known, there is no doubt that carrollite is likely to undergo bio-oxidation at a rate and at an extent which are not known.

When polymetallic sulphides are subjected to bioleaching, the microorganisms could be found either in the liquid phase (refereed as planktonic MO) or attached to the surface (what we called fixed or anchored to the surface MO), the latter ones forming the biofilm. Their importance in view revealing the responsible leaching mechanisms have stimulated numerous studies for counting, characterizing and identification, e.g., Q-PCR, MPN (Beach and Sunner, 2004; Kinzler et al., 2003; Dziurla, 1995, 1998). The ultimate aim has been to design the optimum proportion of microorganism members that a used consortium for bioleaching should contain.

In the studied case having an exploratory character, the initial strategy was to estimate the number of "planktonic"microorganisms met in solution and of those "fixed" or anchored on the carrollite surface in the course of the leaching duration. For the latter, we have chosen a methodology based on physicochemical desorption, which is known by its simplicity and accuracy and at the same time is able to distinguish the two types of anchored bacteria: transiently-bound (reversibly detachable) and strongly-bound. The method was originally developed by Monroy et al. (1993) and used by Dziurla (1995, 1998). In our case, the estimation of the transiently bound (reversibly detachable) microorganisms has been realized by gentle wash-out using 9K solution, while the strongly-adhered ones—by means of a "Tween 80" detergent which is a combination between anionic and non-ionic surfactants. Figure 1 illustrates the experimental protocol used for the estimation of the number of fixed microorganisms.

The findings coming from bacteria counting are complemented by observation on the mineral surface in order to trace the changes resulting from bacterial presence and to link them to the basic bioleach amenability of the mineral.

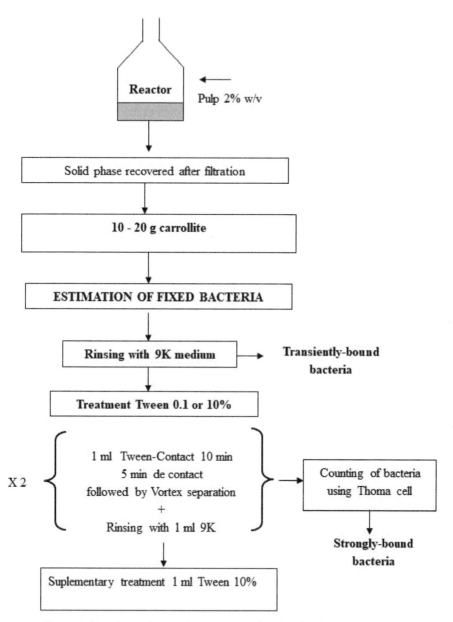

Figure 1. Experimental set-up for estimation of surface-fixed microorganisms.

2. Materials and methods

High purity carrollite samples accompanied by their dolomitic gangue were handpicked from the rich mineralized zones of the Kamoya deposit. The carrollite mono-crystals have been further fragmented and prepared as to render them suitable for bio-leaching. The consortium comprised three different mesophilic iron and sulphur oxidizing bacteria, e.g., *A. ferrooxidans*, *L. Ferrooxidans,* and *A. thiooxidans* isolated in Bulgaria. The mixed culture was adapted to grow on the solid substrate before being used further. The entire bioleaching procedure, together with counting and characterization of methodologies, is described in our previous publications (Nkulu et al., 2013, 2015).

3. Role of microorganisms: experimental results

3.1 Evolution of "fixed" and "planktonic" microorganisms

The grow rate of the "fixed" microorganisms as function of the leaching duration has been followed through implementing the physicochemical desorption protocol described above, while the number of "planktonic" microorganisms was counted directly in the leach solutions. Three distinguished phases characterized by varying number of bacteria in solution and on the surface were detected as follows:

Phase 1—that lasted until approximately day 5, during which the concentration of "planktonic" bacteria remained nearly constant ($\pm 10^7$ cells/mL). It could be argued that this period coincided with the "latent" phase of bacterial activity. During this phase the number of fixed bacteria (both strongly and reversible detachable) increased slightly.

Phase 2—after day 5, we observed strong increase in the "planktonic" bacteria population. We could infer that bacteria started to grow exponentially due to ferric iron reduction taking place continuously at the mineral surface and leading to concomitant increase in ferrous iron concentration in the solution. As a result, the numbers of "planktonic" microorganisms grew exponentially. In parallel, the generated extracellular polymeric substances (EPS) started to form biofilms. At day 15, the number of planktonic microorganisms reached 4.1×10^{10} cell/mL, while the number of the surface-fixed ones approached 5.5×10^8 cells/mL.

After day 20, the third and last phase could be distinguished. This period was characterized by a sharp drop in the number of fixed microorganisms on the expense of just negligible decrease of the planktonic ones. Two reasons could explain these phenomena; (1) either there is lack of nutrient medium being consumed through the jarosite formation or (2) the concentration of ferrous ions is reduced by precipitation or jarosite formation. The latter effect seems plausible since the mineralogical analysis of the leached residue had confirmed jarosite presence—Fig. 2, jarosite marked as J.

In parallel to the estimation of the number and type of microorganisms as functions of the leaching duration, the observations on selected mineral grain surfaces have been useful in defining the role which the fixed microorganisms play during the bioleaching. In the figures which follow, few exemplified situations of microorganisms anchored on the surface of the carrollite grains are illustrated. The perusal of the mineral grain SEM image shown in the left side of Fig. 3, witnesses anchored microorganisms together

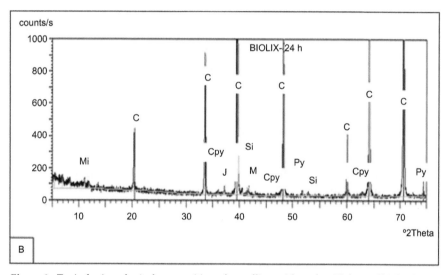

Figure 2. Typical mineralogical composition of carrollite residue after 25 days of bioleaching.

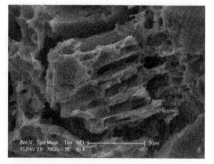

Figure 3. SEM images of mineral grains occupied by bacterial cultures. Left—mineral grain after 12 days leaching; Right—mineral grain after 20 days leaching.

with iron oxides globules and EPS. Aggregates possibly consisting of mineral particles and organic substances are seen. In the right image, mineral grains heavily degraded by the microorganisms met in close proximity do appear. These microorganisms contribute to the generation of visible precipitates as well which most likely do not have chemical origin but rather are formed under bacterial action. The fragments which are composed from mineral particles and organic compounds could be viewed as intermediate sources of energy, thus, supporting the important role of the EPS during the carrollite bioleaching. These observations are in agreement with the model proposed by Schippers et al. (1996) describing the situation where the EPS containing ferric ions have played an important role during pyrite oxidation by *A. ferrooxidans*.

It is known that *L. ferrooxidans* cells tend to accumulate high concentration of Fe^{3+} within the EPS they produce (Rojas-Chapana and Tributsch, 2004). Therefore, it could be argued that cells abundantly seen at the left image in Fig. 4 do belong to this genus. However, due to differences in the adhesion energy between both bacterial types, *A. ferrooxidans* cells are likely to be associated with corrosion pits close to edges (Edwards and Rutenberg, 2001). This situation could be spotted at the right image shown in Fig. 4.

The performed microscopic observations allow us to likewise draw some clues about the life-cycle of the microorganisms. It could be postulated that the evolution of bacterial population follows two distinguished pathways. The first phase, lasting up to day 10, could be considered as maturation stage when bacteria are attaching to the surface and biofilms start emerging. During this period, the anchored bacteria on the surface eventually leave their remnants on it—Fig. 5 on the left. Moreover, visible traces of bacterial attacks toward preferential zones and corrosion pittings could be seen. The second phase, between day 10 and 20, essentially corresponds to a biochemical attack when the mineral surface begins to degrade heavily—Fig. 5 on the right. Physical detachment of mineral's micro-particles takes place followed

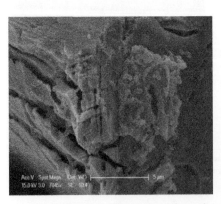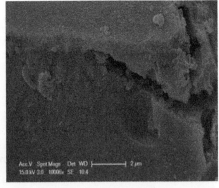

Figure 4. SEM images of carrolite surface after 15 days of bioleaching.

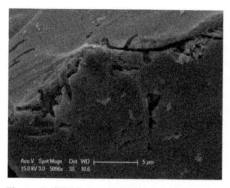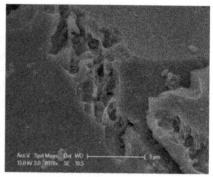

Figure 5. SEM images of mineral grains occupied by bacterial cultures. Left—mineral grain after 6 days bioleaching; Right—mineral grain after 15 days bioleaching.

by the emerging of aggregates containing mineral-organic substances. These phenomena are naturally enhanced by bacterial presence.

3.2 Role of the contact bacteria/mineral

Several studies aiming to define the role of the contact between microorganisms and minerals have pointed out the contribution of the said contact towards efficient bringing of metals in solution. Nevertheless, published results suggest that the leaching efficiency also depends to larger extent on the nature of the metallic sulphides being tested (Konishi et al., 1992; Pistaccio et al., 1994; Porro et al., 1997).

In the mentioned studies, the role of the contact has been often investigated through purposely designed set-ups enabling physical separation between bacteria and mineral. For example, results reported by (Pogliani et al., 1990) for the CuS-*T. ferrooxidans* system were obtained by implementing dialysis sacs. The importance of the direct contact for the studied case has been demonstrated but at the same time the role of the ferrous iron re-oxidation is equally emphasised.

Other studies (Larsson et al., 1993) have shown that in the case of pyrite oxidation by thermophilic archeon *Acidianus brierleyi*, an efficient leaching was only possible if a direct contact between cells and mineral substrate was established.

For the case of carrollite bioleaching in order to experimentally follow the role of bacteria-mineral contact, a double-compartment reactor fitted with microporous membrane has been designed. The objective was to physically separate the microorganisms from the mineral particles and at the same time to allow free exchange of ions and soluble products between both sides of the reactor. This aim was realized by placing a membrane having openings smaller than the size of bacteria. In such a way, the direct action of the fixed bacteria could be identified and compared to the situation where only ferric ions are allowed to enter into contact with the mineral. An intermediate situation simulating semi-contact has been realized as well where half of the mineral mass has been placed in contact while the other one was separated from the microorganisms. Figure 6 shows the methodological set-up and the protocol which was followed to this end.

Figure 7 reports the extraction degree of Co and Cu with time for the cases of: full contact (b); lack of contact (c) and semi-contact (d) between bacteria-mineral as illustrated in Fig. 6. The immediate impression from the cobalt and copper extraction curve trends suggest quite a similar trend for the three cases being studied. In terms of leaching kinetics, one could note three separate zones being function of the leaching duration: the first one between days 0–13, the second one between days 12–20 and the third one between

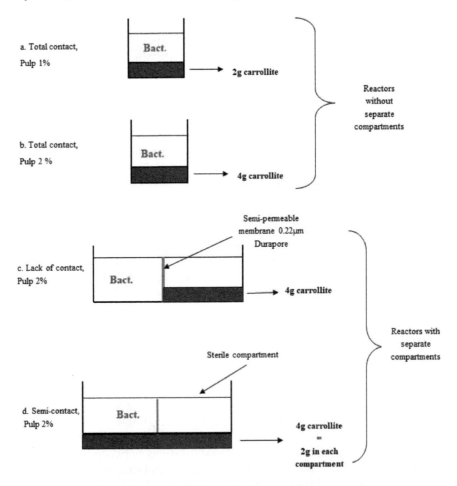

Figure 6. Experimental set-up for simulation of various situation of contact bacteria-carrollite

days 20–30. The first period encompasses both latent and growing phases which are quite evident in the bacteria-mineral total contact mode. Logically, metals recovery in the PLS is much higher in the case of total contact. These findings are in agreement with the published results by Dziurla et al. (1995), where an increased iron solubilization was observed in the case of direct contact between pyrite and *Acidithiobacillus ferrooxidans*, thus confirming the catalytic role of the surface-anchored bacteria. The latent phase in the case of total contact lasts up to day 3 with worth mentioning that during this period the number of fixed bacteria is quite low. Therefore, during this period, the observed leaching of copper and cobalt could be entirely due to the ferric iron accompanying the bacterial inoculum. After day 3 and further to day 13, the rapid raise in metal extraction degree could be attributed to the abundance of ferric iron regenerated by the "planktonic" bacteria. Between days 13–20, a second type of behavior could be identified, where in the case of total contact slight increase in metals recovery was recorded. During this period, the iron oxidation rate (IOR) seems to slow down although the number of "planktonic" bacteria remains constant. However, for the case of non-contact, metal extraction continues to rise being function of the IOR, the latter one being catalyzed by bacteria. It should be noted that the number of bacteria remain nearly constant inside the separate compartment of the reactor where carrollite was absent. This finding suggests that the chemical oxidation of the carrollite continues to progress as long as ferric ions are available in the compartment where only mineral is present, bacteria being absent.

During the third period, covering days 20–30, one could note that in the case of total contact, the metal extraction rate is virtually zero with the sufficient number of "planktonic" bacteria being available. One factor governing this situation could be the excessive formation of jarosite precipitates on the surface of the mineral. These precipitates are limiting the diffusion of the ferric ions towards the surface. For the case of non-contact, the fact that metal leaching continues is supporting the assumption that carrollite solubilization essentially follows chemical route. When comparing

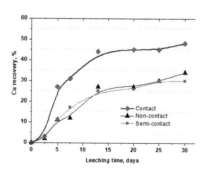 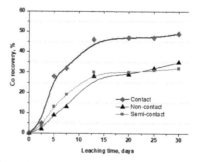

Figure 7. Recovery of Cu and Co as function of leaching duration for the case of total contact, semi-contact and lack of contact.

the situations of total- and semi-contact, one could find the important role which the contact bacteria/mineral plays in relation to the leaching kinetics, especially during the initial period (0–13 days).

These findings corroborate well with the results reported by Konishi et al. (1992), indicating almost equal role of the direct and indirect mechanisms during bioleaching of sphalerite concentrate. However, for the case of molybdenite which is refractory to leaching, Pistaccio et al. (1994) have shown that economically acceptable oxidation is achievable only if sufficient adhesion of *A. ferrooxidans* on the mineral surface takes place. In the same context, results communicated by (Porro et al., 1997) have likewise proved the importance of mineral-bacteria contact during the bioleaching of covellite by *T. ferroox*.

If we compare the carrollite behavior in the tested leaching system with the hypothesis of contact mechanism as proposed by Rohwerder et al. (2003), we could infer that in the case of carrollite bioleaching, the main role of the surface-anchored bacteria will be to catalyze ferrous ions oxidation. Moreover, this catalytic role is supposed to be enhanced by the bacteria-generated EPS, creating an adequate microenvironment for the microorganisms and enabling them to contribute towards mineral oxidation. In such a way, being englobed inside its microenvironment, bacteria appear in "indirect" contact with the mineral surface.

4. Conclusion

This study has allowed following the evolution of bacterial population and its repartitioning between the solid and the liquid phases. The role which bacteria play in carrollite bioleaching system has been thus clarified. SEM examinations have shown pitting patterns for which specific bacterial species present in the consortium are responsible.

Strong adhesion of bacteria to the surface of the carrolite grains was observed during early bioleach stages, manifesting their non-negligible role in the process. The direct contact has favored carrollite oxidation through electrochemical pathway, at the same time being accompanied by release of ferrous ions, elemental sulphur, or sulphur compounds which accumulate on the surface of carrollite. The generated ferrous ions are further used as an energy source by the planktonic bacteria.

Acknowledgment

The authors wish to acknowledge the kind contribution of Prof. Chr. Mustin, LIMOS, Univ Lorraine, France, in studying the role of bacteria-mineral contact. Also special thanks to Prof. St. Groudev from the UMG in Sofia, Bulgaria for providing the initial bacterial consortium.

References

Beach W.B., Sunner J. Biocorrosion: towards understanding interactions between biofilms and metals. Curr. Op. Biotechnol. 15, 181–186, 2004.

Dziurla M.A. Contribution à l'étude de réactions à l'interface bactérie-minéral au cours de la lixiviation de minéraux sulfurés (pyrites) par *Thiobacillus ferrooxidans*. Université Poincaré, 234 p., 1995.

Dziurla M.A., Achouak W., Lam B.T., Heulin T., Berthelin J. Enzyme-linked immunofiltration assay to estimate attachment of *Thiobacilli* to pyrite. Appl. Environ. Microbiol. 64, 2937–2942, 1998.

Edwards K.J., Rutenberg A.D. Microbial response to surface microtopography: the role of metabolism in localized mineral dissolution. Chem. Geol. 180, 19–32, 2001.

Kitobo W. Dépollution et valorisation des rejets miniers sulfurés du Katanga : cas des tailings de l'Ancien Concentrateur de Kipushi, PhD thesis, University of Liege, 2009.

Kinzler K., Gehrke T., Telegdi J., Sand W. Bioleaching—a result of interfacial processes caused by extracellular polymeric substances (EPS). Hydrometallurgy 71, 83–88, 2003.

Konishi Y., Kubo H., Asai S. Bioleaching of zinc sulfide concentrate by *Thiobacillus ferrooxidans*. Biotechnol. Bioengin. 39, 66–74, 1992.

Laurence R. Copper bottomed. Understanding the Central African Copperbelt, Materials World, pp. 24–26, 2005.

Larsson L., Gunnel O., Holst O., Karlsson H.T. Oxidation of pyrite by *Acidianus brierleyi*: Importance of close contact between the pyrite and the microorganisms. Biotechnol. Lett. 15, 99–104, 1993.

Monroy M.G. Biolixiviation—cyanuration de minerais sulfurés aurifères réfractaires en dispositifs de percolation: comportement des populations de *Thiobacillus ferrooxidans* et influence de la minéralogie et des conditions opératoires. PhD thesis, Univ. Nancy, 238 p., 1993.

Nkulu G., Gaydardzhiev S., Mwema E., Compere P. SEM and EDS observations of carrollite bioleaching with a mixed culture of acidophilic bacteria. Min. Eng. 75, 70–76, 2015.

Nkulu G., Gaydardzhiev S., Mwema E. Statistical analysis of bioleaching copper, cobalt and nickel from polymetallic concentrate originating from Kamoya deposit in the Democratic Republic of Congo. Min. Eng. 48, 77–85, 2013.

Pistaccio L., Curutchet G., Donati E., Tedesco P. Analysis of molybdenite bioleaching by *Thiobacillus ferrooxidans* in the absence of iron (II). Biotechnol. Lett. 6, 189–194, 1994.

Porro S., Ramírez S., Reche C., Curutchet G., Alonso-Romanowski S., Donati E. Bacterial attachment: its role in bioleaching processes. Process Biochem. 32, 573–578, 1997.

Pogliani C., Curutchet G., Donati E., Tedesco P.H. A need for direct contact with particle surfaces in bacterial oxidation of covellite in absence of chemical lixiviant. Biotechnol. Lett. 12, 515–518, 1990.

Rohwerder T., Gehrke T., Kinzler K., Sand W. Bioleaching review part A: Progress in bioleaching: fundamentals and mechanisms of bacterial metal sulfide oxidation. Appl. Microbiol. Biotechnol. 63, 239–248, 2003.

Rojas-Chapana J.A., Tributsch H. Interfacial activity and leaching patterns of *Leptospirillum ferrooxidans* on pyrite. FEMS Microbiol. Ecol. 47, 19–29, 2004.

Schippers A., Jozsa P.G., Sand W. Sulfur chemistry in bacterial leaching of pyrite. Appl. Environ. Microbiol. 62, 3424–3431, 1996.

Yager T. The Mineral Industry of Congo (Kinshasa). U.S. Department of the Interior, U.S. Geological Survey, 2014.

CHAPTER 13

Lead Resistance Mechanisms in Bacteria and Co-Selection to other Metals and Antibiotics
A Review

Milind Mohan Naik,[1] *Lakshangy S. Charya*[1,*] and
Pranaya Santosh Fadte[2]

1. Introduction

Environmental lead level has amplified more than 1000-fold over the past few decades as a result of anthropogenic activities in terrestrial as well as aquatic environments. Lead is a persistent environmental pollutant which gradually accumulates leading to biomagnification at different trophic levels in food chain. Environmental contamination by toxic heavy metals and organometals has turned into a foremost global concern as they pose a grave threat to the natural biota along with humans (Nehru and Kaushal, 1992; Hernandez et al., 1998; Nies, 1999; Cerebasi and Yetis, 2001; Hartwig et al., 2002; Dubey and Roy, 2003; Dubey et al., 2006). Lead contaminated sites such as soil, sediments and water create an extreme environment for microbial growth and survival since they are known to cause damage to DNA, protein, and lipid and substitute essential metal ions such as Zn, Ca, and Fe in important enzymes

---
[1] Department of Microbiology, Goa University.
[2] Department of Chemistry, Dnyanprassarak Mandal's college and Research centre, Assagao Goa.
* Corresponding author: lakshangyscharya@gmail.com

(Nies, 1999; Roane, 1999; Asmub et al., 2000; Hartwig et al., 2002). As a result, the U.S. Environmental Protection Agency (EPA) has incorporated lead in the list of hazardous inorganic wastes (Cameron, 1992). Lead is known to inhibit biosynthesis of heme, cause grave neurodegenerative diseases, impede kidney function, and possesses carcinogenic properties (Fowler, 1998; Tong et al., 2000; Watt et al., 2000; Lam et al., 2007). Some natural bacterial isolates have reported to possess potential to colonize sites heavily contaminated with toxic metals by employing various resistance mechanisms which include efflux system, sequestration, oxidation, reduction, bioaccumulation, and biomineralization (Nies, 1999; Roane, 1999; van Hullebusch et al., 2003; Pal and Paul, 2008; Taghavi et al., 2009; Sinha and Khare, 2012). Removal of toxic lead from contaminated terrestrial and aquatic environment is a pressing need and reclamation of lead polluted environments using the lead resistant bacteria has been an effective, affordable, and ecofriendly technological solution. There is also an increasing concern that metal contamination in marine and terrestrial environment may prove to be a selective agent in the proliferation of antibiotic resistance since the same mechanism can confer resistance to lead and antibiotic simultaneously (i.e., cross-resistance) and also due to co-resistance, where different resistance determinants present on the same genetic element (i.e., metal resistance and antibiotic resistance genes present on same plasmid).

2. Various lead resistant mechanisms possessed by terrestrial and aquatic bacteria

Any organism encountering an increased level of metal concentration activates one or more metal resistance mechanisms to combat the stress that they experience under such conditions. For example, lead resistant marine/terrestrial bacteria possess diverse mechanisms viz. intracellular bioaccumulation, extracellular sequestration, cell surface biosorption, biomineralization, modification in cell morphology, siderophore production, and lead efflux pumps to overcome the metal stress as shown in Fig. 1 (Naik and Dubey, 2011; Naik et al., 2012a,b,c; Naik et al., 2013a,b; Sharma et al., 2016).

2.1 Intracellular lead bioaccumulation

Intracellular heavy metal bioaccumulation and homeostasis in bacterial cell cytosol involves low molecular weight cystein-rich metallothionein proteins which vary from 3.5 to 14 kDa (Hamer, 1986). These exclusive proteins were noted to be induced in response to certain particular heavy metals such as Cd, Pb, Zn, and Cu (Gadd, 1990; Turner et al., 1996; Blindauer et al., 2002; Liu et al., 2003). Gram positive *Bacillus megaterium* resists about 0.6 mM lead by sequestering lead intracellularly, by protein resembling metallothionein (Roane, 1999). Aickin and Dean (1977) examined intracellular uptake of lead

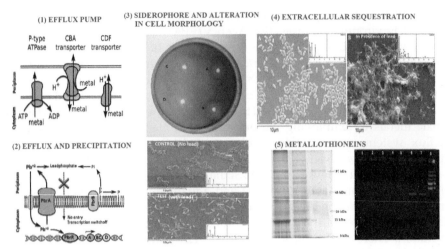

Figure 1. Lead resistance mechanisms in bacteria (Courtesy for 1 & 2: Hynninen et al., 2009; Hynninen, 2010; Naik and Dubey, 2011; Naik et al., 2012a,b,c; Naik et al., 2013a,b).

by microorganisms which have the potential of removing toxic metals from sewage sludge and effluents. *Pseudomonas aeruginosa* strain WI-1 isolated from Mandovi estuary, Goa, India, have been found to possess bacterial metallothionein (BmtA) to alleviate lead toxicity (Naik et al., 2012a). *P. aeruginosa* strain WI-1 resist 0.6 mM lead nitrate and exhibited intracellular accumulation of 26.5 mg.g^{-1} lead/dry weight. Polymerase chain reaction (PCR) amplification of 507 bp internal fragment of smtAB genes, encoding bacterial metallothionein and intracellular bioaccumulation of 19 mg and 22 mg.g^{-1} lead/dry biomass weight in *Salmonella choleraesuis* strain 4A and *Proteus penneri* strain GM10 respectively revealed that metallothionein (SmtA) responsible for lead-resistance is encoded by genomic DNA (Naik et al., 2012c). Lead resistant bacteria such as *B. megaterium*, *P. aeruginosa* strain WI-1, *S. choleraesuis* strain 4A, and *P. penneri* strain GM10 which have potential of bioaccumulating very high amount of lead can be employed for bioremediation of lead present in contaminated environmental sites (Roane, 1999; Naik et al., 2012a, 2012c). Another report by Pedrial et al. (2008) has revealed intracellular sequestration of lead into polyphosphate bodies in bacteria when encountered with the metal in subsurface environments. The same has been confirmed by TEM analysis. Biomineralization of lead in mine tailing by *Bacillus* sp. KK1 into calcite precipitate was revealed by X-ray diffraction studies (Muthusamy et al., 2013). Recently, lead resistant *Providencia vermicola* strain SJ2A (3 mM) isolated from the waste of a battery manufacturing site possessed bacterial metallothionein protein encoded by a plasmid borne *bmtA* gene which is attributed to intracellular sequestration of lead as lead sulfite. TEM analysis of the bacterial cells evidently demonstrated that lead was getting accumulated in the periplasmic space and furthermore

XRD analysis revealed that lead was sequestered in the periplasmic space of *P. vermicola* strain SJ2A as lead sulfite. Furthermore, bacterium demonstrated metallothionein mediated internalization of 155.12 mg.g⁻¹ lead/biomass determined by atomic absorption spectroscopy (Sharma et al., 2016).

2.2 Extracellular lead sequestration

Bacterial exopolysaccharide (EPS) and its possible role in bioaccumulation of Cu and Pb in a marine food chain was investigated using a partially purified and chemically characterized EPS isolated from *Marinobacter* sp. (Bhaskar and Bhosle, 2006). The exopolymer binding to metals is an important process in the downward transport of metals in the ocean environment (Decho and Moriarty, 1990). De et al. (2007) reported that in lead resistant marine *P. aeruginosa* CH07, lead was sequestered extracellularly in EPS signifying it as a possible resistance mechanism. EPS' are high molecular weight polyanionic polymers which bind positively charged metal ions resulting in metal immobilization within the exopolymeric matrix (Roane, 1999; van Hullebusch et al., 2003). *Pseudomonas marginalis* was able to resist 2.5 mM lead by entrapping lead ion (Pb^{+2}) in a negatively charged exopolymer (Roane, 1999). *Enterobacter cloacae* strain P2B isolated from effluent of lead battery manufacturing company of Goa, India, could resist 1.6 mM lead nitrate in Mineral salt medium (MSM) by entraping 17 percent lead (as weight percent) extracellularly by secreting lead enhanced exopolysaccharide as disclosed by SEM-EDX analysis. Considerable increase in exopolysaccharide (EPS) production was observed as the EPS increased from 28 to 108 mg.L⁻¹ dry weight in response to 1.6 mM lead nitrate as compared to control MSM without metal (Naik et al., 2012b). Pb^{+2} could interact with negatively charged carboxyl, hydroxyl, and amide groups and glucuronic acid from different chains of the polyanionic EPS produced by lead resistant *E. cloacae* strain P2B. The feasibility of lead removal from sulfate-rich wastewater through biological sulfate reduction process with hydrogen as electron donor was investigated by Teekayuttasakul and Annachhatre (2008). In this attempt, the sulfide which was the product of sulphate reduction by sulfate-reducing bacteria (SRB) in a gas-lift reactor was used to remove lead as lead sulfide precipitate.

2.3 Cell surface biosorption of lead ions (Pb^{+2})

Bacterial cell surface biosorption of Pb^{+2} ions is due to various negatively charged chemical groups such as carboxyl, hydroxyl, and phosphate present on the bacterial cell surface. The carboxyl (COO⁻) group of the peptidoglycan serve as major metal ion binding site in Gram positive bacteria, whereas phosphate groups contribute considerably in Gram negative bacteria (Gadd and White, 1993). *P. aeruginosa* strain 4EA isolated from soil contaminated with car battery waste from Goa, India resists 0.8 mM lead nitrate by cell surface biosorption (11 percent by weight) of lead as revealed by SEM-EDX (Naik and Dubey, 2011). Biosorption of 97.68% lead ions from a 700 mg.L^{-1} lead aqueous solution by Gram positive bacterium *Bacillus subtilis* at pH 4.5 was reported by Hossain and Anantharaman (2006). Biosorption of lead by bacteria include those physico-chemical mechanisms through which lead ions are removed from an aqueous solution. Lead biosorption by bacteria is mediated by quite a few mechanisms such as ion exchange, adsorption, chelation, and diffusion through bacterial cell walls and membranes.

2.4 Lead biomineralization

The remediation of Pb(II) through biomineralization is observed to be a promising technique as well as an interesting phenomenon for transforming lead from mobile species into very stable minerals in the environment. Aickin et al. (1979) reported precipitation of Pb^{+2} on the cell surface of *Citrobacter* sp. as PbHPO$_4$ which was deduced by electron microscopy and X-ray microanalysis, while Levinson et al. (1996) suggested intracellular bioaccumulation and precipitation of Pb$_3$ (PO$_4$)$_2$ by *S. aureus* grown in the presence of high concentrations of soluble lead nitrate (Aickin et al., 1979; Levinson et al., 1996). *Vibrio harveyi* and *Providentia alcalifaciens* strain 2EA were reported to bioprecipitate soluble Pb^{+2} as unusual phosphate of lead— i.e., Pb$_9$(PO$_4$)$_6$– (Mire et al., 2004; Naik et al., 2013a). *Klebsiella* sp. cultured in phosphate-limited medium has been reported to bioprecipitate lead as black colour lead sulfide (PbS) (Aiking et al., 1985). Lead resistant *Bacillus iodinium* GP13 and *Bacillus pumilus* S3 were reported to precipitate lead as PbS (De et al., 2008). Biomineralization of Pb(II) into nanosized rod-shaped Ca$_{2.5}$Pb$_{7.5}$(OH)$_2$(PO$_4$)$_6$ crystal by *Bacillus cereus* 12–2 has been reported by Chen et al. (2016). XRD and TEM investigation revealed that the Pb(II) loaded on bacteria could be stepwise transformed into rod-shaped Ca$_{2.5}$Pb$_{7.5}$(OH)$_2$(PO$_4$)$_6$ nanocrystal. Another report by Liang et al. (2016) revealed phosphatase-mediated bioprecipitation of lead by soil fungi, *Aspergillus niger*, and *Paecilomyces javanicus* when grown in 5 mM lead nitrate. The minerals were identified as pyromorphite (Pb$_5$(PO$_4$)$_3$Cl), produced only by *P. javanicus* and lead oxalate (PbC$_2$O$_4$) produced by *A. niger* and *P. javanicus*. Biomineralization of Pb, Cu, Ni, Zn, Co, and Cd, by six metal-resistant bacterial strains was investigated using microcosm experiments. Bacterial isolates produced the enzyme urease which hydrolyzed urea and hence soil pH increased and carbonate was produced. This resulted in biomineralization of the soluble lead, copper, nickel, zinc, cobalt, and cadmium present in soil to carbonates (Li et al., 2013). TEM–EDS analysis of lead resistant *Pseudomonas aeruginosa* CHL-004 has shown that lead was transported from the exterior environment, complexed with phosphate, and stored as discrete cellular inclusions (Feldhake et al., 2008).

2.5 Modification in cell morphology and siderophore production

Pseudomonas aeruginosa strain 4EA isolated from lead contaminated soil of battery manufacturing company revealed significant alteration in cell morphology as size reduction when exposed to 0.8 mM lead nitrate suggesting it as a resistance mechanism of bacterial cells against toxic lead (Naik and Dubey, 2011). Significant alteration in cell morphology as reduction in cell size and shrinkage was observed when lead resistant *E. cloacae* strain P2B cells were exposed to 1.6 mM lead nitrate in mineral salt medium (Naik et al., 2012b). Apart from Fe^{+3}, microbial siderophores also form stable complexes with metals such as Cd^{+2}, Pb^{+2}, and Zn^{+2} (Namiranian et al., 1997; Gilis et al., 1998; Hepinstall et al., 2005). *P. aeruginosa* strain 4EA resistant to 0.8 mM lead nitrate revealed lead induced siderophore (pyochelin and pyoverdine) production (Naik and Dubey, 2011). Enhancement of siderophore production by *P. aeruginosa* strain 4EA appears to be an additional mechanism of resistance of bacterial cells in response to toxic level of lead. Pyoverdin siderophore produced by *Pseudomonas aeruginosa* CHL-004, isolated from soil near lead mine was found to bind lead (Feldhake et al., 2008). This proves the potential application of pyoverdin produced by *Pseudomonas aeruginosa* CHL-004 for the bioremediation of lead. Significant increase in extracellular siderophore production in *Bacillus amyloliquefaciens* NAR38.1 isolated from mangrove sediments, observed with Pb and Al at concentrations of 50 µM and above, additionally support the probable role of siderophores in lead resistance (Gaonkar and Bhosle, 2013). The cells of lead resistant *Providencia vermicola* strain SJ2A (3 mM) which showed metallothionein mediated periplasmic sequestration of lead (Pb^{+2}) as lead sulphite also demonstrated a unique alteration in the cell morphology caused due to the exposure of lead (Sharma et al., 2016). In the presence of 0.8 mM lead nitrate, the cells tend to show alteration from rods to filamentous form and appear as long inter-connected chains, approximately 7 to 8 times longer than the usual size revealed by scanning electron microscopy. It was observed that septum formation was inhibited and therefore daughter cells failed to separate resulting in long chains of bacterial cells. Transformation of cells from rods to inter-connected chains of cells reduced the surface area exposed to the toxic lead ions (Pb^{+2}) thus reducing their toxic effect. Additionally, reduction in the total cell surface area exposed to lead (Pb^{+2}) also led to decrease in toxic metal uptake.

2.6 Efflux mechanism in lead resistant bacteria

Releasing excessive metal ions out of cell through efflux pump is one of the main strategies used by bacteria in order to control internal metal ion concentrations and maintain homeostasis (Naik et al., 2013a; Naik et al., 2013b). Bacterial lead resistance via efflux system is a recognized mechanism of resistance. Several efflux systems have been described in bacteria. The two main groups of efflux systems include P-type ATPases, e.g., the Cu(II), Pb (II), Cd(II), and Zn(II) ATPases of Gram-negative bacteria, and chemiosmotic pumps. In order to maintain intracellular heavy metal homeostasis, metal resistant bacteria possesses efflux which effluxes excessive heavy metals outside the cells. Metal resistant bacteria (including lead resistant bacteria) possesses soft metal transporting P_{IB}-type ATPases which are group of proteins involved in transport of heavy metals outside the cell membrane and lead to bacterial heavy metal resistance (Nies and Silver, 1995; Rensing et al., 1999; Coombs and Barkay, 2004). Pb(II) resistant determinant in *Cupriavidus metallidurans* CH34 was investigated and found to be located on pMOL30, one of the two plasmids found in *Cupriavidus metallidurans* CH34 (Diels et al., 1989). It was then more recently described in detail at the molecular level (Borremans et al., 2001; Naik et al., 2013a). From Zuari estuary, two lead-resistant bacteria were isolated, namely *Pseudomonas stutzeri* M-9 and *Vibrio harveyi* M-11, that showed efflux pump mediated lead resistance (Naik et al., 2013b). Lead resistant estuarine bacterial strain M-9 and M-11 exhibited resistance to lead nitrate up to 0.8 mM and 1.2 mM lead nitrate respectively. Nested-PCR using genomic DNA as template clearly demonstrated the presence of *pbrA* gene (amplicon size: 750 bp) which encodes for P-type ATPase efflux pump. Real-time PCR further revealed that 5.4 ± 0.7 and 7.9 ± 0.9 fold expression of *pbrA* gene in *Pseudomonas stutzeri* strain M-9 and *Vibrio harveyi* strain M-11 respectively, when grown in Tris minimal media amended with 0.5 mM lead nitrate, confirmed efflux mediated lead resistance in both bacterial isolates. The complete operon *pbrUTRABCD*, conferring efflux mediated lead resistance, has previously been sequenced in *Ralstonia metallidurans* strain CH34 (Borremans et al., 2001). The lead resistance *pbr* operon were found to contain the structural resistance genes viz: (i) *pbrT*, encodes a Pb(II) uptake protein; (ii) *pbrA*, encodes a P-type lead (II) efflux ATPase pump; (iii) *pbrB*, encodes a integral membrane protein of unknown function; and (iv) *pbrC*, encodes a probable prolipoprotein signal peptidase. Downstream of *pbrC*, the *pbrD* gene, encoding a Pb(II)-binding protein, which was found to be essential for lead sequestration. Pb(II)-dependent inducible transcription of *pbrABCD* from the *PpbrA* promoter is regulated by PbrR, which is lead ion-sensing regulatory protein. This was the first report of a mechanism for specific lead resistance in any bacterial genus. Lead (Pb^{+2}) resistance in bacteria *Cupriavidus metallidurans* CH34, interplay between plasmid and chromosomally-located functions (Taghavi et al., 2009). *pbrUTRABCD* operon is responsible for lead resistance in *Cupriavidus metallidurans* CH34. The defense of *C. metallidurans* CH34 against Pb(II) is by the pMOL30 encoded pbr Pb(II) resistance operon. This was confirmed by several complementary approaches. Pb(II)-induced proteome of *C. metallidurans* CH34 by 2-dimensional gel electrophoresis revealed the induction of proteins PbrT, PbrA, PbrB, PbrC, and PbrD indicating that these proteins are abundant and defend the cell against Pb(II). Transcriptome analysis using quantitative real-time PCR and microarrays also confirmed that the pbr operon is the most induced system under conditions of Pb(II) exposure. Lu et al. (2016) reported that in *Sinorhizobium meliloti* the expression of *zntA* (P_{1B}-type ATPases efflux pump) was induced by Zn, Cd, and Pb whereas *copA1b* (efflux pump) was induced by Cu and Ag. Deletions in *zntA* and *copA1b* led to the increased intracellular concentrations of Zn, Pb and Cd, but not Cu. Complementation of Δ*copA1b* and Δ*zntA* mutants demonstrated a restoration of tolerance to Zn, Cd, and Pb to a certain extent. The results advocate an important role of *copA1b* and *zntA* in Zn homeostasis and Cd and Pb detoxification in *S. meliloti* CCNWSX0020. Efflux of Pb(II) has also been reported for the CadA ATPase of the *S. aureus* and the ZntA ATPase of *Escherichia coli* (Rensing et al., 1998). CadC is a metal-responsive repressor that responds to soft metals in the order Pb > Cd > Zn. Also both CadA and ZntA bestow resistance to Pb(II). Transport of Zn(II) in everted membrane vesicles of *E. coli* catalyzed by either of these two P-type ATPase superfamily members is inhibited by Pb(II). An efflux transporter PbrA and a phosphatase PbrB cooperate in a lead-resistance mechanism in bacteria as reported by Hynninen et al. (2009). As a model of action for PbrA and PbrB, they proposed a mechanism where Pb^{2+} is exported from the cytoplasm by PbrA and then sequestered as a phosphate salt with the inorganic phosphate produced by PbrB. P_{1B}-type ATPases can be divided into two subgroups: (i) Cu(I)/Ag(I)-translocating ATPases encoded by gene *copA* in *Enterococcus hirae*, *Helicobacter pylori,* and *E. coli*; (ii) Zn(II)/Cd(II)/Pb(II)-translocating ATPases encoded by gene *zntA* in *E. coli,* and gene *cad A* in *Staphylococcus aureus* plasmid, pI258 (Nies and Silver, 1995; Rensing et al., 1999). Genes encoding P_{1B}-type ATPases are found in majority of sequenced bacterial and archaeal genomes (Coomb and Barkay, 2004, 2005). *Achromobacter xylosoxidans* A8 have been reported to carry a plasmid, pA81 harbouring *pbtTFYRABC* gene cluster that is responsible for lead resistance (Hlozkova et al., 2013). Elimination of *pbtTFYRABC* from *Achromobacter xylosoxidans* A8 resulted in increased sensitivity toward Pb and Cd. It has been observed that pbtTRABC products share strong similarities with Pb uptake transporter PbrT, transcriptional regulator PbrR, metal efflux P1-ATPases PbrA and CadA, undecaprenyl pyrophosphatase PbrB and its signal peptidase PbrC from *Cupriavidus metallidurans* CH34.

Three major families of efflux transporters involved in $Zn^{2+}/Cd^{2+}/Pb^{2+}$ resistance include: (1) P-type ATPase, e.g., the Cu(II), Pb(II), Cd(II), and Zn(II) ATPases of Gram-negative bacteria, (2) cation diffusion facilitator (CDF) and (3) CBA. CBA transporters are three-component trans-envelope pumps of Gram negative bacteria that operate as chemiosmotic antiporters. The three-component divalent-cation efflux systems *cnr*, *ncc*, and *czc* of *Ralstonia metallidurans* (formerly *Alcaligenes eutrophus* CH34) (Borremans et al., 2001). Cation diffusion facilitator (CDF) family transporters act as chemiosmotic ion-proton exchangers. P-type ATPases and CDF transporters export metal ions from the cytoplasm to the periplasm; whereas CBA transporters chiefly detoxify periplasmic metals (outer membrane efflux), i.e., CBA transporters further eliminate periplasmic ions transported there by ATPases or CDF transporters, a way before ions enter the cytoplasm. P-type ATPases and CDF transporters can functionally substitute each other but they cannot substitute CBA transporter and vice versa (Hynninen et al., 2009; Hynninen, 2010).

3. The co-selection mechanisms (co-resistance and cross resistance) in metal resistant bacteria

Pseudomonas stutzeri strain M-9 and *Vibrio harveyi* strain M-11 resistant to lead (0.8 mM and 1.2 mM lead nitrate, respectively) via efflux pump P_{IB}-type ATPase encoded by gene *pbrA* gene also possess *mdrL* gene for multi-drug resistance suggesting probable contamination of Zuari estuary with heavy metals/antibiotics. MIC of antibiotics for lead-resistant bacterial strain M-9 was chloramphenicol (30 µg per disc), ampicillin (50 µg per disc), norfloxacin (10 µg per disc), cephalexin (30 µg per disc), and co-trimoxazole (25 µg per disc); whereas for strain M-11 it was ampicilin (50 µg per disc), nalidixic acid (30 µg per disc), erythromycin (15 µg per disc), cep-halexin (30 µg per disc), chloramphenicol (30 µg per disc), and co-trimoxazole (25 µg per disc) (Naik et al., 2013b). Comparative genomics of multi-drug resistance *Acinetobacter bauman* revealed that it also possesses efflux pumps for antibiotics, heavy metals, and metalloids viz. Pb, Cd, Hg, As, and Antimony (Fournier et al., 2006). Lead-resistant *P. aeruginosa* strain WI-1 isolated from Mandovi estuary possesses bacterial metallothionein (BmtA) and exhibited bioaccumulation of Pb to alleviate Pb^{+2} toxicity. The isolate also demonstrated cross-tolerance to cadmium, mercury, and tributyltin chloride (TBTCl) along with resistance to multiple antibiotics (Naik et al., 2012a). *Providencia alcalifaciens* strain 2EA resisted lead nitrate up to 0.0014 mol l^{-1} by precipitating soluble lead (Pb^{+2}) as insoluble light brown solid. This bacterial strain also crossed tolerated cadmium and mercury with MIC values 0.0002 and 0.00003 $mol.L^{-1}$ respectively and showed resistance to several antibiotics viz. ampicillin (25 µg per disc), cephalexin (30 µg per disc), oleondamycin (15 µg per disc), ciphaloridine (30 µg per disc), erythromycin (15 µg per disc), and amikacin (10 µg per disc) (Naik et al., 2013a). *Aeromonas caviae* strain KS-1 isolated from Mandovi estuary, Goa, showed tolerance to lead nitrate up to 1.4 and with MIC values of 1.6. Cross tolerance to other metals was also observed as MIC values were 1.2 mM, 30 µM, 0.4 mM, and 0.9 mM for $ZnSO_4$, $HgCl_2$, $CdSO_4$, and $CuSO_4$, respectively (Shamim et al., 2012).

There is a pressing concern considering that anthropogenic levels of heavy metals including lead are currently several times greater than levels of antibiotics (Stepanauskas et al., 2005). Importantly, a considerable number of reports recommend that metal pollution in natural environments could have a significant role in the maintenance and propagation of antibiotic resistance (Summers et al., 1993; Alonso et al., 2001; Summers, 2002). It has been identified for several decades that metal and antibiotic resistance genes are linked, predominantly on plasmid DNAs, because the proof for co-resistance as a mechanism of antibiotic-metal co-selection came from investigations that used plasmid DNA transformation, plasmid curing, and plasmid sequencing techniques (Novick and Roth, 1968; Foster, 1983). The genetic linkage of antibiotic and mercury Soil microbes isolated from a copper spiked field (21 months later copper spiking) were more resistant to copper and antibiotics as compared to strains isolated from control plots (Berg et al., 2005). Here, copper-resistant strains were considerably more resistant to ampicillin and sulfonamide as compared to copper-sensitive isolates which support the dispute that the traits are co-selected. Mercury present in dental amalgams has been found to be responsible for co-selection of antibiotic-resistant bacteria by investigation that examined bacterial isolates from intestinal and oral bacterial communities (Summers et al., 1993; Osterblad et al., 1995; Wireman et al., 1997). Ampicillin resistance pattern in bacterial isolates observed after dental amalgam installation pointed that resistance to ampicillin amplified relative to pre-installation levels.

Cross resistance has also been seen in Some few agricultural based investigations wherein same resistance mechanism can confer resistance to other metals and antibiotics. Huysman et al. (1994) observed greater incidence of resistance to a range of metals and antibiotics (nickel, zinc, cadmium, cobalt, ampicillin, olaquindox, streptomycin, and spiramycin) in copper-resistant bacterial isolates as compared to copper sensitive isolates obtained from agricultural fields in which copper-contaminated pig manure had been used. Also, Lu et al. (2016) reported that in *Sinorhizobium meliloti* the expression of *zntA* (P_{1B}-type ATPases efflux pump) was induced by Zn, Cd, and Pb, whereas *copA1b* (efflux pump) was induced by Cu and Ag. Efflux of Pb(II) has also been reported for the CadA ATPase of the *S. aureus* and the ZntA ATPase of *Escherichia coli* (Rensing et al., 1998). Here CadC is a metal-responsive repressor that responds to soft metals in the order Pb > Cd > Zn. Also, as reported earlier, both CadA and ZntA confer resistance to Pb(II) (Rensing et al., 1998). ATP7B is a P-type ATPase that mediates the efflux of Cu and recent studies have demonstrated that ATP7B regulates the cellular efflux of cisplatin (DDP) and controls sensitivity to the cytotoxic effects of this drug (Safaei et al., 2008).

Heavy metal pollution (viz. Cd, Hg, Zn, Ag, Cu) has been reported to work as a selective agent in the propagation of antibiotic resistance (Baker-Austin et al., 2006). The co-selection mechanisms of heavy metals and antibiotics include co-resistance (different resistance genes present on the same genetic element) and cross resistance (the same genetic determinant responsible for resistance to multiple antibiotics and heavy metals). Co-resistance occurs when the genes responsible for resistance are present together on the same genetic element viz. transposons, plasmid, or integron (Chapman, 2003). There is a rising worry that metal contamination in soil and aquatic environment may co-select multiple drug-resistant pathogens. Therefore, biomonitoring and bioremediation of marine and estuarine environment is a prime need.

References

Aickin R.M., Dean A.C.R. Lead accumulation by microorganisms. Microbios Lett. 5, 129–133, 1977.

Aickin R.M., Dean A.C.R., Cheetham A.K., Skarnulis A.J. Electron microscope studies on the uptake of lead by a *Citrobacter* species. Microbios Lett. 9, 7–15, 1979.

Aiking H., Govers H., van'tRiet J. Detoxification of mercury, cadmium and lead in *Klebsiella aerogenes* NCTC 418 growing in continuous culture. Appl. Environ. Microbiol. 50, 1262–1267, 1985.

Alonso A., Sanchez P., Martinez J.L. Environmental selection of antibiotic resistance genes. Environ. Microbiol. 3, 1–9, 2001.

Asmub M., Mullenders L.H.F., Hartwig A. Interference by toxic metal compounds with isolated zinc finger DNA repair proteins. Toxicol. Lett. 112, 227–231, 2000.

Baker-Austin C., Wright M.S., Stepanauskas R., McArthur N.J.V. Co-selection of antibiotics and metal resistance. Trends Microbiol. 14, 176–182, 2006.

Berg J., Tom-Petersen A., Nybroe O. Copper amendment of agricultural soil selects for bacterial antibiotic resistance in the field. Lett. Appl. Microbiol. 40, 146–151, 2005.

Bhaskar P.V., Bhosle N.B. Bacterial extracellular polymeric substances (EPS) a carrier of heavy metals in the marine food-chain. Environ. Int. 32, 191–198, 2006.

Blindauer C.A., Harrison M.D., Robinson A.K., Parkinson J.A., Bowness P.W., Sadler P.J., Robinson N.J. Multiple bacteria encode metallothioneins and SmtA-like fingers. Mol. Microbiol. 45, 1421–1432, 2002.

Borremans B., Hobman J.L., Provoost A., Brown N.L., van der Lelie D. Cloning and functional analysis of the pbr lead resistance determinant of *Ralstonia metallidurans* CH34. J. Bacteriol. 183, 5651–5658, 2001.

Cameron R.E. Guide to site and soil description for hazardous waste characterization. Volume 1: Metals. Environmental Protection Agency EPA/600/4–91/029. 250, 1992.

Cerebasi I.H., Yetis U. Biosorption of Ni (II) and Pb (II) by *Phanerochaete chrysosporium* from a binary metal system-kinetics. Water Res. 24, 15–20, 2001.

Chapman J.S. Disinfectant resistance mechanisms, cross-resistance and co-resistance. Int. Biodeterior. Biodegr. 51, 271–276, 2003.

Chen Z., Pan X., Chen H., Guan X., Lin Z. Biomineralization of Pb(II) into Pb-hydroxyapatite induced by *Bacillus cereus* 12–2 isolated from Lead–Zinc mine tailings. J. Hazard. Mater. 301, 531–537, 2016.

Coombs J.M., Barkay T. Molecular evidence for the evolution of metal homeostasis genes by lateral gene transfer in bacteria from the deep terrestrial subsurface. Appl. Environ. Microbiol. 70, 1698–1707, 2004.

Coombs J.M., Barkay T. New findings on evolution of metal homeostasis genes: evidence from comparative genome analysis of bacteria and archaea. Appl. Environ. Microbiol. 71, 7083–7091, 2005.

De J., Ramaiah N., Bhosle N.B., Garg A., Vardanyan L., Nagle V.L., Fukami K. Potential of mercury resistant marine bacteria for detoxification of chemicals of environmental concern. Microbes Environ. 22, 336–345, 2007.

De J., Ramaiah N., Vardanyan L. Detoxification of toxic heavy metals by marine bacteria highly resistant to mercury. Mar. Biotechnol. 10, 471–477, 2008.

Decho A.W., Moriarty D.J.W. Bacterial exopolymer utilization by a harpacticoid copepod: A methodology and results. Limnol. Oceanogr. 35, 1039–1049, 1990.

Diels L., Sadouk A., Mergeay M. Large plasmids governing multiple resistance to heavy metals: a genetic approach. Toxicol. Environ. Chem. 23, 79–89, 1989.

Dubey S.K., Roy U. Biodegradation of tributyltins (organotins) by marine bacteria. Appl. Organomet. Chem. 17, 3–8, 2003.

Dubey S.K., Tokashiki T., Suzuki S. Microarray mediated transcriptome analysis of the tributyltin (TBT)-resistant bacterium *Pseudomonas aeruginosa* 25 W in the presence of TBT. J. Microbiol. 44, 200–205, 2006.

Feldhake D.J., Grosser R.J., Davis-hoover W.J. A pyoverdin siderophore produced by *Pseudomonas aeruginosa* chl-004 binds lead and other heavy metals. United States Environmental Protection Agency, 2008.

Foster T.J. Plasmid-determined resistance to antimicrobial drugs and toxic metal ions in bacteria. Microbiol. Rev. 47, 361–409, 1983.

Fournier P.-E., Vallenet D., Barbe V., Audic S., Ogata H., Poirel L., Richet H., Robert C., Mangenot S., Abergel C., Nordmann P., Weissenbach J., Raoult D., Claverie J.-M. Comparative genomics of multidrug resistance in *Acinetobacter baumannii*. PLoS Genet. 2, e7, 2006.

Fowler B.A. Roles of lead binding proteins in mediating lead bioavailability. Environ. Health Perspect. 106, 1585–1587, 1998.

Gadd G.M., White C. Microbial treatment of metal pollution—a working biotechnology. Trends Biotechnol. 11, 353–359, 1993.

Gaonkar T., Bhosle S. Effect of metals on a siderophore producing bacterial isolate and its implications on microbial assisted bioremediation of metal contaminated soils. Chemosphere 93, 1835–1843, 2013.

Gilis A., Corbisier P., Baeyens W., Taghavi S., Mergeay M., van der Lelie D. Effect of the siderophore alcaligin E on the bioavailability of Cd to *Alcaligenes eutrophus* CH34. J. Ind. Microbiol. Biotechnol. 20, 61–68, 1998.

Hamer D.H. Metallothioneins. Annu. Rev. Biochem. 55, 913–951, 1986.

Hartwig A., Asmuss M., Ehleben I., Herzer U., Kostelac D., Pelzer A., Schwerdtle T., Burkle A. Interference by toxic metal ions with DNA repair processes and cell cycle control: molecular mechanisms. Environ. Health Perspect. 110, 797–799, 2002.

Hepinstal S.E., Turner B.F., Maurice P.A. Effect of siderophores on Pb and Cd adsorption to kaolinite. Clays Clay Miner. 53, 557–563, 2005.

Hernandez A., Mellado R.P., Martinez J.L. Metal accumulation and vanadium-induced multidrug resistance by environmental isolate of *Escherichia hermannii* and *Enterobacter cloacae*. Appl. Environ. Microbiol. 64, 4317–4320, 1998.

Hlozkova, K., Suman J., Strnad H., Ruml T., Paces V., Kotrba P. Characterization of *pbt* genes conferring increased Pb^{2+} and Cd^{2+} tolerance upon *Achromobacter xylosoxidans* A8. Res. Microbiol. 164, 1009–1018, 2013.

Hossain S.M., Anantharaman N. Studies on bacterial growth and lead biosoption using *Bacillus subtilis*. Indian J. Chem. Technol. 13, 591–596, 2006.

Huysman F., Verstraete W., Brookes P.C. Effect of manuring practices and increased copper concentrations on soil microbial populations. Soil Biol. Biochem. 26, 103–110, 1994.

Hynninen A., Touze T., Pitkanen L., Mengin-Lecreilx D., Virta M. An efflux transporter PbrA and a phosphatase PbrB cooperate in a lead-resistance mechanism in bacteria. Mol. Microbiol. 74, 384–394, 2009.

Hynninen A. Zinc, cadmium and lead resistance mechanisms in bacteria and their contribution to biosensing. Academic Dissertation in Microbiology. Department of Food and Environmental Sciences, Faculty of Agriculture and Forestry, University of Helsinki, 2010.

Lam T.V., Agovino P., Niu X., Roche L. Linkage study of cancer risk among lead exposed workers in New Jersey. Sci. Total Environ. 372, 455–462, 2007.

Levinson H.S., Mahler I., Blackwelder P., Hood T. Lead resistance and sensitivity in *Staphylococcus aureus*. FEMS Microbiol. Lett. 145, 421–425, 1996.

Levinson H.S., Mahler I. Phosphatase activity and lead resistance in *Citrobacter freundii* and *Staphylococcus aureus*. FEMS Microbiol. Lett. 161, 135–138, 1998.

Li M., Cheng X., Guo X. Heavy metal removal by biomineralization of urease producing bacteria isolated from soil. Int. Biodeterior. Biodegradation 76, 81–85, 2013.

Liang X., Kierans M., Ceci A., Hillier S., Gadd G.M. Phosphatase-mediated bioprecipitation of lead by soil fungi. Environ. Microbiol. 18, 219–231, 2016.

Liu W.X., Li X.D., Shen Z.G., Wang D.C., Wai O.W., Li Y.S. Multivariate statistical study of heavy metal enrichment in sediments of the Pearl river estuary. Environ. Pollut. 121, 377–388, 2003.

Lu M., Li Z., Liang J., Wei Y., Rensing C., Wei G. Zinc resistance mechanisms of P1B-type ATPases in Sinorhizobium meliloti CCNWSX0020. Scientific Reports 6, doi:10.1038/srep29355, 2016.

Mire C.E., Tourjee J.A., O'Brien W.F., Ramanujachary K.V., Hecht G.B. Lead precipitation by *Vibrio harveyi*: evidence for novel quorum–sensing interactions. Appl. Microbiol. Biotechnol. 70, 855–864, 2004.

Muthusamy G., Lee K.-J., Cho M., Kim J.S., Kamala-Kannan S., Oh B.-T. Significance of autochthonous *Bacillus* sp. KK1 on biominerallization of lead in mine tailings. Chemosphere 90, 2267–2272, 2013.

Naik M.M., Dubey S.K. Lead-enhanced siderophore production and alteration in cell morphology in a Pb-resistant *Pseudomonas aeruginosa* strain 4EA. Curr. Microbiol. 62, 409–414, 2011.

Naik M.M., Pandey A., Dubey S.K. *Pseudomonas aeruginosa* strain WI-1 from Mandovi estuary possesses metallothionein to alleviate lead toxicity and promotes plant growth. Ecotoxicol. Environ. Saf. 79, 129–133, 2012a.

Naik M.M., Pandey A., Dubey S.K. Biological characterization of lead-enhanced exopolysaccharide produced by a lead resistant *Enterobacter cloacae* strain P2B. Biodegradation 23, 775–783, 2012b.

Naik M.M., Shamim K., Dubey S.K. Biological characterization of lead resistant bacteria to explore role of bacterial metallothionein in lead resistance. Curr. Sci. 103, 1–3, 2012c.

Naik M.M., Khanolkar D.S., Dubey S.K. Lead resistant *Providentia alcalifaciens* strain 2EA bioprecipitates Pbþ2 as lead phosphate. Lett. Appl. Microbiol. 56, 99–104, 2013a.

Naik M.M., Dubey S.K., Khanolkar D., D'Costa B. P-type ATPase and MdrL efflux pump-mediated lead and multi-drug resistance in estuarine bacterial isolates. Curr. Sci. 105, 1366–1372, 2013b.

Namiranian S., Richardson D.J., Russell D.A., Sodeau J.R. Excited state properties of siderophore pyochelin and its complex with zinc ions. Photochem. Photobiol. 65, 777–782, 1997.

Nehru B., Kaushal S. Effect of lead on hepatic microsomal enzyme activity. J. Appl. Toxicol. 12, 401–405, 1992.

Nies D., Silver S. Ion efflux systems involved in bacterial metal resistance. J. Ind. Microbiol. 14, 186–199, 1995.

Nies D. Microbial heavy-metal resistance. Appl. Microbiol. Biotech. 51, 730–750, 1999.

Novick R.P., Roth C. Plasmid-linked resistance to inorganic salts in *Staphylococcus aureus*. J. Bacteriol. 95, 1335–1342, 1968.

Osterblad M., Leistevuo J., Leistevuo T., Jarvinen H., Pyy L., Tenovuo J., Huovinen P. Antimicrobial and mercury resistance in aerobic Gram–negative bacilli in fecal flora among persons with and without dental amalgam fillings. Antimicrob. Agents Chemother. 39, 2499–2502, 1995.

Pal A., Paul A.K. Microbial extracellular polymeric substances: central elements in heavy metal bioremediation. Indian J. Microbiol. 48, 49–64, 2008.

Pedrial N., Liewig N., Delphin J.E., Elsass F. TEM evidence for intracellular accumulation of lead by bacteria in subsurface environments. Chem. Geol. 253, 196–204, 2008.

Rensing C., Sun Y., Mitra B., Rosen B.P. Pb(II)-translocating P-type ATPases. J. Biol. Chem. 273, 32614–32617, 1998.

Rensing C., Ghosh M., Rosen B.P. Families of soft metal ion transporting ATPases. J. Bacteriol. 181, 5891–5897, 1999.

Roane T.M. Lead resistance in two bacterial isolates from heavy metal-contaminated soils. Microb. Ecol. 37, 218–224, 1999.

Safaei R., Otani S., Larson B.J., Rasmussen M.L., Howell S.B. Transport of cisplatin by the copper efflux transporter ATP7B. Mol. Pharmacol. 73, 461–468, 2008.

Shamim K., Naik M.M., Pandey A., Dubey S.K. Isolation and identification of *Aeromonas caviae* strain KS-1 as TBTC- and lead resistant estuarine bacteria. Environ. Monit. Assess. 185, 5243–5249, 2013.

Sharma J., Shamim K., Dubey S.K., Meena R.M. Metallothionein assisted periplasmic lead sequestration as lead sulfite by *Providencia vermicola* strain SJ2A. Sci. Total Environ. 579, 359–365, 2017.

Sinha A., Khare S.K. Mercury bioremediation by mercury accumulating *Enterobacter* sp. cells and its alginate immobilized application. Biodegradation 23, 25–34, 2012.

Stepanauskas R., Glenn T.C., Jagoe C.H., Tuckfield R.C., Lindell A.H., McArthur J.V. Elevated microbial tolerance to metals and antibiotics in metal-contaminated industrial environments. Environ. Sci. Technol. 39, 3671–3678, 2005.

Summers A.O., Wireman J., Vimy M.J., Lorscheider F.L., Marshall B., Levy S., Bennett S., Billard L. Mercury released from dental "silver" fillings provokes an increase in mercury-resistant and antibiotic resistant bacteria in oral and intestinal floras of primates. Antimicrob. Agents Chemother. 37, 825–834, 1993.

Summers A.O. Generally overlooked fundamentals of bacterial genetics and ecology. Clin. Infect. Dis. 34, S85–S92, 2002.

Taghavi S., Lesaulnier C., Monchy S., Wattiez R., Mergeay M., vanderLelie D. Lead (II) resistance in *Cupriavidus metallidurans* CH34: interplay between plasmid and chromosomally located functions. Anton. Leeuw. 96, 171–182, 2009.

Teekayuttasakul P., Annachhatre A.P. Lead removal and toxicity reduction from industrial wastewater through biological sulfate reduction process. J. Environ. Sci. Health A 43, 1424–1430, 2008.

Tong S., von Schirnding Y.E., Prapamontol T. Environmental lead exposure: a public health problem of global dimensions. Bull. World Health Organ. 78, 1068–1077, 2000.

Turner J.S., Glands P.D., Samson A.C.R., Robinson N.J. Znþ2-sensing by the cyanobacterial metallothionein repressor SmtB: different motifs mediate metal-induced protein-DNA dissociation. Nucleic Acids. Res. 24, 3714–3721, 1996.

van Hullebusch E.D., Zandvoort M.H., Lens P.N.L. Metal immobilisation by biofilms: mechanisms and analytical tools. Rev. Environ. Sci. Biotechnol. 2, 9–33, 2003.

Watt G.C.M., Britton A., Gilmour H.G., Moore M.R., Murray G.D., Robertson S.J. Public health implications of new guidelines for lead in drinking water: a case study in an area with historically high water lead levels. Food Chem. Toxicol. 38, 73–79, 2000.

Wireman J., Liebert C.A., Smith T., Summers A.O. Association of mercury resistance with antibiotic resistance in the Gram-negative fecal bacteria of primates. Appl. Environ. Microbiol. 63, 4494–4503, 1997.

CHAPTER 14

Mercury Toxicity
The Importance of Microbial Diversity for Improved Environmental Remediation

Mohammed H. Abu-Dieyeh, Kamal Usman, Haya Alduroobi
and *Mohammad Al-Ghouti**

1. Introduction

Mercury is one of the naturally occurring elements found in the environment. This heavy metal occurs naturally in minute amounts and is the sixteenth rarest element on earth. It is ranked third among the most toxic biosphere occurring elements and when present in high concentration in a given soil, sediment and/or air compartments of the environment, it leaches and deposits in water bodies, eventually ending up in the food chain (Sinha and Khare, 2012), thereby causing a great health risks. Mercury pollution is a serious global environmental problem attracting the attention of many stakeholders around the world (Mason et al. 2012). Traditionally, chemical and physical methods were employed in the reduction or removal of mercury from contaminated aqueous and soil environments, however, owing to their labor intensiveness, high operational cost and low efficiency among other limitations, these remediation measures are adjudged unsatisfactory and hence the search for alternative means of clean-up of mercury contaminated environments becomes imperative (Wuana et al., 2011). Accordingly, natural remediation processes and technologies are being explored, among which bioremediation proved promising. Although a wide range of groups of organisms are used to degrade mercury, many challenges still persist, and much of the successes recorded are limited to laboratory scales (Xu et al., 2015). It will therefore be

Department of Biological and Environmental Sciences, College of Arts and Sciences, Qatar University, P.O. Box: 2713, Doha–Qatar.
* Corresponding author: mohammad.alghouti@qu.edu.qa

of interest to demonstrate the efficiency of these technologies at a much larger scale and optimize performance where necessary. This chapter overview the diversity of microorganisms, and prospects in the improvement of mercury contaminated environment remediation process.

2. Mercury toxicity and adverse effects

Mercury is one of the naturally occurring elements found in the environment. It is ranked number three among the most toxic biosphere occurring elements and when present in high concentration in a given soil, sediment and/or air compartments of the environment, it leaches and deposits in water bodies, eventually ending up in the food chain (Sinha and Khare, 2012). Common natural processes leading to mercury deposition in the environment include volcanoes and forest fires as well as geothermal activities. Anthropogenic sources include power plants, mining, and refining industries (Li et al., 2009; Xu et al., 2015). Once emitted, mercury is immediately transported from point sources to other environmental compartments and phases alike. Although it gets into the biogeochemical cycle in terrestrial or aquatic habitats, it is however reported to persist more in soil and sediments (Tangahu et al., 2011). The environmental persistence of mercury permits long range transport, and often, it partitions to areas with very less or no human activities, and hence fast becoming a major health concern (Kabata-Pendias, 2010; Mahbub et al., 2016). Anthropogenic sources appear to aggravate mercury deposition into the environment and various industries contribute up to 1960 tons of total emission every year (AMAP, 2013). In a recent review, Li et al. (2015) summarize regional and sectoral sources of mercury globally, with Asia and Africa recording the highest air emission of this heavy metal.

2.1 Sources of mercury contamination

Mercury is found widely distributed throughout the world. Common natural processes leading to deposition in the environment include volcanoes and forest fires as well as geothermal activities. Anthropogenic activities are estimated to release approximately 1960 tons of mercury per year (UNEP, 2013). This indicates the huge effect of human activities to the spate of global mercury pollution. Mercury dispersal by human activities started before the industrial revolution when it was commonly used in gold extraction, medical antiseptics, pharmaceutical products, and agricultural fungicides (Eisler, 2006). While this heavy metal is still applied in gold mining, other activities emerged after the industrial revolution, which greatly contributed to the global pollution of this metal element (Mason et al., 2012). Additionally, the production of various mercury-containing products such as thermometers, paints, dental amalgams, batteries, and fluorescent lamps and their eventual disposal are major sources of mercury contamination in all biosphere compartments of air, soil and water. In North America, for example, waste materials containing mercury including discarded thermometers, batteries, as well as fluorescent lamps are the reason behind 40% mercury emissions in the region (Xu et al., 2015). Moreover, the incineration of sewage sludge, medical, and industrial wastes is also considered as one of the major source of mercury pollution (Eisler, 2006).

Although release from certain industries such as coal-operated power plants is being regulated, thereby reducing its emission, other industrial activities such as cement production is increasing mercury emission. The emissions from cement industries was estimated to have increased by almost 30% between 2005 and 2009 (USGS, 2012), in addition to global emissions associated with "Artisanal and small scale gold mining" (ASGM) operations, which is significantly higher than previously reported (UNEP, 2013). Environmental monitoring equipments such as barometers commonly used in airports, wind tunnels, onshore and offshore installations, and mechanical manufacturing processes are also reported to release mercury into the atmosphere (Hutchison and Atwood, 2003). Recent data indicate worldwide mercury emissions in the atmosphere amounting to around 3000 t as at 2005 (Branch, 2008), with more than 2000 t arising from human activities (Li et al., 2009). However, oceanic emissions are considered to be the largest contributors from all natural sources (Lollar, 2005). It was estimated that the total oceanic mercury emission worldwide is 2600 tons/yr, but a part of this is recycled back to the ocean surface through photochemical processing, leaving only 1500 tons/year as net flux out of the marine waters (Bank, 2012). By 2050, it is estimated that global emission may reach 4,800 mg or more (Streets et al., 2009). Looking at individual nations and their contribution to mercury pollution worldwide, China comes first as the largest contributor to the global mercury emissions (about 700 tons/year) and the next top five emitters are South Africa, India, Japan, Australia and the US, respectively (Wu et al., 2006). This may be attributed to the strong industrial growth and over-reliance on coal combustion for power generation purposes in these countries. Indeed, the background level of natural mercury is already high, additionally, the large amount of anthropogenic mercury adds to the serious pollution problems and could prove dangerous to human health if not properly reduced and managed.

2.2 *Mercury and environmental toxicity*

Mercury is increasingly getting concentrated in sediments and soil becoming tenfold, largely owing to the continued burning of fossil based fuels and air transport. Although natural depletion of accumulated pollutants occur in the soil environment through various means which includes transformation, diffusion, and dissipation to reduce contamination levels (Trasande et al., 2005). However, the concentration of heavy metals in the soil including mercury often exceeds the attenuation level which will unavoidably result to a high level pollution in the soil and present a potential health hazard to humans and other living organisms (Xu et al., 2015). To reduce, remove, or convert mercury to a less toxic form in the soil, effective remediation strategies are thus proposed (Cui et al., 2011), and accordingly other physical, chemical, and biological techniques are employed to thwart its contagion in the soil (Assad et al., 2016; Rodríguez et al., 2012). Mercury toxicity could also adversely affect the balance of the ecosystem. For instance, it was found to negatively affect birds "flight" behavior and reproduction, it causes liver and kidney damages to arctic ringed seals, beluga whales and other mammals as well (UNEP, 2013). Similarly, increased mercury concentrations in the blood of black–legged kittiwakes (*Rissa tridactyla*) have been implicated in missed breeding and irregular reproductive hormone responses (Overjordet et al., 2015), yet in another study, some Minnesota loons found to have accumulated high mercury concentrations demonstrate impaired ability to reproduce (Ensor et al., 1992). Other organisms such as minks, otters, and several other avian species are susceptible to the effects of methylmercury (from a fish diet) according to U.S. EPA (1997).

2.3 Mercury and human health

The environmental persistence of mercury permits long range transport. It often partitions to areas with less or no human activities, and hence increasingly becomes a major health concern (Kabata-Pendias, 2010; Mahbub et al., 2016). Volatilization of inorganic mercury and eventual transformation into methyl–mercury in water present an even worst scenario. In this form, it becomes easily bio-accumulated in living organisms (Humphries, 2012). This could dramatically raise the bio-concentration factor relative to water in aquatic organisms and enters into the food chain. Indeed, this poses a serious threat to global health and in particular to high tropic level species (Gabriel and Williamson, 2004). Humans largely get mercury from the ingestion of aquatic organisms, and fish is adjudged to be the main source (Selid et al., 2009). It is a well-known fact that the beginning of awareness of the health impacts of mercury pollution dates back to 1950s when mercury-containing chemical waste was released into the nearby Minamata Sea by the Chisso Corporation in Minamata, Japan. This was a consequence of the large amount of bioavailable methyl mercury accumulated by fishes and other aquatic organisms. Being the main source of diet to the local population in the area, continued ingestion of these contaminated aquatic food led to the detrimental health effects witnessed (Kurland et al., 1960). Ever since then, many studies have demonstrated that pre-natal or post-natal exposure to methylmercury results in serious neurological impacts among adults and children, now known as "Minamata Disease". Distal paresthesia of the extremities and the lips were the main symptoms of chronically-exposed patients even 30 years after termination of exposure (Ekino et al., 2007). Furthermore, recent evidence suggest even general population exposed to methyl mercury in

Minamata who were not legitimate "Minamata Disease" patients suffered from "psychiatric symptoms including impairment of intelligence and mood and behavioral disruption" (Yorifuji et al., 2011). It negatively affects the central nervous system in humans even in low exposure conditions (Nance et al., 2012). Similarly, mercury is also implicated in physiological disorders by directly binding to the cysteine amino acid residues and nitrogen atoms in proteins and nucleic acids respectively. It is considered dangerous for the unborn babies and children especially (Holmes et al., 2009). In addition to loss of memory and damaged brain, grave exposure to any form of elemental mercury at high concentration could result in the loss of internal organs such as the liver and heart. For instance, Zuber and Newman (2011) reported that one gram dosage of mercury is capable of causing death to humans. Bose-O'Reilly et al. (2010) also demonstrated that long term exposure to mercury could lead to renal failure, interfere with reproductive organs, and cause neuronal disorders such as Parkinson's and Alzheimer's. However, due to increased awareness on the detrimental health effects of mercury pollution, incidents such as the Minamata are not common at present time.

3. Remediation of mercury and benefits of bioremediation

The increase in mercury contamination, especially from wastewaters in addition to other sources necessitated the exploration of many technologies to reduce this contamination level (Sinha and Khare, 2012). Conventional treatments such as carbon adsorption, precipitation, reverse osmosis, and the ion exchange are routinely used to reduce mercury contamination (Akpor and Muchie, 2010). Additional physicochemical means of mercury remediation involve thermal treatment, soil washing, and acid extraction. More recent strategies saw the use of manganese oxide and gold nanoparticles as viable alternatives (Zhang et al., 2013). While these processes are in the early developmental stages, additional concerns are their high costs and requirement of large concentration of metal contamination for use, generally not specific to targets and often resulting in the large production of by-products (Wagner-Döbler, 2013). Hence, the use of biological based remediation processes such as bioremediation using microorganisms and phytoremediation employing plants species are gaining popularity for their host of advantages. They are relatively inexpensive, more environment-friendly and are more efficient under low mercury contamination (Mahbub et al., 2016). Remediation wise, in the soil, trace elements are usually not degraded in a similar fashion as organic contaminants. Rather, the process involves relocation from place of primary contamination to other sites such as the landfill (Xu et al., 2015).

Common strategies involve the stabilization of the element right in the area of primary occurrence (Karami et al., 2011). In the majority of cases, mercury extraction is commonly employed for the separation of the element from soil or lowering its bioavailability to non-toxic level. Another strategy is immobilization, this helps to protect humans and other organisms from exposure by encapsulating mercury in the soil (Petruzzelli et al., 2013). The following sub-sections highlight some mercury chemical and biological remediation methods based on recent technologies.

3.1 Chemical remediation methods

Mercury can be converted into chemical form by solidification or stabilization process. It is usually transformed to a stable and non-soluble form even at varying pH concentration and various reduction-oxidation reaction (redox) conditions (Figueiredo et al., 2016). Solidification is the enclosure of stable mercury in a protective matrix as a way of reducing the bioavailability, emission, and deposition. This way, leachable mercury is prevented from partitioning (Svensson, 2006) and upto 90% of the element can be sequestrated in the process (Bolan et al., 2014). The *in situ* stabilization/solidification is an emerging technology that utilizes minerals, fly ashes, phosphates, and aluminum silicates as agents of mercury to stabilization.

Although yet to be popularly practiced on a large scale capacity, the most studied among chemical remediation methods involve the use of phosphatase agents (Randall and Chattopadhyay, 2013). The success of this technology largely depends on mercury transport, which in turn depends on the elements properties, especially the state of oxidation. Specific mercury present and the pH of the environment also contribute in this regard (Xu et al., 2015).

3.2 Phytoremediation

Conventional remediation methods of metals including mercury are recognized for a number of disadvantages or limitations such as high cost and labor intensiveness. Chemical processes create yet another pollution source and are especially costly since they generate heaps of sludge (Tangahu et al., 2011). In view of this context, new and more efficient techniques to reduce or remove metal contaminants became imperative and hence the exploration of various biological remediation techniques. The use of biological agents are considered cheap, safe, and have limited or no negative impact on the environment (Doble and Kumar, 2005). Among the various biological remediation technologies, phytoremediation assures to be a viable and promising alternative, which over the years has gained increased attention (Ullah et al., 2015; Witters, 2011). Phytoremediation refers to a process where plants are employed to reduce or clean-up organic and inorganic contaminants from the environment with or without the aid of associated microbes (Visioli and Marmiroli, 2013). The processes by which contaminants are remediated differs; it may be in the form of removal, transfer, degradation, and immobilization from either soil or water (Ahmadpour et al., 2012). It is a unique approach capitalizing on plants' roots ability for the initial uptake of pollutants and eventually accumulating them into their shoot tissue by translocation across the stem. In comparison with other conventional treatment techniques, phytoremediation could be considered a novel method with a great potential to provide the much-needed green technology solution to our deteriorating environment.

3.3 Microbial bioremediation

Considering the fact that mercury is a toxic contaminant that cannot be treated as a non-toxic material, it can be either transformed to a less toxic form or removed from the contaminated matrix by microbial remediation. Bioremediation could be either in the form of biosorption or volatilization. Volatilization is popularly known to involve the use of mercury resistant bacteria that bears the *mer*-operon, which binds, transports, and detoxifies mercury (II) and organic mercury to elemental form. This was reported to help in preventing the accumulation of metal in the food chain (Wagner-Döbler, 2013). However, microbial volatilization other than those mediated by *mer*-operon was also found. The dissimilation of mercury (II) to mercury (Hg^0) by the reducing bacterial species *Shewanella oneidensis* was not *mer*A (mercuric reductase) mediated but MR-1 (Wiatrowskim et al., 2006).

The biomass of both living and non-living microbes inclusive of algae, bacteria and fungi have been used to reduce mercury solubility and toxicity via biosorption (Adeniji, 2004) involving various immobilization processes and systems including exchange of ions and adsorption (François et al., 2012). Although a wide range of groups of organisms are able to degrade soil organic mercury, many challenges still persist. These include the activity of co-contaminants which affect the activity of mercury degrading bacteria, less bioavailability, and insufficient nutrients supply (Krämer and Chardonnens, 2001; Xu et al., 2015). To overcome these challenges, genetic modification of soil microbes are being explored (He et al., 2011).

3.3.1 Role of microbial transformations in the mercury geochemical cycle

The transport, distribution and bioavailability of chemicals do not only depend on their concentration, but most importantly on their form of natural occurrence. Transport and bioavailability depends much on the reactivity of trace elements in the sediments such as mercury (Violante et al., 2007). Many factors inclusive of thermodynamic solubility affect the movement and bioavailability of mercury and associated compounds in aquatic habitats. Mercury may be adsorbed to suspended particles and organic matter, precipitate in water, spread in sediments, and host of other substrates that can sequestrate mercury from the mobile aqueous compartment (Reduction, 2012). Speciation involves localizing mercury in different environmental phase components and such phenomenon involving mercury and coordination processes in an aqueous phase were detailed in many studies (Mason et al., 2012; Wang et al., 2012a).

4. Biochemical basis of bacterial mercury resistance

Microorganisms have developed an extensive defense mechanism against mercury toxicity (Fig. 1). "Mercuric reductase" is an essential bacterial cytoplasmic enzyme that transforms ionic mercury (Hg^{2+}) to elemental mercury (Hg^0), which is then diffused out of the bacterial cell (Wagner-Döbler, 2003). The most widely studied defense mechanism is the degradative enzymatic pathways of organic to inorganic and ionic form (Hg^{2+}). It is reduced to metallic mercury (Hg^0), removed from the cells, and eventually volatized to the atmosphere (Chien et al., 2012). However, Wagner-Döbler (2003) reports that instead of releasing metallic mercury back to the atmospheric environment, it can be accumulated in a mechanical bioreactor. The sequestration of mercury by microbial cell surface components and dead cells is also a common occurrence (François et al., 2012). Adsorption of ionic mercury (Hg^{2+}) by the secretion of exo-polymers under suitable condition have been reported in some resistant microbes (Mahbub et al., 2016).

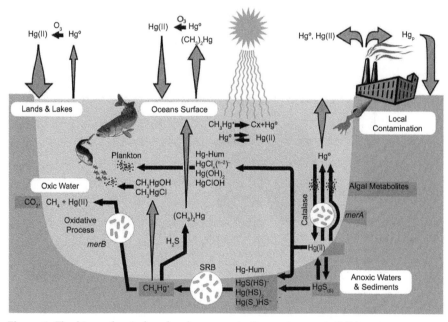

Figure 1. The biogeochemical cycle of mercury in the environment. Solid arrows represent transformation or uptake of mercury. Hollow arrows indicate flux of mercury between different compartments in the environment. The width of the hollow arrows is approximately proportional to the relative importance of the flux in nature. The speciation of Hg(II) in oxic and anoxic waters is controlled by chloride and hydroxide, and by sulfide respectively. Transformations known to be mediated by microorganisms are represented by circles depicting bacterial cells. SRB stands for sulfate-reducing bacteria and merB and merA refer to the activity of genes encoding the enzymes organomercurial lyase and mercuric reductase respectively. A group of dots indicate the involvement of unicellular algae. Light-mediated water column transformations are positioned below the sun. Photodegradation of CH 3 Hg þ results in mostly Hg‡ and an unknown C 1 species depicted as C x (D. Krabbenhoft, personal communication) (Adapted from Barkay et al., 2003).

4.1 Role of mercury-resistant bacteria mer operon

The process of degradative enzymatic pathways of organic to inorganic and ionic form (Hg^{2+}), and eventual reduction to metallic mercury (Hg^0) rely on the expression of the *mer* operon (Fig. 2), a cluster of mercury resistance genes that transforms extremely toxic soluble (ionic) mercury to an insoluble form (metallic). In a reduced form, the metallic mercury is easily immobilized from microbial cells and volatilized (Chien et al., 2012). Many studies suggest the ubiquity of such bacterial functions in various environments, which indicates their worldwide distribution and evolution (Osborn et al., 1997).

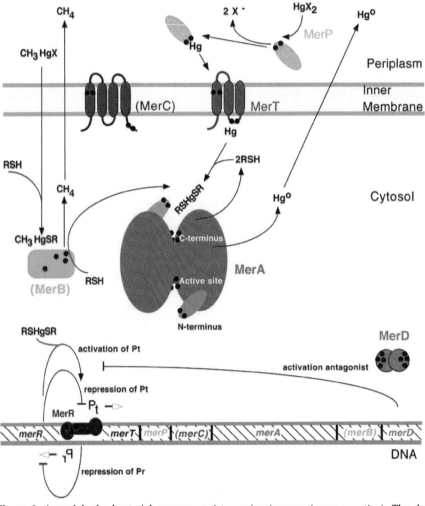

Figure 2. A model of a bacterial mercury resistance *(mer)*operon (gram negative). The dot symbol is a cysteine residue. X represents generic solvent nucleophile. RSH, a low molecular weight cytosolic thiol redox buffer. Parentheses around gene/proteins are proteins/genes that are not common to all operon (Adapted from Barkay et al., 2003).

The *mer* operon of gram negative bacteria is the most studied instance of mercury resistance by microorganisms. Similar set of genes with the same pattern are also found in the gram positives bacteria (Wiatrowski et al., 2006). Typically, resistance mechanism involves the reduction of ionic mercury to a volatile form by regulator merA, a mercuric reductase "cytosolic Flavin disulfide oxidoreductase" and uses NADPH as reductant in the process. *Mer*P, a homologue of *Mer*A, is relatively small mercury biding peptide and replaces nucleophiles such as Cl⁻ to which mercury (II) likely coordinates in an oxygenated bacterial growth medium. It is understood to exchange mercury (II) ions for cysteines in MerT which is another protein found in the cytosolic membrane. Others are *Mer*C, *Mer*F, *Mer*B, *Mer*G, and *Mer*D which together coordinate the resistance mechanism conferred by bacteria via the famous *mer* operon family and the details of their roles and regulatory functions could be found here (Wiatrowski et al., 2006).

4.2 Genetic engineering of bacteria for bio-sorption of mercury

Although various groups of organisms are able to degrade soil organic mercury, many challenges still persist. These include the activity of co-contaminants which influence the activity of mercury degrading bacteria, less bioavailability, and insufficient nutrients supply (Krämer, 2005). To overcome these challenges, genetic modification of soil microbes are being explored (He et al., 2011). Metal binding proteins could be expressed in bacteria as a strategy to improve metal bioremediation. In an experiment to improve heavy metals accumulation in bacteria, Kotrba et al. (1999) demonstrated that engineering cysteine or histidine rich peptides in LamB and expressed in *E. coli* significantly improved metal uptake by two and four folds respectively. However, novel strategies with variable arrangement and expression pattern of the key genes such as the *mer* operon are desired to combat the increasing mercury pollution in the environment (Sangvanich et al., 2014).

4.3 Sources of mercury-resistant microorganisms

Mercury-resistant bacterial strains are found in both aquatic and terrestrial ecosystems in various studies. Their resistance to mercury is usually confirmed by the presence of the *mer* operon. More specifically, resistant strains are especially found in stressful environmental conditions whether in aquatic or soil habitats.

4.3.1 Marine microorganisms

Many bacterial strains showing resistance to mercury isolated from marine environments were reported in different studies. Examples of identified species are listed in Table 1 below along with their mercury resistance level expressed as "Minimum Inhibitory Concentration" (MIC).

Table 1. Marine mercury-resistant bacteria and their tolerance level.

Strain Name	MIC (mg/L)	Reference
Alcaligenes faecalis	75 mg/L	
Bacillus pumilus	25 mg/L	(De and Ramaiah, 2007)
Brevibacterium iodinium	75 mg/L	
Pseudomonas aeruginosa	75 mg/L	
Aeromonas media	9.87 mg/L	
Citrobacter freundii	19.3 mg/L	(Figueiredo et al., 2016)
Vibrio metschnikovii	9.87 mg/L	
Acinetobacter sp.	20 mg/L	
Bacillus sp.	20 mg/L	
Citrobacter sp.	30 mg/L	
Escherchia coli	40 mg/L	
Klebsiella sp.	40 mg/L	(Zeng et al., 2010)
Micrococcus sp.	40 mg/L	
Proteus sp.	30 mg/L	
Pseudomonas spp.	70 mg/L	
Serratiamarcescens	30 mg/L	
Staphylococcus aureus	30 mg/L	

4.3.2 Soil microorganisms

Soil microorganisms were also repeatedly isolated in a wide range of studies. Some of those identified species are summarized in Table 2 below along with their mercury resistance level expressed as "Minimum Inhibitory Concentration" (MIC).

Table 2. Soil mercury-resistant bacteria and their tolerance level.

Strain name	MIC (mg/L)	Reference
Pseudoxanthomonas sp.	23.1 mg/L	(Mahbub et al., 2016)
Sphingobium sp.	48.48 m/L	(Mahbub et al., 2016)
Bacillus sp.	500 mg/L	(Kotala et al., 2014)
Aeromonas hydrophila	500 mg/L	(Kotala et al., 2014)
Pseudomonas pseudoalcaligenes	600 mg/L	(Sorkhoh et al., 2010)
Pseudomonas stutzeri	600 mg/L	(Sorkhoh et al., 2010)
Citrobacter freundii	600 mg/L	

4.3.3 Microorganisms under environmental stress

Stressful environmental conditions could result in several reactions at the "morphological, physiological and biochemical levels" (Gustavs et al., 2009). The ability of bacterial strains to handle sudden changes in the surrounding environment guarantees their ecological control under stressful situations. Studies suggest a positive correlation between extreme levels of pH and salinity of a certain environment and the availability of mercury-resistant bacterial strains. For instance, *Bacillus* sp. and *Vibrio* sp. were two isolates found in the brackish, moderately acidic sediments of Bhitarkanika mangrove ecosystem in Odisha, India (Dash and Das, 2014). Moreover, hot springs microorganisms particularly were forced to evolve tactics to cope with the toxic effects of high mercury concentrations in their habitats, and thus, the *Mer*A protein first evolved in thermophilic microbes in geothermal environments before it spread to widely distributed mesophilic microbes via evolutionary processes (Boyd and Barkay, 2012).

5. Strategies of mercury-resistant bacteria against mercury

Mercury and its salts are released into the environment in bioavailable forms both from natural and anthropogenic sources. Environmental persistence of this heavy metal triggers the development of resistance mechanisms by a large group of bacterial species, especially in the soil, which are critical to the reduction and eventual removal of toxic mercury from the environment (Jan et al., 2009). Bacterial species, *Staphylococcus aureus* was first reported to have demonstrated mercury resistance (Richmond and John, 1964). The most reported mechanism of resistance of mercury by bacteria is via the *"mer* operon", a cluster of genes commonly found in different bacterial species and in different kinds of environments but most commonly in the soil, and till date, it has the most outstanding bacterial characteristics utilized to remediate toxic mercury from contaminated areas (Jan et al., 2009).

5.1 Mobilization processes

5.1.1 Enzymatic oxidation and reduction

Biological oxidation of Hg (0) received the least attention in the biogeochemical cycle of mercury (Barkay et al., 2003). Bacterial oxidation was demonstrated to be facilitated by peroxidases, however, even when mutant strains of the enzyme were subjected, low level oxidation was observed and hence the suggestion that other bacterial oxidases exists (Smith et al., 1998). The same authors report high oxidation activity against mercury by soil bacteria, *Streptomyces* and *Bacillus* as a recurring event in the element's environmental cycling.

Interestingly, mercuric oxidation by these bacterial species were more efficient when compared with *Mer*A facilitated reduction demonstrated by *mer* operon bearing species (Summers and Silver, 1972). The resistance mechanism also involves the reduction of ionic mercury to a volatile form by regulator *Mer*A, a mercuric reductase "cytosolic Flavin disulfide oxidoreductase" and it uses NADPH as reductant in the process. *Mer*P, a homologue of *Mer*A, is relatively small mercury biding peptide and replace nucleophiles such as Cl⁻ to which mercury (II) is likely coordinate in an oxygenated bacterial growth medium. It is understandable to exchange mercury (II) ions for cysteines in *Mer*T, which is another protein found in the cytosolic membrane. Others are *Mer*C, *Mer*F, *Mer*B, *Mer*G, and *Mer*D which together coordinate the resistance mechanism conferred by bacteria via the famous *mer* operon and an in-depth explanation of their roles and regulatory functions could be found here (Barkay et al., 2003).

5.1.2 Complexation

Complexation is crucial to mercury cycle in the environment and the most common complexing agent is methylation of mercury (Gabriel and Williamson, 2004). Methylation of mercury is generally believed to be facilitated by microorganisms, while abiotic factors are more likely to influence organic contaminants (Zhang et al., 2012). The production of methyl mercury is not as simple as a factor of total concentration mercury in the entire system, but a function of several environmental determinants; diversity of bacterial community, temperature, redox potential and persistence, as well as organic and inorganic complexing substances which together interact to determine methyl mercury formation (Randall and Chattopadhyay, 2013). Naturally occurring organic matter are known to be composed of a heterogeneous mixture of organic compounds that are widely distributed in the environment (Wang et al., 2012b). The natural organic matter binds strongly to heavy metals and influences their speciation, solubility, transport, and subsequent environmental toxicity (Buffle, 1988).

5.1.3 Siderophores

Almost all microorganisms require iron as an essential nutrient, however, it usually exists in an insoluble form, and hence unavailable for uptake by the microbes. Therefore, these microorganisms have to develop a mechanism through which iron can be solubilized to enable uptake. To do this, bacteria produce siderophores, which are small phytochelatins with high affinity to the iron metal (Rajkumar et al., 2010). Siderophores act as solubilizing agents under limited iron condition. There exist about 500 siderophores, however, despite their vast number, they are not as diverse in terms of their functions (Boukhalfa and Crumbliss, 2002). Generally, siderophores form a complex with iron (III), after which it is then taken by the bacterial cell membrane. Inside the cell membrane, iron (III) is reduced to iron (II) and then released into the cell from the siderophores. This mechanism of iron uptake has been identified and reported in both gram positive and gram negative bacterial species (Krewulak and Vogel, 2008).

5.2 Immobilization processes

5.2.1 Bio-uptake (bio-sorption)

Neutral mercury is assimilated by bacteria through passive diffusion but actively assimilated when in both charged and non-charged form (Kelly et al., 2003; Wiatrowski et al., 2006). Sediments found in wetland areas generally have low redox potential which serially reduce mercury(II); low redox potential stimulate bacteria facilitated sulfur-reduction which also promotes the methylation of mercury. However, when present in large quantity, sulfide results in the formation of soluble mercury in complex with sulfur (King et al., 2002). Natural organic matter dissolved in water can promote or hinder the formation of dangerous and bio-accumulative methyl mercury species. The dissolved organic matter is critical to the bio-concentration and bio-magnifications of mercury (Cormack, 2001).

Methylation and bioaccumulation constitutes important components in the mercury geo-chemical cycle found in different aquatic environmental phases, water, the sediment, and biota. In fishes, for instance, mercury is majorly found in methyl mercury (CH_3Hg) form and conversion from ionic to mono methyl mercury form is critical to fish bioaccumulation, exposure, and toxicity to living organisms including humans (Randall and Chattopadhyay, 2013). Complexation with dissolved organic matter in bacterial mediated methylation is expressed and measured as dissolved organic carbon and is crucial to the bioavailability of mercury. The more the dissolved organic matter, the less is the availability of inorganic mercury thereby limiting organismic uptake. In sulfur limited environment, the dissolved organic matter stimulate microbial growth and hence improve the rate of mercury methylation in sediments and aqueous phases (Graham et al., 2012; Kelly et al., 2003).

5.2.2 Precipitation

Precipitation as HgS is a potential remediating strategy for the immobilization of mercury by sulfate-reducing bacteria. For example, a bacterium that produces dimethyl sulfide called *Klebsiella pneumoniae* was found to produce mercury-containing precipitates when the off-gas, produced from the culture, was passed through a mercury contaminated solution (Essa et al., 2006). This bacterium removed nearly 99% of the bioavailable mercury in the solution, thereby, indicating that bio-precipitation can be performed even separately from the bacterial growth medium. It was also reported that cyanobacteria, "*Limnothrixplanctonica* (Lemm.), *Synechoccus leopoldiensis* (Racib.) Komarek, and *Phormidium limnetica* (Lemm.)" were able to convert considerable amounts of Hg(II) into β–HgS precipitates (Lefebvre et al., 2007).

5.3 Marine mercury reduction in contaminated soil

To remediate soil contaminated with mercury, the trace element in a contaminated site is transferred to another location, say a landfill or better still, *in situ* immobilization of a stable form of the toxic element (Kumpiene et al., 2008; Mukherjee et al., 2015). In the majority of cases, mercury is extracted from soil or reduced to an acceptable bioavailable form by reducing its concentration or the amount of contaminated soil. When immobilized, mercury in the soil is encapsulated and stabilized to eliminate toxic effects to humans and other living species (Petruzzelli et al., 2013). Several remediation techniques exist for the removal of soil mercury, including "soil washing, thermal treatment, stabilization/solidification, nanotechnology, vitrification, electro-remediation, and phytoremediation" however, this chapter focuses on bioremediation based treatment method. Several authors have extensively reported the above techniques, detailing underlying principles, as well as the comparison of the pros and cons of each strategy (Wang et al., 2012b; Xu et al., 2015).

The biomass of both living and non-living microbes inclusive of algae, bacteria, and fungi have been used to reduce mercury solubility and toxicity via bio-sorption (Adeniji, 2004) involving various immobilization processes and systems including exchange of ions and adsorption (François et al., 2012). Although several species of microorganisms are able to degrade soil organic mercury, many challenges still persist. These include the activity of co-contaminants which affect the activity of mercury degrading bacteria, less bioavailability, and insufficient nutrients supply (Krämer and Chardonnens, 2001; Xu et al., 2015). To overcome these challenges, genetic modification of soil microbes are being explored (He et al., 2011).

6. Treatment applications

Mercuric environmental persistence suggests the need for the exploration of many remedial methods to effectively decontaminate polluted environments. Capping and dredging are commonly employed to actively remediate polluted sediments in water (Randall and Chattopadhyay, 2013). Dredging results in the recovery of water circulation and removal of bottom sediments (Barbosa and de Almeida, 2001). Capping may be *in situ*, which involves the placement of a layer of separating sand between contaminated sediment and the water (Palermo, 1998).

However, in contaminated soil, dredging is not suitable, instead, the trace element in contaminated sites are transferred to another location, say a landfill or better still, *in situ* immobilization of stable form of the toxic element (Kumpiene et al., 2008; Mukherjee et al., 2015). In the majority of cases, mercury is extracted from the soil or reduced to an acceptable bioavailable form by reducing its concentration or the amount of contaminated soil. When immobilized, mercury in the soil is encapsulated and stabilized to eliminate toxic effects to humans and other living species (Dermont et al., 2008).

7. Conclusion

Mercury pollution is a serious global environmental problem attracting the attention of many stakeholders around the world. This heavy metal occurs naturally in minute amounts in the environment and is the sixteenth rarest element on earth. Anthropogenic activities could also result in increased mercury deposition into the environment, and although widespread globally, this is especially true in countries with the highest industrial activities where emission could reach up to 1960 t annually. Famous since the Minamata incidence, increased mercury pollution pose great health risks to the all important higher tropic level species, threatens the diversity of natural ecosystem, and consequently endangers environmental sustainability. Traditionally, chemical and physical methods were employed in the reduction or removal of mercury from contaminated aqueous and soil environments, however, owing to their labor intensiveness, high operational cost and low efficiency among other limitations, these remediation measures are adjudged unsatisfactory and hence the search for alternative means of clean-up of mercury contaminated environments is ongoing. Accordingly, natural remediation processes and technologies are being explored, among which bioremediation is included. Microbes assisted remediation of mercury contaminated sites mostly involve the use of diverse groups of mercury resistant bacteria that bears the *mer*-operon gene, which binds, transports, and detoxifies mercury (II) and organic mercury to elemental form. These have so far been assuring in comparison to the conventional treatment technologies. While several studies that have been conducted and the others that are still ongoing demonstrated that the efficiency of mercury resistant bacteria in the remediation process could further be improved by the genetic manipulation of genes involved in this process, however, it is yet to achieve the desired performance level and process evaluation is still limited to laboratory scale experiments. It will therefore be of interest to prospect for new bacterial strains, with improved mercury degradation potential. Genetic manipulation of metabolic pathways/genes involved in mercury resistance and transformation should also be afforded priority. It is also important that efforts should be made towards the scale up of current laboratory successes.

References

Adeniji A. Bioremediation of arsenic, chromium, lead, and mercury. US Environmental Protection Agency, Washington DC, 2004.

Ahmadpour P., Ahmadpour F., Mahmud T., Abdu A., Soleimani M., Tayefeh F.H. Phytoremediation of heavy metals: A green technology. African J. Biotechnol. 11, 14036–14043, 2012.

Akpor O., Muchie M. Remediation of heavy metals in drinking water and wastewater treatment systems: Processes and applications. Int. J. Phys. Sci. 5, 1807–1817, 2010.

AMAP A. Assessment 2013: Arctic Ocean Acidification. Arctic Monitoring and Assessment Programme (AMAP), Oslo, Norway, 2013.

Assad M., Parelle J., Cazaux D., Gimbert F., Chalot M., Tatin-Froux F. Mercury uptake into poplar leaves. Chemosphere 146, 1–7, 2016.

Bank M.S. Mercury in the Environment: Pattern and Process: University of California Press, 2012.

Barbosa M.C., de Almeida M. Dredging and disposal of fine sediments in the state of Rio de Janeiro, Brazil. J. Hazard. Mater. 85, 15–38, 2011.

Barkay T., Miller S.M., Summers A.O. Bacterial mercury resistance from atoms to ecosystems. FEMS Microbiol. Rev. 27, 355–384, 2003.

Bolan N., Kunhikrishnan A., Thangarajan R., Kumpiene J., Park J., Makino T., Scheckel K. Remediation of heavy metal (loid) s contaminated soils-to mobilize or to immobilize? J. Hazard. Mater. 266, 141–166, 2014.

Bose-O'Reilly S., McCarty K.M., Steckling N., Lettmeier B. Mercury exposure and children's health. Curr. Probl. Pediatr. Adolesc. Health Care 40, 186–215, 2010.

Boukhalfa H., Crumbliss A.L. Chemical aspects of siderophore mediated iron transport. BioMetals 15, 325–339, 2002.

Branch U.C. The global atmospheric mercury assessment: Sources, emissions and transport. UNEP–Chemicals, Geneva, 2008.

Buffle J. Complexation reactions in aquatic systems. An analytical approach J. Wiley & Sons Ltd./Ellis Horwood Ltd., 1988.

Chien M., Nakahata R., Ono T., Miyauchi K., Endo G. Mercury removal and recovery by immobilized *Bacillus megaterium* MB1. Front. Chem. Sci. Eng. 6, 192–197, 2012.

Cormack R. Sediment Quality Guideline Options for the State of Alaska: Alaska Department of Environmental Conservation, 2001.

Cui W., Wang X., Duan J., Yang M., Zi W., Liu X. Advances on remediation techniques of heavy metal Cd and Hg in contaminated soil. Guizhou Agric. Sci. 7, 67–70, 2011.

Dash H.R., Das S. Assessment of mercury pollution through mercury resistant marine bacteria in Bhitarkanika mangrove ecosystem, Odisha, India. IJMS 43, 1103–1115, 2014.

De J., Ramaiah N. Characterization of marine bacteria highly resistant to mercury exhibiting multiple resistances to toxic chemicals. Ecological Indicator 7, 511–520, 2007.

Dermont G., Bergeron M., Mercier G., Richer-Laflèche M. Metal-contaminated soils: remediation practices and treatment technologies. Pract. Periodical Hazard. Toxic. Radioact. Waste Manage. 12, 188–209, 2008.

Doble M., Kumar A. Biotreatment of industrial effluents: Butterworth-Heinemann, 2005.

Eisler R. Mercury Hazards to Living Organisms: CRC Press, 2006.

Ekino S., Susa M., Ninomiya T., Imamura K., Kitamura T. Minamata disease revisited: an update on the acute and chronic manifestations of methyl mercury poisoning. J. Neurol. Sci. 262, 131–144, 2007.

Essa A.M.M., Creamer N.J., Brown N.L., Macaskie L.E. A new approach to the remediation of heavy metal liquid wastes via off-gases produced by *Klebsiella pneumoniae* M426. Biotechnol. Bioeng. 95, 574–583, 2006.

Figueiredo N.L., Canário J., O'Driscoll N.J., Duarte A., Carvalho C. Aerobic mercury-resistant bacteria alter mercury speciation and retention in the Tagus Estuary (Portugal). Ecotoxicol. Environ. Saf. 124, 60–67, 2016.

François F., Lombard C., Guigner J.-M., Soreau P., Brian-Jaisson F., Martino G., Pignol D. Isolation and characterization of environmental bacteria capable of extracellular biosorption of mercury. Appl. Environ. Microbiol. 78, 1097–1106, 2012.

Gabriel M.C., Williamson D.G. Principal biogeochemical factors affecting the speciation and transport of mercury through the terrestrial environment. Environ. Geochem. Health 26, 421–434, 2004.

Graham A.M., Aiken G.R., Gilmour C.C. Dissolved organic matter enhances microbial mercury methylation under sulfidic conditions. Environ. Sci. Technol. 46, 2715–2723, 2012.

Gustavs L, Eggert A, Michalik D, Karsten U. Physiological and biochemical responses of green microalgae from different habitats to osmotic and matric stress. Protoplasma. 243: 3–14, 2009.

He Z., Siripornadulsil S., Sayre R.T., Traina S.J., Weavers L. Removal of mercury from sediment by ultrasound combined with biomass (transgenic *Chlamydomonas reinhardtii*). Chemosphere 83, 1249–1254, 2011.

Holmes P., James K., Levy L. Is low-level environmental mercury exposure of concern to human health? Sci. Total Environ. 408, 171–182, 2009.

Humphries M. Rare earth elements: The global supply chain. Congressional Research Service, 2012.

Hutchison A.R., Atwood D.A. Mercury pollution and remediation: the chemist's response to a global crisis. J. Chem. Crystallogr. 33, 631–645, 2003.

Jan A.T., Murtaza I., Ali A., Haq Q.M.R. Mercury pollution: an emerging problem and potential bacterial remediation strategies. World J. Microbiol. Biotechnol. 25, 1529–1537, 2009.

Kabata-Pendias A. Trace elements in soils and plants: CRC Press, 2010.

Karami N., Clemente R., Moreno-Jiménez E., Lepp N.W., Beesley L. Efficiency of green waste compost and biochar soil amendments for reducing lead and copper mobility and uptake to ryegrass. J. Hazard. Mater. 191, 41–48, 2011.

Kelly C., Rudd J.W., Holoka M. Effect of pH on mercury uptake by an aquatic bacterium: implications for Hg cycling. Environ. Sci. Technol. 37, 2941–2946, 2003.

King J.K., Harmon S.M., Fu T.T., Gladden J.B. Mercury removal, methylmercury formation, and sulfate-reducing bacteria profiles in wetland mesocosms. Chemosphere 46, 859–870, 2002.

Kotala S., Kawuri R., Gunam I.B.W. The presence of mercury resistant bacteria in sediment of gold processing plant at Waekerta village of Buru district, Maluku province and their activity in reducing mercury. Curr. World Environ. 9, 271–278, 2014.

Kotrba P., Pospisil P., de Lorenzo V., Ruml T. Enhanced metallosorption of *Escherichia coli* cells due to surface display of β- and α-domains of mammalian metallothionein as a fusion to LamB protein. J. Recept. Signal Transduct. 19, 703–715, 1999.

Krämer U. Phytoremediation: novel approaches to cleaning up polluted soils. Curr. Opin. Biotechnol. 16, 133–141, 2005.

Krämer U., Chardonnens A. The use of transgenic plants in the bioremediation of soils contaminated with trace elements. Appl. Microbiol. Biotechnol. 55, 661–672, 2001.

Krewulak K.D., Vogel H.J. Structural biology of bacterial iron uptake. Biochim. Biophys. Acta (BBA)-Biomembranes 1778, 1781–1804, 2008.

Kumpiene J., Lagerkvist A., Maurice C. Stabilization of As, Cr, Cu, Pb and Zn in soil using amendments—a review. Waste Manage. 28, 215–225, 2008.

Kurland L.T., Faro S.N., Siedler H. Minamata disease. The outbreak of a neurologic disorder in Minamata, Japan, and its relationship to the ingestion of seafood contaminated by mercuric compounds. World Neurol. 1, 370–395, 1960.

Lefebvre D.D., Kelly D., Budd K. Biotransformation of Hg(II) by cyanobacteria. Appl. Environ. Microbiol. 73, 243–249, 2007.

Li P., Feng X., Qiu G., Shang L., Li Z. Mercury pollution in Asia: a review of the contaminated sites. J. Hazard. Mater. 168, 591–601, 2009.

Li X., Wasila H., Liu L., Yuan T., Gao Z., Zhao B., Ahmad I. Physicochemical characteristics, polyphenol compositions and antioxidant potential of pomegranate juices from 10 Chinese cultivars and the environmental factors analysis. Food Chem. 175, 575–584, 2015.

Lollar B.S. Environmental Geochemistry: Elsevier, 2005.

Mahbub K.R., Krishnan K., Megharaj M., Naidu R. Bioremediation potential of a highly mercury resistant bacterial strain *Sphingobium* SA2 isolated from contaminated soil. Chemosphere 144, 330–337, 2016.

Mahbub K.R., Krishnan K., Naidu R., Megharaj M. Mercury resistance and volatilization by *Pseudoxanthomonas* sp. SE1 isolated from soil. Environ. Technol. Inn. 6, 94–104, 2016.

Mason R.P., Choi A.L., Fitzgerald W.F., Hammerschmidt C.R., Lamborg C.H., Soerensen A.L., Sunderland E.M. Mercury biogeochemical cycling in the ocean and policy implications. Environ. Res. 119, 101–117, 2012.

Mukherjee S., Mukhopadhyay S., Hashim M.A., Sen Gupta B. Contemporary environmental issues of landfill leachate: assessment and remedies. Crit. Rev. Environ. Sci. Technol. 45, 472–590, 2015.

Nance P., Patterson J., Willis A., Foronda N., Dourson, M. Human health risks from mercury exposure from broken compact fluorescent lamps (CFLs). Regul. Toxicol. Pharmacol. 62, 542–552, 2012.

Osborn A.M., Bruce K.D., Strike P., Ritchie D.A. Distribution, diversity and evolution of the bacterial mercury resistance (mer) operon. FEMS Microbiol. Rev. 19, 239–262, 1997.

Overjordet I.B., Kongsrud M.B., Gabrielsen G.W., Berg T., Ruus A., Evenset A., Jenssen B.M. Toxic and essential elements changed in black-legged kittiwakes (*Rissa tridactyla*) during their stay in an Arctic breeding area. Sci. Total Environ. 502, 548–556, 2015.

Palermo M.R. Design considerations for *in situ* capping of contaminated sediments. Water Sci. Technol. 37, 315–321, 1998.

Petruzzelli G., Pedron F., Rosellini I., Barbafieri M. Phytoremediation towards the future: focus on bioavailable contaminants Plant-based remediation processes. Springer, 2013.

Rajkumar M., Ae N., Prasad M.N., Freitas H. Potential of siderophore-producing bacteria for improving heavy metal phytoextraction. Trends Biotechnol. 28, 142–149, 2010.

Randall P.M., Chattopadhyay S. Mercury contaminated sediment sites—an evaluation of remedial options. Environ. Res. 125, 131–149, 2013.

Reduction M. Oxidation by reduced natural organic matter in anoxic environments zheng, wang; Liang, Liyuan; Gu, Baohua. Environ. Sci. Technol. 46, 292–299, 2012.

Richmond M., John M. Co-transduction by a staphylococcal phage of the genes responsible for penicillinase synthesis and resistance to mercury salts. Nature 202, 1360–1361, 1964.

Rodríguez O., Padilla I., Tayibi H., López-Delgado A. Concerns on liquid mercury and mercury-containing wastes: A review of the treatment technologies for the safe storage. J. Environ. Manage. 101, 197–205, 2012.

Sangvanich T., Morry J., Fox C., Ngamcherdtrakul W., Goodyear S., Castro D., Yantasee W. Novel oral detoxification of mercury, cadmium, and lead with thiol-modified nanoporous silica. ACS Appl. Mat. Interf. 6, 5483–5493, 2014.

Selid P.D., Xu H., Collins E.M., Striped Face-Collins M., Zhao J.X. Sensing mercury for biomedical and environmental monitoring. Sensors 9, 5446–5459, 2009.

Sinha A., Khare S.K. Mercury bioremediation by mercury accumulating *Enterobacter* sp. cells and its alginate immobilized application. Biodegradation 23, 25–34, 2012.

Smith T., Pitts K., McGarvey J.A., Summers A.O. Bacterial oxidation of mercury metal vapor, Hg (0). Appl. Environ. Microbiol. 64, 1328–1332, 1998.

Sorkhoh N.A., Ali N., Al-Awadhi H., Dashti N., Al-Mailem D.M., Eliyas M., Radwan S.S. Phytoremediation of mercury in pristine and crude oil contaminated soils: contributions of rhizobacteria and their host plants to mercury removal. Ecotoxicol. Environ. Saf. 73, 1998–2003, 2010.

Streets D.G., Zhang Q., Wu Y. Projections of global mercury emissions in 2050. Environ. Sci. Technol. 43, 2983–2988, 2009.

Summers A.O., Silver S. Mercury resistance in a plasmid-bearing strain of *Escherichia coli*. J. Bacteriol. 112, 1228–1236, 1972.

Svensson M. Mercury immobilization: a requirement for permanent disposal of mercury waste in Sweden, Örebro Universität, 2006.

Tangahu B.V., Sheikh Abdullah S.R., Basri H., Idris M., Anuar N., Mukhlisin M. A review on heavy metals (As, Pb, and Hg) uptake by plants through phytoremediation. Int. J. Chem. Eng. 2011, 1–31, 2011.

Trasande L., Landrigan P.J., Schechter C. Public health and economic consequences of methyl mercury toxicity to the developing brain. Environ. Health Perspect. 113, 590–596, 2005.

Ullah A., Heng S., Munis M.F.H., Fahad S., Yang X. Phytoremediation of heavy metals assisted by plant growth promoting (PGP) bacteria: a review. Environ. Exp. Bot. 117, 28–40, 2015.

UNEP (United Nations Environment Programme) Mercury: Time to Act. Geneva, Switzerland: Division of Technology, Industry and Economics, UNEP Chemicals Branch, 2013a.

UNEP. Global Mercury Assessment 2013: Sources, Emissions, Releases and Environmental Transport. Geneva, Switzerland: UNEP Chemicals Branch, 2013b.

U.S. geological Survey (USGS). Mineral commodity summaries, 2012.

USEPA (US Environmental Protection Agency). Mercury Study Report to Congress. Vol. VI: An Ecological Assessment for Anthropogenic Mercury Emissions in the United States. Office of Research and Development, Washington, DC, USA, 1997.

Violante A., Krishnamurti G., Pigna M. Factors affecting the sorption-desorption of trace elements in soil environments. Biophysico-Chemical Processes of Heavy Metals and Metalloids in Soil Environments. New Jersey, NJ: John Wiley & Sons, 169–213, 2007.

Visioli G., Marmiroli N. The proteomics of heavy metal hyperaccumulation by plants. J. Proteomics 79, 133–145, 2013.

Wagner-Döbler I. Pilot plant for bioremediation of mercury-containing industrial wastewater. Appl. Microbiol. Biotechnol. 62, 124–133, 2003.

Wagner-Döbler I. Bioremediation of mercury: current research and industrial applications: Horizon Scientific Press, 2013.

Wang J., Feng X., Anderson C.W., Wang H., Zheng L., Hu T. Implications of mercury speciation in thiosulfate treated plants. Environ. Sci. Technol. 46, 5361–5368, 2012a.

Wang S., Zhang M., Li B., Xing D., Wang X., Wei C., Jia Y. Comparison of mercury speciation and distribution in the water column and sediments between the algal type zone and the macrophytic type zone in a hypereutrophic lake (Dianchi Lake) in Southwestern China. Sci. Total Environ. 417, 204–213, 2012b.

Wiatrowski H.A., Ward P.M., Barkay T. Novel reduction of mercury (II) by mercury-sensitive dissimilatory metal reducing bacteria. Environ. Sci. Technol. 40, 6690–6696, 2006.

Witters N. Phytoremediation: an alternative remediaton technology and a sustainable marginal land management option. Hasselt University, 2011.

Wu Y., Wang S., Streets D.G., Hao J., Chan M., Jiang J. Trends in anthropogenic mercury emissions in China from 1995 to 2003. Environ. Sci. Technol. 40, 5312–5318, 2006.

Xu J., Bravo A.G., Lagerkvist A., Bertilsson S., Sjöblom R., Kumpiene J. Sources and remediation techniques for mercury contaminated soil. Environ. Int. 74, 42–53, 2015.

Yorifuji T., Tsuda T., Inoue S., Takao S., Harada M. Long-term exposure to methylmercury and psychiatric symptoms in residents of Minamata, Japan. Environ Int. 37, 907–913, 2011.

Zeng X.-X., Tang J.-X., Jiang P., Liu H.-W., Dai Z.-M., Liu X.-D. Isolation, characterization and extraction of mer gene of Hg^{2+} resisting strain D2. Trans. Nonferr. Met. Soc. China 20, 507–512, 2010.

Zhang H., Feng X., Zhu J., Sapkota A., Meng B., Yao H., Larssen T. Selenium in soil inhibits mercury uptake and translocation in rice (*Oryza sativa* L.). Environ. Sci. Technol. 46, 10040–10046, 2012.

Zhang Y., Liu J., Zhou Y., Gong T., Wang J., Ge Y. Enhanced phytoremediation of mixed heavy metal (mercury)-organic pollutants (trichloroethylene) with transgenic alfalfa co-expressing glutathione S-transferase and human P450 2E1. J. Hazard. Mater. 260, 1100–1107, 2013.

Zuber S.L., Newman M.C. Mercury Pollution: A Transdisciplinary Treatment: CRC Press, 2011.

PART IV
FIELD APPLICATIONS

CHAPTER 15

Potential Application of an Indigenous Actinobacterium to Remove Heavy Metal from Sugarcane Vinasse

Verónica Leticia Colin,[1,][*] *Macarena María Rulli,*[2] *Luciana Melisa Del Gobbo*[2] and *María Julia del Rosario Amoroso*[1,2]

1. Introduction

Vinasse is a dark brown acid effluent result of the ethyl alcohol industry which represents a serious problem for this sector due to the large quantities that are produced. In fact, a production volume of 9–14 L per liter of alcohol obtained was reported (España-Gamboa et al., 2012). The physicochemical characteristics of vinasses largely depend on the raw material used in the production process of ethanol, the use of sugarcane being predominant in the northwest of Argentina. Only in the province of Tucumán, a planted area of sugarcane of approximately 260,000 hectares is reported. Ethanol can be produced from sugarcane juice rich in sucrose or from molasses which is a by-product of raw sugar production, or from molasses, a by-product of raw sugar. Interestingly, an almost identical yield of ethanol from fermentable sugars of both feedstock's was reported (Troiani, 2009).

[1] Planta Piloto de Procesos Industriales y Microbiológicos (PROIMI), CONICET, Av. Belgrano y Pje. Caseros, 4000, Tucumán, Argentina.
[2] Facultad de Bioquímica, Química y Farmacia, Universidad Nacional de Tucumán, 4000 Tucumán, Argentina.
[*] Corresponding author: veronicacollin@yahoo.com.ar

Vinasse, resulting from ethanol production, has been classified as a class II residue, not inert and potentially dangerous (Ahmed et al., 2013). Despite their variable composition, raw effluent is mainly composed of water, mineral elements, and a high concentration of organic matter; with a chemical oxygen demand (COD) between 50 and 150 g L^{-1}, and a biological oxygen demand (BOD) about 30–70% of the COD (Pant and Adholeya, 2007).

Final disposition of vinasse in lakes and rivers without prior treatment may cause eutrophication, leading to highly undesirable changes in ecosystem's structure and functions. Emission of disagreeable odors caused by the putrefaction of organic matter is also a widely reported effect (Belhadj et al., 2013). However, the high content of colored compounds including phenols and polyphenols can lead to reduced sunlight penetration in rivers and lakes, thereby reducing the dissolved oxygen concentration and causing hazardous conditions for aquatic life (Prasad et al., 2008). In fact, during November 2011 it was reported that more than 30 tons of fish died as a direct consequence of the large volumes of vinasse discharged into the Rio Hondo Lake, located in the deadline area between the provinces of Santiago del Estero and Tucumán (Rocha, 2012).

High contents of nitrogen and phosphorous are typical characteristics of raw vinasse (Colin et al., 2016). The total nitrogen in wastewater is composed of organic nitrogen, ammonia, nitrite, and nitrate; while usual forms of phosphorus found include orthophosphate (e.g., PO_4^{3-}, HPO_3^{2-}, H_2PO_4, and H_3PO_4), condensed phosphates (pyro-, meta-, and other polyphosphates), and organic phosphate. Potassium is also a predominant element since vinasse accumulates more than 70% of the potassium exported by sugarcane. Besides, variable concentrations of diverse heavy metals were also reported in raw vinasse (Camiloti et al., 2007; Colin et al., 2016). Thus, acidic nature of vinasse could provide a favorable environment for metals to react with water and other elements (Habeeb et al., 2015). However, hypoxic conditions provided by effluents with high organic content may favor the metals' accumulation until extremely toxic levels (Haiyan et al., 2013).

Based on this background, it is assumed that vinasse requires a conditioning treatment prior to its discharge in the environment. Besides, different alternatives for vinasse re-use as a by-product of the alcohol industry have been purposed in order to decrease their volume (Fig. 1). For example, agricultural use of vinasse as irrigation water or composting has been widely assayed because of the high potassium, nitrogen, and phosphorus concentrations that make it particularly attractive as a soil amendment or fertilizer (de Mello Prado et al., 2013). However, there are practical and legal restrictions on potassium content in the irrigation water; and these limits are often exceeded in raw vinasse (Soler da Silva et al., 2013). The presence of heavy metals in vinasses is also a major concern for agriculture when this effluent is used without a prior conditioning (Srivastava and Jain, 2010; Jain

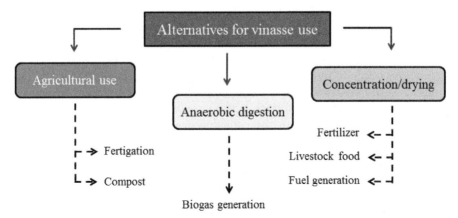

Figure 1. Main alternatives of treatment/re-use of the vinasse.

and Srivastava, 2012; Tchounwou et al., 2012). It is important to remark that small amounts of some metals such as cobalt (Co), copper (Cu), chromium (Cr), iron (Fe), magnesium (Mg), manganese (Mn), molybdenum (Mo), nickel (Ni), selenium (Se), and zinc (Zn) are common in our environments being nutritionally essential for a healthy life (WHO/FAO/IAEA, 1996). However, some metals such as lead (Pb) can show cellular toxicity even at low levels (Saidi, 2010).

Aerobic and anaerobic treatments are usually effective strategies to neutralize the pH of the vinasse, to balance C/N ratio, and to remove a wide range of toxins from it (Juarez Cortes, 2016). With respect to this, anaerobic digestion is a process of high biotechnological value where about 90% of the biodegradable organic matter from vinasse is removed, which implies a similar reduction of their effluence. Additionally, this process has an economic return due to the production of methane biogas which is used as an alternative fuel (Janke et al., 2015) (Fig. 1). Concentration/drying of vinasse is also a conventional practice and may be used as fertilizer or as an additive for feed supplement of ruminant and non-ruminant animals. Total dehydration of vinasse and its subsequent burning in boilers for fuel generation is also a routine practice (Mornadini and Quaia, 2013) (Fig. 1).

A variety of technologies based upon the recovery of heavy metals from industrial wastes via microbial pathways have been evaluated (Wasiullah et al., 2015). Although microorganisms cannot destroy metals, they can alter their chemical properties via surprising array of mechanisms. Because of the wide variety of pollutants found in vinasse, various types of microorganisms could be required for their effective remediation. However, actinobacteria have been reported as biological agents of interest for the bioremediation because they have relatively rapid growth rates and are metabolically versatile

(Ravel et al., 1998). In fact, during the last couple of years, actinobacteria from our collection culture have been highlighted for their ability to remove heavy metals from artificially contaminated liquid systems (Colin et al., 2012; Álvarez et al., 2017). Recently, we began to study the potential of an indigenous strain to remove metals from a regionally important effluent like vinasse. In the next sections, the first advances on this topic are reviewed. Main aspects related to the impact of heavy metals present in vinasse on soils and aquatic environments are also considered.

2. Environmental impact of heavy metals present in vinasse

2.1 Impact on agricultural soils

As distillery vinasse contains all the essential elements for growth, in countries such as India, Brazil, and other Latin American countries, this residue of the sugar-alcohol industries has been traditionally disposed by using it as a fertilizer or disposed as irrigation water for sugarcane and other crops (Mornadini and Quaia, 2013). This can produce short-term benefits because the vinasse contains nutrients such as potassium, magnesium, and calcium which are needed for crops such as sugarcane. However, over a long term, such disposal on the crops without prior treatment of the vinasse can cause severe deterioration of soil and ground water due to the high content of organic material, heavy metals' excessive levels, etc. As mentioned before, certain metals can inhibit diverse biological processes, even in low concentrations (Colussi et al., 2009; Sa'idi, 2010). Even more, some heavy metals can enhance the toxicity of others by synergistic effects (Colussi et al., 2009). Thereby, dose of vinasse applied in agriculture must follow certain guidelines, which vary according to the characteristics of each soil. There are specific recommendations for each region that must be followed in order to prevent excessive use and consequent mineral lixiviation, and contamination of subterranean waters.

Diverse biomarkers can be used to predict the biological responses and the future harms involving hazardous agents present in industrial wastes such as heavy metals. In the field of environmental mutagenesis, assays of chromosome aberrations in plants have been reported as one of the simplest and least expensive biomarkers (Abdel Migid et al., 2010). Interestingly, chromosome aberrations in mitotic cells reflect a rapid response of organisms against environmental toxins, providing early warning signs. With respect to this, Srivastava and Jain (2010) evaluated the effects of either crude or digested vinasse used as irrigation water on *de novo* cytomorphological changes in sugarcane. As expected, higher concentrations of K, P, S, Fe, Mn, Zn and Cu, and heavy metals were detected in crude vinasse as compared to the digested one. Root meristem assay of settlings grown showed higher number of chromosomal abnormalities with both vinasses as compared to those under control conditions. Mitotic index of root-meristems for different sugarcane genotypes showed a decrease that ranged 62.65 to 100% for treatment with crude vinasse while this index varied from 36.94 to 90.33% for treatments with digested vinasse. Inhibitory effects on the bud sprouting and settling height was also associated with the application of crude and digested vinasse. However, these authors noted an improvement in the mitotic activity of root-meristems treated with diluted vinasse in water (1:5 v/v).

A more recent study reported the nutrient composition of crude and digested vinasse as well as their effects on growth and biochemical attributes of the sugarcane (Jain and Srivastava, 2012). Higher content of essential nutrients and heavy metals such as Cd, Cr, Ni, and Pb, were present in crude vinasse compared to the digested effluent. These authors detected a stimulatory effect of low wash rate with crude vinasse (5 ml kg^{-1} soil) on root and shoot growth of sugarcane; while higher dose (100 ml kg^{-1} soil) of both crude and digested vinasse caused inhibitory effect. Therefore, they concluded that a judicious application of vinasse could improve crop productivity, while the environmental pollution problems would be partially resolved.

In our research group, the effects of raw and treated vinasse with the actinobacterium *Streptomyces* sp. MC1 on the attributes of *Lactuca sativa* seedlings were recently evaluated (Colin et al., 2016). We noted a significant reduction in the content of certain metals present in the effluent after being subjected to the microbial action. Interestingly, application of metal-free treated vinasse increased the germinated seed number and improved the development characteristics of the seedlings with respect to the use of raw effluent (Fig. 2) (Juárez Cortes, 2016). These findings could be attributed to the reduction of heavy metals' levels and other toxins by microbial action, enhancing the effluent quality.

Figure 2. Germination and development characteristics of *Lactuca sativa* subjected to the effects of the running water—(A) raw vinasse and (B) treated vinasse with the actinobacterium *Streptomyces* sp. MC1 (C) (Modified from Juárez Cortes, Undergraduate thesis, National University of Tucumán, 2016).

2.2 Impact on aquatic life

The vinasses' disposal into the water bodies without prior treatment affects the quality of aquatic life. The effect depends on the dose of vinasses discharged. Although some countries have made stringent amendments, many countries still have a long way to go about it. If the concentration of the metals in industrial wastes is not within permissible limits, they can dissolve in water and can be easily absorbed by fish and other aquatic organisms. Metal toxicity produces adverse biological effects on survival, activity, growth, metabolism, and/or reproduction of aquatic organisms; thereby they can be lethal or harm the organism without killing it directly (Solomon, 2008). There are several pathways by which each organism can be exposed to the metals' action. One of them is by diffusion in the bloodstream via the gills and skin of the fishes. Besides, animal or plants can be exposed by drinking water or eating sediments which have been contaminated with metal.

Marinho et al. (2014) conducted bioassays on the fish *Oreochromis niloticus* to evaluate the toxicity of different dilutions of vinasse, containing metals such as Cu, Cd, Cr, Pb, Ni, Zn. In the histological analysis, the groups treated with vinasse exhibited significant cytoplasmic alterations, with loss of cell limit and tissue disorganization. A recent study investigated the spatial variability of diverse elements as a consequence of the vinasse disposal into the main canal that runs through a sugar factory located in Guayana (Clementson et al., 2016). Samples collected from four locations along the canal and at five different time periods were analyzed in order to determine the physical and chemical parameters including metals such as Mg, Al, Mn, Cu, Cd, Cr, Co, Ni, Pb, and Zn. The authors concluded that continuous disposal of vinasse in the waterway could be detrimental to aquatic life, due to the depletion of the oxygen supply and the accumulation of heavy metals at toxic levels along the entire length of the canal.

Several studies demonstrated that the bioaccumulation of heavy metals carried by industrial wastes can affect fish histopathology, which is a sensitive biomarker of the overall health and ecology of water bodies. It has been reported, for instance, that the essential metals such as Cu, Zn, and Fe are commonly accumulated in the fish's liver, while Pb and Mn can exhibit their highest concentrations in gills (El-Moselhy et al., 2014). In the same study, it was reported that the accumulation pattern of metals such as Cd can differ among species with the highest concentrations varying between the liver and gills. Other studies conducted on diverse fish species also revealed histopathological effects as a consequence of the heavy metals exposure. Besirovic et al. (2010) performed an exposure study of brown trout to the Cd and Zn effects and showed that the tissues in kidney and gills were the most affected. Effect of heavy metals mixture on fishes from Dam Lake (Bulgaria) used as a model, revealed alterations in tissues in gills and liver (Velcheva et al., 2010). However, Osman et al. (2010) reported that both liver tissues as enzymatic activities of a fish species from Nyle River (Egypt) were susceptible to the effects of diverse metals including Pb, Cd, Zn, Cu, Cr, Fe, Hg, and Mn. A study of the exposure of copper sulfate conducted by Wani et al. (2011) revealed histopathological alterations in African catfish's multiple tissues; while Cr effects on *Labeo rohita* fish used as a model species, produced histopathological changes in specific tissues of gills and liver (Muthukumaravel and Rajaraman, 2013). Multiple tissues of fishes recollected from Kor River (Iran), including liver, kidney, muscle, gonad, and brain were highly susceptible to the Hg, As, Cd, and Pb action (Ebrahimi and Taherianfard, 2011), while fish species from Indus River (Pakistan) showed histopathological alterations in tissues in gills and liver after Mn, Pb, Cu, Zn, Hg, and Cr exposure (Jabeen and Chaudhry, 2013). However, histopathological alteration in fish species from Yamuna River (India) after their exposure to Cr, Ni, and Pb was reported (Fatima and Usmani, 2013). Based on this background, it is concluded that the specific occurrence of heavy metals on fishes could vary depending on diverse factors such as their species, age and development state.

3. Heavy metals removal from vinasse using an indigenous actinobacterium

In order to avoid and/or attenuate the detrimental effects of heavy metals on both soil and aquatic life, different ways of treating industrial wastes are being developed. A variety of physicochemical methods are available for metal recovery, some of them being: electrochemical treatments, ion exchange, precipitation, osmosis, evaporation, and sorption (Fu and Wang, 2011; Aman et al., 2015). Selection of the more adequate method depends on wastewater type, depends on wastewater type. In this connection, the precipitation of metals in forms of hydroxides is considered as an optimal method (Zabochnicka-Świątek and Krzywonos, 2014). However, the presence of high content of organic compounds, as is the case of vinasse, can have negative effect on the efficiency of the metal chemical precipitation. In addition, many of the physicochemical methods have significant disadvantages such as incomplete removal of pollutants and high-energy requirements, many of them not being environment-friendly as well.

The use of microorganisms to remove heavy metals is useful even when they are present in very low concentrations. As shown in Fig. 3, two metal-removal processes have been identified: biosorption and bioaccumulation (Chojnacka, 2009). These methods differ in terms of the mechanisms involved in binding to the metal. Biosorption is usually a passive process that allows for binding contaminants mostly on the surface of the microbial cell wall. In this process, both living and dead biomass can be used. On the contrary, bioaccumulation is an active process where the metals are transported into microbial cells. This process occurs when microbial cells are alive because removal of metals requires their metabolic activity. It is interesting to

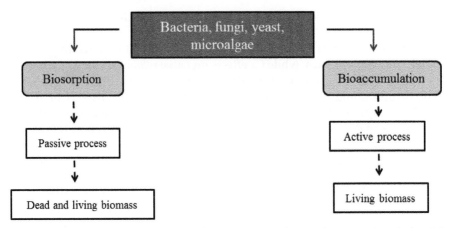

Figure 3. Processes involved in the use of microorganisms for metal recovery from industrial wastes.

highlight that both techniques offer their advantages on physicochemical methods such as a low operating cost and high efficiency for removing metals. However, the high concentration of heavy metals in wastewater can be toxic for living microorganisms. In fact, heavy metals tend to adhere to the bacteria and produce their complexation and death (Sa'idi, 2010). Therefore, the selection of heavy metal-resistant microorganisms would be the first step in order to ensure the success of the microbial remediation (Colin et al., 2012).

The phylum Actinobacteria is one of the most diverse phyla within the domain of Bacteria. Although this phylum encompasses six classes, 19 orders, 50 families, and 221 genera, new taxa are being continuously discovered (Goodfellow et al., 2012). They exhibit a cosmopolitan distribution since their members are distributed in varied ecosystems, both aquatic and terrestrial. In the environment, they play relevant ecological roles including recycling of several substances.

The Pilot Plant of Industrial and Microbiological Processes (PROIMI) has a collection of more than 50 taxa, confirming the placement of most of these strains within the class of Actinobacteria. During the last years, our research group has provided extensive information on the ability of these strains to remove pollutants, majorly heavy metals, from diverse systems (Colin et al., 2012; Álvarez et al., 2017). Actinobacterium *Streptomyces* sp. MC1 isolated from sugarcane grown on a polluted area of the province of Tucuman (Polti et al., 2007), is till date, one of the most studied strains of our collection. Mostly studies provide significant advances regarding the Cr(VI)-removal mechanisms operating in this strain (Polti et al., 2007, 2009, 2010, 2011a,b, 2014). However, these researches are limited to a sole metal, always using artificially-contaminated systems. In order to evaluate the performance of this strain to remove diverse heavy metals from an actual effluent, the first assays using sugarcane vinasse were conducted (Colin et al., 2016). Presence of metals in a sample of the effluent (Cr, Mn, Fe, Cu, Zn, As, Cd, Hg, and Pb) was determined using inductively coupled plasma mass spectrometry, before and after bacterial growth. A strong decline in Mn and Fe levels in raw vinasse was observed after 4 days of treatment, with reduction percentage of 95% and 62% respectively. Metals such as Zn, As, and Pb were also detected which were completely removed by microbial action towards the end of culture period (Colin et al., 2016).

As mentioned throughout of the current chapter, irrigation with vinasse is a promising practice since its return to the soil could be desirable and productive. This practice has been widely accepted in countries such as Brazil, which is the main producer of fuel alcohol, as a result of the pro-alcohol program that generates very large vinasse volume per harvest (Goldemberg et al., 2008). In our region (Tucumán, Argentina) the agricultural use of vinasse is also recognized as a sustainable practice being regulated by Resolution N°030 of the Secretary of State for the Environment. However, due to practical and legal restrictions on content of certain elements in the irrigation water, vinasse has been used in diluted form with water (1:10 to 1:30, v/v). This practice has an important disadvantage since water resources are consumed, increasing so the handled volume (Colin et al., 2016). As an alternative, a significant reduction of the organic matter, heavy metals, and other toxins could be achieved via microbial pathways.

Based upon the presumption that microbiological treatment could enhance the agricultural quality of vinasse, Colin et al. (2016) evaluated the effects of raw and treated effluent with *Streptomyces* sp. MC1 for 4 days on growth parameters of *Lactuca sativa*. Ecotoxicological tests using plants are relatively recent as compared to those with aquatic species. They are also an important tool to assess the toxicity of chemical substances toward different species. So, after 5 days of exposure of the *Lactuca sativa* seeds to the two effluents as well as to running water (used as control), the number of germinated seeds (G), hypocotyl length (HL), root length (RL), and the vigor index (VI) were determined (Fig. 4). Under the three assayed conditions, G parameter was similar, and not significant differences could be found (data not shown). Relevant differences could also not be found in RL parameter related to raw or treated vinasse exposure (Fig. 4A). However, the authors noted a significant increase in HL and VI parameters of the seedlings exposed to treated vinasse compared to seedlings exposed to the raw effluent (Fig. 4A). This positive effect could be attributed to the reduction of the concentration of heavy metals and other toxins by microbial action. It is important to remark that the exposure of seeds to the treated microbiologically vinasse did not improve the growth parameters when it was compared with the running water exposure (control condition) (Fig. 4B). These findings suggest the presence of some toxins even after growth of *Streptomyces* sp. MC1, which

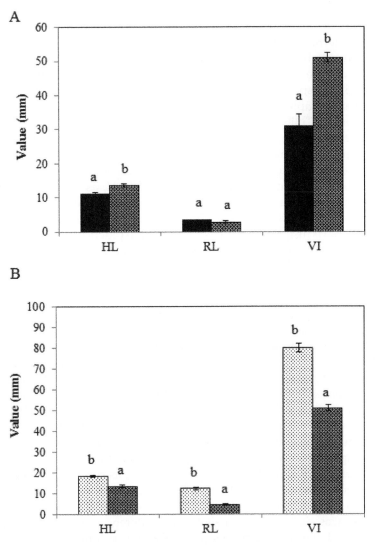

Figure 4. Growth parameters of *Lactuca sativa* determined after 5 days of incubation in the presence of raw vinasse (■), vinasse subjected to the microbiological treatments with *Streptomyces* sp. MC1 (▨), and running water (▦).

would inhibit the normal development of the seedlings. Based upon these findings, it was concluded that the microbial treatment of vinasse within a time period as short as 4 days could be useful, but insufficient for effluent's total conditioning. Thus, further studies will be required in order to optimize the microbial treatment-time.

4. Remarked conclusions and perspectives

Although the treatment of vinasse does not usually represent an economic benefit, this practice is closely related to the sustainability of the sugar-alcohol industries. This question is of vital importance in regions such as the province of Tucuman; where concentration of sugar industries combined with autonomous alcohol distilleries generate large vinasse volume, endangering the environmental preservation. Although new alternatives for the integral use of the vinasse are continually being evaluated in order to mitigate their detrimental effects, none of them solves pollution problem alone, and two or more of these alternatives are commonly combined to improve the management of this waste. Irrigation with vinasses has been recognized as a routine practice in many agricultural countries. However, this practice is controversial because of some negative effects. Therefore, experts in this matter advise carrying out conditioning treatments of vinasse prior to their disposal in agricultural soil. The effluents recovery via microbial pathways is a feasible approach because high concentration of biodegradable organic carbon. In addition, heavy metals' removal by microbial action can be effective even when they are present in extremely low concentrations where conventional physicochemical methods fail to operate. Studies reviewed here emphasize on the potential of an indigenous actinobacterium to remove heavy metals from sugarcane vinasse sample in a short time period. Toxicity mitigation of the treated effluent compared to crude vinasse was reflected in the significant increase in the vigor index of *Lactuca sativa*. It is important to remark that the response to toxicity testing is highly dependent on the organism used as bioindicator. Thereby, a variety of crops of regional interest could be used as potential bioindicators to evaluate the agricultural quality of effluent subjected to the microbial action.

Findings represented in this chapter, are the first advances in the recovery and re-evaluation of an actual effluent by using *Streptomyces* sp. MC1 from our collection of cultures. However, further studies will be required to evaluate the effectiveness of the microbial treatment. Elucidating the detoxification mechanisms operating in this strain during the heavy metals' removal from vinasse is also a desirable aim which is an ongoing study.

References

Abdel Migid H.M., Haroun S.A., Abdelwahed A.A., Younis H.N. Detection of genetic damage induced by pesticides using cytogenetic and biomarkers assays in *Allium* and *Pisum*. Egypt. J. Genet. Cytol. 39, 271–289, 2010.

Ahmed O., Sulieman A.M.E., Elhardallou S.B. Physicochemical, chemical and microbiological characteristics of vinasse, a by-product from ethanol industry. Amer. J. Biochem. 3, 80–83, 2013.

Alvarez A., Saez J.M., Dávila Costa J.S., Colin V.L., Fuentes M.S., Cuozzo S.A., Benimeli C.S., Polti M.A., Amoroso M.J. Actinobacteria: current research and perspectives for bioremediation of pesticides and heavy metals. Chemosphere 166, 41–62, 2017.

Aman A., Saraswathi M., Viswanath B. The trends in removal of heavy metals contaminants from the waste water by using biomaterials. I.J.D.R. 5, 4660–4669, 2015.

Belhadj S., Karouach F., El Bari H., Joute Y. The biogas production from mesophilic anaerobic digestion of vinasse. IOSR–JESTFT 5, 72–77, 2013.

Besirovic H., Alic A., Prasovic S., Drommer W. Histopathological effects of chronic exposure to cadmiun and zinc on kidneys and gills of Brown trout (*Salmo trutta m. fario*). Turk. J. Fish. Aquat. Sci. 10, 255–262, 2010.

Camiloti F., Marques M.O., Andrioli I., Da Silva A.R., Tasso Junior L.C., de Nobile F.O. Heavy metals accumulation in sugarcane after application in sewage sludge and vinasse. Eng. Agríc. Jaboticabal 27, 284–293, 2007.

Chojnacka K. Biosorption and bioaccumulation in practice, Nova Science Publishers Inc., New York, 2009.

Clementson C., Abrahim B.N., Homenauth O. An investigation of the spatial variability of elements due to vinasse disposal in waterways at the albion bioethanol plant, Berbice Guyana. G.J.E.M.P.S. 5. 74–87, 2016.

Colin V.L., Villegas L.B., Abate C.M. Indigenous microorganisms as potential bioremediators for environments contaminated with heavy metals. Int. Biodeter. Biodegr. 69, 28–37, 2012.

Colin V.L., Juárez Cortes A.A., Aparicio J.D., Amoroso M.J. Potential application of a bioemulsifier-producing actinobacterium for treatment of vinasse. Chemosphere 144, 842–847, 2016.

Colussi I., Cortesi A.L., Vedova D.V., Gallo V., Robles F.K.C. Start-up procedures and analysis of heavy metals inhibition on methanogenic activity in EGSB reactor. Bioresour. Technol. 100, 6290–6294, 2009.

de Mello Prado R., Caione G., Silva Campos C.N. Filter cake and vinasse as fertilizers contributing to conservation agriculture. Appl. Environ. Soil Sci. 1–8, 2013.

Ebrahimi M., Taherianfard M. The effects of heavy metals exposure on reproductive systems of cyprinid fish from Kor River. Iran. J. Fish. Sci. 10, 13–24, 2011.

El-Moselhy Kh. M., Othman A.I., Abd El-Azem H., El-Metwally M.E.A. Bioaccumulation of heavy metals in some tissues of fish in the Red Sea. Turk. J. Fish. Aquat. Sci. 1, 97–105, 2014.

España-Gamboa E., Mijangos-Cortés J.O., Hernández-Zárate G., Domínguez Maldonado J.A., Alzate-Gaviria L.M. Methane production by treating vinasses from hydrous ethanol using a modified UASB reactor. Biotechnol. Biofuels 5, 82–91, 2012.

Fatima M., Usmani N. Histopathology and bioaccumulation of heavy metals (Cr, Ni and Pb) in fish (Channa striatus and Heteropneustes fossilis) tissue: A study for toxicity and ecological impacts. Pak. J. Biol. Sci. 16, 412–420, 2013.

Fu F., Wang Q. Removal of heavy metal ions from wastewaters: A review. J. Environ. Manage. 92, 407–418, 2011.

Goldemberg J., Teixeira Coelho S., Guardabassi P. The sustainability of ethanol production from sugarcane. Energy Policy 36, 2086–2097, 2008.

Goodfellow M., Kämpfer P., Busse H.-J., Trujillo M.E., Suzuki K.-I., et al. Bergey's Manual of Systematic Bacteriology, Springer, New York, 2012.

Habeeb M., Al-Berman A.-K., Salman J. Environmental study of water quality and some heavy metals in water, sediments and aquatic macropytas in lotic ecosystem, Iraq. Mesopotamia. Environ. J. 1, 66–84, 2015.

Haiyan L., Shi A., Li M., Zhang X. Effect of pH, temperature, dissolved oxygen, and flow rate of overlying water on heavy metals release from storm sewer sediments. J. Chem. 1–11, 2013.

Jabeen F., Chaudhry A.S. Metal uptake and histological changes in gills and liver of *Oreochromis mossambicus* inhabiting Indus River. Pak. J. Zool. 45, 9–18, 2013.

Jain R., Srivastava S. Nutrient composition of spent wash and its impact on sugarcane growth and biochemical attributes. Physiol. Mol. Biol. Plants 18, 95–99, 2012.

Janke L., Leite A., Nikolausz M., Schmidt T., Liebetrau J., Nelles M., Stinner W. Biogas production from sugarcane waste: assessment on kinetic challenges for process designing. Int. J. Mol. Sci. 16, 20685–20703, 2015.

Juárez Cortes A.A. Aplicación de procesos microbiológicos para el tratamiento de vinaza y la producción de bioemulsificantes por *Streptomyces* sp. MC1. Undergraduate thesis, Faculty of Biochemistry, Chemistry and Pharmacy, National University of Tucumán, Argentina, 2016.

Marinho J.F., Correia J.E., Marcato A.C., Pedro-Escher J., Fontanetti C.S. Sugar cane vinasse in water bodies: impact assessed by liver histopathology in tilapia. Ecotoxicol. Environ. Saf. 110, 239–245, 2014.

Mornadini M., Quaia E. Alternativas para el aprovechamiento de la vinaza como subproducto de la actividad sucroalcoholera. Avance Agroindustrial 34, 1–12, 2013.

Muthukumaravel K., Rajaraman P. A study on the toxicity of chromium on the histology of gill and liver of freshwater fish *Labeo rohita*. J. Pure Appl. Zool. 1, 122–126, 2013.

Osman A.G.M., Abd El reheem A.E.B.M., Abuel Fadl K.Y., El-Rab A.G.G. Enzymatic and histopathologic biomarkers as indicators of aquatic pollution in fishes. Nat. Sci. 2, 1302–1311, 2010.

Pant D., Adholeya A. Biological approaches for treatment of distillery wastewater: a review. Bioresour. Technol. 98, 2321–2334, 2007.

Polti M.A., Amoroso M.J., Abate C.M. Intracellular chromium accumulation by *Streptomyces* sp. MC1. Water Air Soil Pollut. 214, 49–57, 2011a.

Polti M.A., Amoroso M.J., Abate C.M. Chromate reductase activity in *Streptomyces* sp. MC1. J. Gen. Appl. Microbiol. 56, 11–18, 2010.

Polti M.A., Amoroso M.J., Abate C.M. Chromium (VI) resistance and removal by actinomycete strains isolated from sediments. Chemosphere 67, 660–667, 2007.

Polti M.A., Aparicio J.D., Benimeli C.S., Amoroso M.J. Simultaneous bioremediation of Cr(VI) and lindane in soil by Actinobacteria. Int. Biodeterior. Biodegrad. 88, 48–55, 2014.

Polti M.A., Atjian M.C., Amoroso M.J., Abate C.M. Soil chromium bioremediation: synergic activity of actinobacteria and plants. Int. Biodeterior. Biodegrad. 65, 1175–1181, 2011b.

Polti M.A., García R.O., Amoroso M.J., Abate C.M. Bioremediation of chromium (VI) contaminated soil by *Streptomyces* sp. MC1. J. Basic Microbiol. 49, 285–292, 2009.

Prasad K.R., Kumar R.R., Srivastava S.N. Design of optimal response surface experiments for electro-coagulation of distillery spend wash. Water Air Soil Poll. 191, 5–13, 2008.

Ravel J., Amoroso M.J., Colwell R.R., Hill R.T. Mercury resistant actinomycetes from Chesapeake Bay, FEMS Microbiol. Lett. 162, 177–184, 1998.

Rocha L. Rio Hondo, amenazado por vinaza, La Nación Ecológico, Noticias ambientales, 2012.

Sa'idi M. Experimental studies on effect of heavy metals presence in industrial wastewater on biological treatment. Int. J. Environ. Sci. 1, 666–676, 2010.

Soler da Silva M.A., Kliemann H.J., De-Campos A.B., Madari B.E., Borges J.D., Gonçalves J.M. Effects of vinasse irrigation on effluent ionic concentration in Brazilian Oxisols. Afr. J. Agric. Res. 8, 5664–5672, 2013.

Solomon F. Impacts of metals on aquatic ecosystems and human health. Environ. Communities 14–19, 2008.

Srivastava S., Jain R. Effect of distillery spent wash on cytomorphological behaviour of sugarcane settlings. J. Environ. Biol. 31, 809–812, 2010.

Tchounwou P.B., Yedjou C.G., Patlolla A.K., Sutton D.J. Heavy metals toxicity and the environment. Experimental Supplementum 101, 133–164, 2012.

Troiani E. Distillery Efficiency on Molasses and Cane Juice personal communication, A R Gopal, Berkeley, CA, 2009.

Velcheva I., Tomova E., Arnaudova D., Arnaudov A. Morphological investigation on gills and liver of freshwater fish drom Dam Lake Studen Kladenets. Bulg. J. Agric. Sci. 16, 364–368, 2010.

Wani A.A., Sikdar-Bar M., Borana K., Khan H.A., Andrabi S.S.M., Pervaiz P.A. Histopathological alterations induced in gill epitelium of african Catfish, *Clarias gariepinus*, exposed to cooper sulfate. Asian J. Exp. Biol. Sci. 2, 278–282, 2011.

Wasiullah R.D., Malaviya D., Pandiyan K., Singh U.B., Sahu A., Shukla R., Singh B.P., Rai J.P., Sharma P.K., Lade H., Paul D. Bioremediation of heavy metals from soil and aquatic environment: an overview of principles and criteria of fundamental processes. Sustainability 7, 2189–2212, 2015.

WHO/FAO/IAEA.World Health Organization. Switzerland: Geneva; 1996. Trace Elements in Human Nutrition and Health.

Zabochnicka-Świątek M., Krzywonos M. Potentials of biosorption and bioaccumulation processes for heavy metal removal. Pol. J. Environ. Stud. 23, 551–561, 2014.

CHAPTER 16

Bioremediation of Polluted Soils in Uranium Deposits

Stoyan Groudev,* Plamen Georgiev, Irena Spasova and
Marina Nicolova

1. Introduction

The pollution of soils and waters by radionuclides and toxic heavy metals is
a serious environmental problem in many countries, especially in those with
intensive industrial development and/or with a large-scale recovery of such
elements from the relevant mineral deposits. The pollution is due to different
mechanisms, some of which are acting under natural conditions but others
are directly connected with the human activity. In some cases, this activity
is connected with the recovery of valuable components from the relevant
natural sources, mainly the ore deposits, but in some cases the recovery is
connected with the processing of some mineral wastes or even of industrial
products such as metal-bearing concentrates. Since a relatively long period
of time, some of these technologies have been connected with the application
of different microorganisms which are able to solubilize or to precipitate
the relevant metals under suitable conditions. Apart from the recovery of
different metals, some microorganisms are used to prevent pollution or even
to participate in the cleaning of ecosystems polluted by toxic elements.

In Bulgaria, for a long period of time, uranium was leached commercially
in a large number of deposits using mainly different *in situ* technologies. Most

University of Mining and Geology, Sofia 1700, Bulgaria.
* Corresponding author: groudev@mgu.bg

of these commercial-scale operations were connected with the acid leaching of uranium due to the presence of pyrite and the negative net neutralization potential of the relevant uranium ores. In some of these operations, the leaching was connected with the action of some acidophilic chemolithtrophic bacteria which were able to oxidize the tetravalent uranium to the soluble hexavalent form and to generate sulfuric acid and ferric ions by the oxidation of pyrite present in such deposits (eq. 1–4):

$$U(IV)O_2 + 0.5\ O_2 + H_2SO_4 \rightarrow bacteria \rightarrow \quad U(VI)O_2SO_4 + H_2O \tag{1}$$

$$4\ FeS_2 + 15\ O_2 + 2\ H_2O \rightarrow bacteria \rightarrow \quad 2\ Fe_2(SO_4)_3 + 2\ H_2SO_4 \tag{2}$$

$$U(IV)O_2 + Fe_2(SO_4)_3 \rightarrow \quad U(VI)O_2SO_4 + 2\ FeSO_4 \tag{3}$$

$$FeSO_4 + 15\ O_2 + 2\ H_2SO_4 \rightarrow bacteria \rightarrow \quad 2\ Fe_2(SO_4)_3 + H_2O \tag{4}$$

Several years ago, all commercial-scale operations for uranium leaching in the country were stopped due to a complex of different political, economical, and environmental reasons. Regardless of some preventive and remedial actions during the uranium recovery, many natural ecosystems were heavily polluted with radioactive elements and several toxic metals, mainly through the seepage of acid drainage waters. Soils around the water flowpath were polluted with these toxic elements and some of them are still unsuitable for agriculture use. It is known that different methods for assessment and remediation of soils contaminated with uranium and toxic heavy metals are available (Knox et al., 2008; Park et al., 2011; Malavija and Sing, 2012; Clean et al., 2013; Sing et al., 2014; Romero et al., 2016; Groudev et al., 2001, 2008, 2010, 2014). However, only few of them have been applied under real large-scale conditions. The excavation and transportation of the heavily polluted soils to specific depositories is still a common practice in most countries. In some cases, the disposal is followed by off-site treatment of the relevant soils. The *in situ* monitored natural attenuation or passive capping using the installation of clean, inert material over the contaminated soil is also largely applied. The application of methods for remediation *in situ* of soils contaminated with toxic elements (such as heavy metals, uranium, and arsenic) is still limited but can be very attractive especially from the economical point of view. The *in situ* leaching is connected with solubilization of uranium and heavy metals by means of different chemical lixiviants (bicarbonate, mineral acids, and some organic complexing agents) or by means of different microorganisms, mainly acidophilic chemolithotrophic bacteria are able to oxidize the insoluble tetravalent uranium to the soluble hexavalent form. The bacterial leaching is especially efficient in the cases when the metals are present in the form of relevant sulfide minerals and in some cases the soil remediation can be connected with the recovery of the dissolved metals from the relevant pregnant leach solutions.

Some *in situ* bioremediation methods are connected with the immobilization of uranium and heavy metals inside the soils by converting them into their least soluble or toxic forms or by encapsulation in solid products of high structural integrity (Mulligan et al., 2001). The anaerobic sulfate-reducing bacteria are especially effective in this respect since in the presence of suitable electron donors, mainly different biodegradable organic compounds and also hydrogen, they are able to, by the process of dissimilatory sulfate reduction, precipitate the dissolved heavy metals as relevant insoluble sulfides and the dissolved hexavalent uranium as the insoluble tetravalent form.

Another group of *in situ* bioremediation methods is based on the ability of some plants to accumulate uranium and heavy metals from contaminated soils via their root systems (Hinchee et al., 1995). Considerable portions of the bioaccumulated contaminants are then transported to the plant biomass located above ground. The biomass is periodically removed and burned to ashes. These ashes are suitable for disposal or can be used for recovering some valuable components.

Different variants of the above methods have been applied in several countries, mainly in Northern America and Europe. Some data about the application of such methods for remediation of contaminated soils in two uranium deposits in Bulgaria are shown in this publication.

The uranium deposit Curilo, located in Western Bulgaria, for a long period of time was a site of intensive mining activities including both open-pit and underground techniques as well as *in situ* leaching of uranium. The mining operations were ended in 1990 but until some years ago both the surface and ground waters and soils within and near the deposit were heavily contaminated with radionuclides (mainly uranium and radium) and toxic heavy metals (mainly copper, zinc, and cadmium). Some soil plots located in this area were treated *in situ* by means of methods based on the activity of the indigenous soil microflora. This activity was enhanced by suitable changes in the levels of some essential environmental factors such as pH and water, oxygen, and nutrient contents in the soil.

The polymetallic Rossen deposit is located in the Vromos Bay area, near the Black Sea coast, Southeastern Bulgaria. Some agricultural lands located in this area have been contaminated with radionuclides (mainly uranium and radium) and heavy metals (mainly copper, cadmium and lead) as a result of mining and processing of polymetallic ores. Several *in situ* methods were applied to treat contaminated soils in this area (chemical leaching with different reagents such as bicarbonate, mineral and organic acids, phytoremediation by herbaceous plants able to accumulate some of the above-mentioned contaminants). Enhanced natural attenuation consisting of periodic ploughing and addition of organic substrates and sources of N and P to stimulate the growth and activity of some indigenous microorganisms (mainly sulfate and iron-reducing bacteria) are able to reduce and precipitate uranium as uraninite. To precipitate the non-ferrous metals as the relevant insoluble sulfides is also applied.

Some of these methods (mainly the bicarbonate leaching) were partially efficient and decreased the concentrations of some contaminants and the toxicity of the soil but caused a negative effect on its structure and composition, as well as on its microflora. The best results were achieved by an *in situ* bioremediation method based on the enhanced activity of the indigenous microflora.

The two field operations (in the Curilo deposit and in the Vromos Bay area) are typical examples for *in situ* bioremediation of acidic and alkaline soils respectively.

2. Bioremediation of contaminated soil in the Curilo deposit

Two soil plots located in the Curilo deposit consisted of a leached cinnamonic forest soil and were 180 m² in size each. The soil profile was 100 cm deep (horizon A, 30 cm; horizon B, 50 cm; horizon C, 20 cm) and was underlined by intrusive rocks with a very low permeability. Data about the chemical composition and some essential geo-technical parameters of the soil are shown in Table 1. The contaminants were located mainly in the upper soil layers (in the horizon A) (Groudev et al., 2010).

The soil treatment in both plots was connected with the initial solubilization of contaminants. Water acidified with sulfuric acid to pH in the range of 2.8–3.5 was used as leach solution. Periodically, this solution was supplemented with ammonium and phosphate ions in concentrations sufficient to maintain their concentrations in the soil pore solution in the range of about 20 and 10 mg/L respectively. The irrigation rate and acidity of the leach solutions were adjusted in connection with the levels of the local natural rainfall and temperature to maintain the water-filled porosity in the soil's upper layers (mainly in the horizon A) at about 60%, and the pH of the soil pore solution within the range of about 3.0–3.5.

This level of the soil moisture was optimum for the activity of the aerobic microorganisms inhabiting these soil layers. The pH of the soil pore solution was higher than the optimal values for the acidophilic chemolithotrophic bacteria but was still suitable for their growth and activity. Lower pH values had a negative effect on the soil structure and composition. The upper soil layers were ploughed up periodically to enhance the natural aeration. The contaminants dissolved in the first plot were removed from the soil profile through the soil effluents. Periodically, usually once per 1–3 weeks, higher irrigation rates were applied for flushing these contaminants from the soil profile. The contaminants dissolved in the second plot were transferred to the deeply located soil subhorizon B_2 where they were immobilized as a result of the activity of the indigenous sulfate-reducing bacteria. Water solutions of dissolved organic compounds (lactate and acetate) and ammonium

Table 1. Characteristics of the soil in the plots in the Curilo deposit before and after the treatment.

Parameters	Before treatment	After treatment
Chemical composition (%)		
SiO_2	77.4	78.1
Al_2O_3	12.5	12.0
Fe_2O_3	2.35	1.58
P_2O_5	0.14	0.10
K_2O	2.12	1.70
N total	0.10	0.08
S total	1.72	0.79
S sulfidic	1.54	0.71
Carbonates	0.14	0.01
Humus	2.10	1.41
pH (H_2O)	4.40	3.21
Net neutralization potential (kg $CaCO_3.t^{-1}$)	−44.8	−21.4
Bulk density (g.cm^{-3})	1.32	1.27
Specific density (g.cm^{-3})	2.68	2.62
Porosity (%)	51	46
Permeability (cm.h^{-1})	10.4	8.2
Particle size (mm) (%):		
1.00 – 0.25	18.9	18.1
0.25 – 0.01	49.5	50.9
< 0.01	31.6	31.0

and phosphate ions were injected through vertical boreholes to this soil subhorizon to enhance the bacterial activity.

The flowsheet also included a system to collect the soil effluents and to avoid the migration of contaminants into the environment. The system consisted of several ditches, boreholes and wells located in suitable sites in the plots. The soil effluents collected by this system were then treated by constructed wetlands located near the plots to remove the dissolved contaminants. The wetlands were characterized by a mixed (surface/subsurface) water flowpath and by abundant water, emergent vegetation and a diverse microflora. *Typha latifolia* and *Phragmites australis* were the prevalent plant species in the wetlands but representatives of the genera *Scirpus, Juncus, Eleocharis, Carex,* and *Poa* were also present, as well as several algae.

The leaching of contaminants in the horizon A in the two plots was efficient and within a period of 20 months (including a 2-month pause in irrigation during the cold winter months) their concentrations were decreased below the relevant permissible levels (Table 2). The analysis of microflora in this soil horizon revealed that it was characterized by a rich diversity of microorganisms (Groudev et al., 2010).

The mesophilic acidophilic chemolithotrophic bacteria related to the species *Acidithiobacillus ferrooxidans, At. thiooxidans,* and *Leptospirillum ferrooxidans* were the prevalent microorganisms in the horizon A. These bacteria were able to oxidize the sulfide minerals present in the soil and to solubilize their metal components. The non-ferrous metals (Cu, Zn, and Cd) were solubilized mainly in this way and as the relevant free cations were removed from the soil profile of the first plot by drainage effluents. The hexavalent uranium was readily solubilized by the sulfuric acid present in the irrigating solutions or generated in the soil as a result of the oxidation of sulfides, mainly of pyrite.

Table 2. Contents of contaminants in the horizon A of the soil from the plots in the Curilo deposit.

Parameters	U	Ra	Cu	Zn	Cd
Contents of contaminants (mg.L^{-1})					
- before treatment	68	510	190	215	4.6
- after treatment	8.0	65	35	48	0.4
Permissible levels for soils					
with pH 4.1–5.0	10	65	40	60	1.5
Permissible levels for soils					
with pH < 4.1	10	65	20	30	0.5
Bioavailable fractions (mg.L^{-1})					
(a) by DTPA leaching					
- before treatment	12	105	35	28	0.5
- after treatment	1.4	10	4.1	3.5	0.04
(b) by EDTA leaching					
- before treatment	5.3	41	25	15	0.35
- after treatment	0.7	5.0	1.9	1.7	0.01
Easily leachable fractions (mg.L^{-1})					
- before treatment	19	120	68	62	1.0
- after treatment	1.2	10	3.7	4.8	0.02
Inert fractions (mg.L^{-1})					
- before treatment	5.1	50	21	28	0.35
- after treatment	4.6	45	18	28	0.30

Notes: The contents of radium are shown in Bq/kg dry soil or in Bq/L.

The tetravalent uranium was solubilized as a result of its prior bacterial oxidation to the hexavalent state. The bacterial oxidation of sulfides and tetravalent uranium was carried out by well-known direct and indirect (via the oxidation of Fe^{2+} to Fe^{3+}) mechanisms. The oxidative activity of these chemolithotrophic bacteria and their ability to fix CO_2 as a carbon source depended on some essential environmental factors such as pH and temperature (Table 3), and the availability of substrates, oxygen, and nutrients (mainly sources of N and P). High numbers of bacteria and efficient oxidation rates were achieved during the first 5–6 months of treatment when most of the sulfide sulfur and soluble iron were still present in the soil (in the horizon A) and considerable portions of sulfides were well exposed. The maintenance of concentrations of dissolved oxygen higher than 5–6 $mg.L^{-1}$ in the pore solution (by means of periodic ploughing and control of irrigation) resulted in relatively fast growth and oxidation, while at concentrations lower than 2 $mg.L^{-1}$ these processes were much slower. Concentrations of ammonium and phosphate ions in the range of about 15–25 $mg.L^{-1}$ for each of these ions were needed for the optimum growth and activity of these bacteria. A portion of the complex uranyl sulfate anions formed in the acidic leach solutions was absorbed on the positively

Table 3. Microbial activity *in situ* at different environmental conditions in the plots in the Curilo deposit.

Sample and conditions of testing	Fe^{2+} oxidized for 5 days $(g.L^{-1})$	$^{14}CO_2$ fixed for 5 days, $(counts.min^{-1}.mL^{-1})$ (g)
Soil effluents with pH of 3.8 + Fe^{2+} (10 $g.L^{-1}$) at 9 – 11°C	0.51–1.40	1500–4100
Soil effluents with pH of 3.2 + Fe^{2+} (10 $g.L^{-1}$) at 9 – 11°C	0.91–2.84	2600–7700
Soil effluents with pH of 3.2 + Fe^{2+} (10 $g.L^{-1}$) at 16 – 18°C	1.54–4.21	4400–11,200
Soil effluents with pH of 3.2 + Fe^{2+} (10 $g.L^{-1}$) at 21 – 23°C	1.90–6.44	5000–16,200
Soil suspensions in K nutrient medium (with 10 $g.L^{-1}$ Fe^{2+} and pH 3.8) at 9 – 11°C	0.60–1.61	1500–4400
Soil suspensions in K nutrient medium (with 10 $g.L^{-1}$ Fe^{2+} and pH 3.2) at 9 – 11°C	0.99–3.05	2800–7900
Soil suspensions in K nutrient medium (with 10 $g.L^{-1}$ Fe^{2+} and pH 3.2) at 16 – 18°C	1.45–4.85	4400–12,500
Soil suspensions in K nutrient medium (with 10 $g.L^{-1}$ Fe^{2+} and pH 3.2) at 21 – 23°C	2.05–7.11	5500–18,100

charged surface of the soil minerals, mainly iron oxides and alumosilicates. Regardless of this, a considerable portion of uranium was also removed from the soil profile through the drainage effluents. This was connected with the saturation and passivation of the active sites on the mineral surfaces.

The pregnant soil effluents from the first plot were treated efficiently in the constructed wetland. The dissolved non-ferrous metals and iron were precipitated mainly as the relevant insoluble sulfides by the sulfate-reducing bacteria inhabiting the wetland. Uranium was precipitated mainly as uraninite (UO_2) as a result of the reduction of U^{6+} to U^{4+} carried out also by sulfate-reducing bacteria as well as by some Fe^{3+} reducing bacteria. Portions of iron and manganese were precipitated as $Fe(OH)_3$ and MnO_2 as a result of the prior bacterial oxidation of Fe^{2+} and Mn^{2+} to Fe^{3+} and Mn^{4+} respectively. Radium and portions of the heavy metals and uranium were removed by sorption on the living and dead plant biomass and on the clay minerals present in the wetland.

The contaminants dissolved in the horizon A of the second soil plot were also transferred by the drainage waters to the deeply located soil layers (in the subhorizon B_2) but here they were precipitated as the relevant insoluble forms (sulfides and UO_2) as a result of the enhanced activity of the indigenous sulfate-reducing bacteria. The concentrations of contaminants in the effluents from this plot were lower than the relevant permissible levels for water intended for use in the agriculture and/or industry.

The monitoring of the soil toxicity revealed that it was connected with the concentrations of contaminants in the soil pore solutions and of their bioavailable fractions in the soil. The toxicity during the treatment was initially increased in comparison with that before the treatment. However, after reaching the maximum during the period from the 3rd to the 8th month since the start of the treatment, the toxicity then steadily decreased and at the end of the treatment was considerably lower than that before the treatment (Table 4).

The data from Table 4 revealed that the toxicity of this soil was due not only to the radionuclides and heavy metals but also to high acidity which was further increased during the treatment.

The clean-up of the heavily contaminated soil in another plot in this deposit was connected with an efficient treatment of a small portion of the effluents from the subhorizon B_2 by means of a microbial fuel cell generating electricity (Groudev et al., 2014). It must be noted that the effluents from the horizon B_2 were efficiently treated by means of a constructed wetland located near the soil plot. The initial flowpath of these effluents through the surface aerobic water layer in the wetland was connected with the removal of the dissolved bivalent forms of iron and manganese as a result of their oxidation by different heterotrophic bacteria to Fe^{3+} and Mn^{4+}, followed by precipitation as $Fe(OH)_3$ and MnO_2 respectively. After the aerobic zone of the wetland, the waters passed through a thick (~ 50 cm) soil layer in which the lower part was a typical anaerobic zone. This zone was inhabited by various anaerobic

Table 4. Toxicity of soils from the plots in the Curilo deposit before and after the treatment.

Test-organisms	Toxicity		
	Before treatment (pH 4.4)	After treatment (pH 3.2)	After treatment (pH 4.4 by lime)
Bacillus cereus	40	40	80
Pseudomonas putida	30	40	100
Lactuca sativa	40	40	NOEC at 100
Trifolium repens	40	50	NOEC at 100
Avena sativa	30	40	90
Lumbricus terrestris	20	10	60

Notes: The toxicity was exposed as the lowest observed effect concentration (LOEC) at different contents (in wt. %) of contaminated soil in a mixture with a clean soil of the relevant type; NOEC = no observed effect concentration.

microorganisms, including sulfate-reducing bacteria and other metabolically connected microorganisms. The effluents from this zone practically contained no oxygen, had negative redox potential and pH about the neutral point but still contained large amounts of biodegradable organic compounds and sulfates. The effluents were rich in heterotrophic microorganisms (> 10^9 cells.mL^{-1}), included such related to the genera *Geobacter, Shewanella* and *Clostridium*, which, in principle, contained some electrochemically active representatives. These effluents were used in experiments for producing electricity in a microbial fuel cell, constructed near the wetland and soil plot. Samples from the effluents from the subhorizon B$_2$, without the above-mentioned prior cleaning for decreasing the residual concentrations of iron, manganese, and other heavy metals and radionulides, were also used in such experiments.

The results from the experiments for treatment of these waters by the microbial fuel cell revealed that the prior decrease of the concentrations of heavy metals and radionuclides facilitated the removal of organic compounds from the waters and increased the electricity production. It was also found that the increase in the concentration of biodegradable organic compounds in the waters increased the amount of active heterotrophic microorganisms as well as the rate of biodegradation of the organic compounds and the efficiency of the electricity production. Furthermore, several mixed microbial populations present in anaerobically digested sludges taken from different wastewater treatment plants were also included in the formations of biofilms in the anodic section of the relevant microbial fuel cell.

The formation of active and stable biofilms was a slow process and several months were needed for obtaining the maximum efficiency of wastewater treatment and power density generation by the microbial fuel cell. It is difficult to predict the future role of the microbial fuel cells as systems of electricity generation connected with the wastewater and soil treatment. At present, it is clear that such systems are promising since a considerable part of the chemical energy of organic contaminants is converted to electricity, reducing in this way the generation of excess amounts of sludge.

In any case, after the end of the operation for removal or detoxification of the contaminants, the treated soils are subjected to some conventional melioration procedures such as liming (if necessary), grassing, mulching, addition of fertilizers, and animal manure as well as periodic ploughing and irrigation.

3. Bioremediation of contaminated soil in the Vromos Bay area

Several soil plots consisted of cinnamonic soil were tested in this area. The size of these plots varied from 24 to 400 m^2. The most typical soil profile was 95 cm deep (horizon A, 30 cm; horizon B_2, 50 cm; horizon C, 15 cm). Data about the chemical composition and some essential geotechnical parameters of this soil are shown in Table 5.

The system established in the horizon A of the soil plot in the Vromos Bay area was favorable for the growth and activity of several aerobic and facultatively aerobic microorganisms, including cellulose-degrading bacteria and fungi (Groudev et al., 2010). This resulted in an efficient removal of radionuclides and heavy metals from this soil horizon within 20 months of treatment (Table 6).

The biodegradation of biopolymers resulted in increasing the concentrations of monomers (mainly monosaccharides and organic acids) in the pore solution of the soil and maintained a relatively stable pH in the system (in the range of about 7.5–8.0), regardless of the irrigation with hydrocarbonates and the solubilization of carbonates from the soil. In this system, U^{4+} was oxidized to U^{6+} by some heterotrophic bacteria producing peroxides. The hexavalent uranium was solubilized as different complexes— mainly with carbonate [$UO_2(CO_3)_2^{2-}$ and $UO_2(CO_3)_3^{4-}$] but also with carbonate and calcium [$CaUO_2(CO_3)_3^{2-}$ and $Ca_2UO_2(CO_3)_3$] (Bernhard et al., 2001) and with organic monomers such as carbonic andhumic acids (Finch and Murakami, 1999).

Radium also was solubilized as such complexes. The non-ferrous metals (Cu, Zn, Cd, and Pb) were solubilized mainly as complexes with organic acids. Small portions of these metals, mainly of the lead, were solubilized as complexes with chloride ions. It must be noted that small portions of the non-ferrous metals in the soil were present as the relevant sulfides. The decrease in the content of sulphidic sulfur and the increase in the number of basophilic chemolithotrophs in the soil were indications of the role played by these bacteria in the solubilization of metals.

Table 5. Characterization of the soil in the plot in the Vromos Bay area before and after the treatment.

Parameters	Before treatment	After treatment
Chemical composition (%)		
SiO_2	73.8	74.3
Al_2O_3	14.0	13.7
Fe_2O_3	3.29	3.02
P_2O_5	0.10	0.08
K_2O	3.25	2.37
N total	0.19	0.15
S total	1.40	0.99
S sulfidic	0.82	0.51
Carbonates	4.51	2.21
Humus	3.50	2.82
pH (H_2O)	7.52	7.81
Net neutralization potential (kg $CaCO_3 \cdot t^{-1}$)	49.8	21.3
Bulk density (g.cm^{-3})	1.50	1.43
Specific density (g.cm^{-3})	2.80	2.73
Porosity (%)	48	44
Permeability (cm.h^{-1})	8.2	7.3
Particle size (mm) (%)		
> 1.00	6.0	5.5
1.00–0.25	11.1	10.4
0.25–0.01	34.3	36.1
< 0.01	48.6	48.0

Most of them were related to the genus *Thiobacillus*, mainly to the species *T. thioparus* and *T. denitrificans*, as well as to the species *Halothiobacillus neapolitanus*. These bacteria oxidized the elemental sulfur which was generated as a result of the chemical and elemental sulfur which was generated as a result of the chemical and electrochemical oxidation of sulfides and was precipitated on the surface of these minerals as passivation films (Groudev et al., 2008). The removal of these passivation films enhanced the sulfide oxidation. The metal ions liberated during this oxidation were mantained in solution as the above-mentioned complexes.

The dissolved radionuclides and heavy metals were transported by the drainage solutions into the more deeply located soil layers (in the horizon B). In the top layer of this horizon, under anaerobic conditions and a negative redox potential at levels relatively close to zero, rich and diverse populations of denitrifying and iron-reducing bacteria were present. The ability of iron-

Table 6. Contents of contaminants in the horizon A of the soil from the plot in the Vromos Bay area.

Parameters	U	Ra	Cu	Zn	Cd	Pb
Contents of contaminants (mg.L⁻¹)						
- before treatment	41	280	611	251	7.3	268
- after treatment	7.1	55	242	99	1.7	109
Permissible levels for soils with pH > 7.0	10	65	280	370	3.0	80
Bioavailable fractions (mg.L⁻¹)						
(a) by DTPA leaching						
- before treatment	18	60	64	32	0.8	28
- after treatment	1.4	5	12	5.5	0.1	10
(b) by EDTA leaching						
- before treatment	7.1	25	41	17	0.44	46
- after treatment	0.9	3.0	3.5	2.8	0.02	10
Easily leachable fractions (mg.L⁻¹)						
- before treatment	18	73	109	62	1.5	53
- after treatment	1.2	8.2	9.9	7.7	0.02	6.8
Inert fractions (mg.L⁻¹)						
- before treatment	4.6	30	104	62	1.4	52
- after treatment	4.1	23	99	59	1.2	48

Notes: The contents of radium are shown in Bq/kg dry soil or in Bq/L.

reducing bacteria to reduce the hexavalent uranium to the tetravalent state is well known (Anderson et al., 2003). On the contrary, denitrification is a process which maintains uranium in solution. This is due to the fact that the nitrogen intermediates formed during this process (NO_2^-, NO and N_2O) oxidize U^{4+} to U^{6+} both directly and indirectly (Senko et al., 2002). The indirect oxidation is connected with the oxidation of Fe^{2+} to Fe^{3+} by these intermediates. As a result of this oxidation, poorly crystalline Fe^{3+}-oxide minerals are formed and they are able to oxdize U^{4+} to U^{6+} more efficiently than the nitrogen intermediates. The rate of U^{4+} oxidation coupled to microbial denitrification is increased when aqueous Fe^{2+} is added to a solution containing U^{4+} and nitrate. The Fe^{3+}-oxide minerals produced during the chemical oxidation of Fe^{2+} by nitrite are poorly crystalline but oxidize U^{4+} efficiently.

Fe²⁺ ions were solubilized in the upper layers of the horizon B as a result of the reduction of Fe^{3+}-oxide minerals by the iron-reducing bacteria. The injection of nitrite ions to these soil layers stimulated the microbial denitrification and in this way decreased the uranium precipitation (Elias et al., 2003). However, more intensive precipitation of uranium occured in the deeply located soil layers inhabited by iron and mainly by sulfate-reducing bacteria, regardless of the fact that the Ca-U^{6+}-CO_3 and some U^{6+}-organics complexes are quite refractory to both chemical and biological reduction. In some cases, sulfate-reducing bacteria were able to reduce the U^{6+} present in some complexes but these complexes were not degraded and U^{4+} remained in the solution.

In most cases, the microbial sulfate reduction was enhanced by injecting dissolved organic compounds such as lactate and acetate to the deeply located soil layers. However, it must be noted that the high contents of dissolved organic carbon (higher than 25–30 mg/l) inhibited the precipitation of uranium. It is well-known that long-term amendments of uranium contaminated soils with organic electron donors result in re-oxidation of the initially bioreduced uranium (U^{4+}) under reducing conditions (Mulligan et al., 2001). In such cases, Ca-U-CO_3 complexes are formed due to the increased concentrations of some microbial metabolites, including hydrocarbonate ions. Furthermore, the reduction of U^{6+} decreases when sulfate-reducing bacteria not able to carry out an enzymatic reduction of this ion are dominant over the iron-reducing bacteria (Vrionis et al., 2005).

Considerable portions of the non-ferrous metals were precipitated in the deeply located soil layers (mainly in the subhorizon B_2) as the relevant insoluble sulfides as a result of the activity of sulfate-reducing bacteria. Most of the uranium was precipitated as uraninite also by these bacteria. It must be noted that relatively small portions of uranium and non-ferrous metals were removed from the percolating drainage waters by means of sorption on the soil particles and by the hydroxides and oxides of iron and manganese present in the soil. However, the sorption was an essential mechanism for the removal of radium and lead from these waters. Small amounts of the insoluble (Pb, Ra)SO_4 were also detected in the soil. There was a clear tendency for the initially adsorbed uranium and non-ferrous metals to be subjected to reduction in the course of time.

In any case, the effluents from the soil profile still contained dissolved radionuclides and heavy metals but in concentrations lower than the relevant permissible levels.

The soil toxicity initially increased during the treatment but then steadily decreased and at the end of the operation was much lower than that of the contaminated soil before treatment (Table 7).

Further investigations on the bioremediation of soil of this type revealed that the addition of the soluble Mn^{2+} to the leach solution considerably increased the efficiency of the uranium removal from the soil. This was connected with the presence of several microorganisms in the soil that are able to oxidize the Mn^{2+} to the solid Mn^{4+} present in the form of MnO_2, an efficient oxidant of U^{4+} (Groudev et al., 2015).

It was found that at least five different groups of microorganisms able to oxidize the Mn^{2+} to MnO_2 were present in the treated soil (Groudev et al., 2015). Some of the microorganisms (mainly bacteria related to the genera *Bacillus*, *Pseudomonas*, and *Arthrobacter*) were able to oxidize Mn^{2+}

Table 7. Toxicity of the soil from the plot in the Vromos Bay area before and after the treatment.

Test-organisms	Toxicity	
	Before treatment (pH 7.52)	After treatment (pH 7.81)
Bacillus cereus	40	NOEC at 100
Pseudomonas putida	40	100
Lactuca sativa	50	NOEC at 100
Trifolium repens	50	NOEC at 100
Avena sativa	30	90
Lumbricus terrestris	30	90

Notes: The toxicity was exposed as the lowest observed effect concentration (LOEC) at different contents (in wt. %) of contaminated soil in a mixture with a clean soil of the relevant type; NOEC = no observed effect concentration.

enzymatically using the molecular oxygen (O_2) as the terminal electron acceptor and were able to converse energy (as ATP) from the oxidation. Some bacteria of the genus Bacillus were also able to use this type of oxidation but only in the case when soluble Mn^{2+} was pre-bound to some solid substrates, mainly to the hydrated manganese oxide ($MnO_2.H_2O$). The third group of Mn oxidizers consisted of bacteria (mainly of the genera *Bacillus* and *Leptothrix*) oxidizing the Mn^{2+} to Mn^{4+} in the presence of oxygen but without the conversion of energy from this process. The fourth group included bacteria (mainly of the genera *Metallogenium*, *Siderobacter* and *Crenothrix*) using the hydrogen peroxide (H_2O_2) as oxidant catalyzed by the enzyme catalase. The last group consisted of different bacteria and fungi oxidizing the Mn^{2+} non-enzymaticlly, by means of secreted hydroxycarboxylic acids, mainly by citrate and lactate. The growth of these heterotrophic microorganisms was possible only in the presence of sufficient amount of biodegradable organic matter. Chemolithotrophic bacteria able to use the energy liberated from the oxidation of the bivalent manganese to the tetravalent form (MnO_2) fixing organic carbon (CO_2) for producing a primary organic matter have not been detected as yet.

The oxidation of the solid U^{4+} by the solid MnO_2 requires physical contact or close proximity of these two solids. Without a direct contact, the dissolved U^{6+} is adsorbed on the solid MnO_2 and maintains a driving force for U^{4+} dissolution by keeping the dissolved uranium concentrations low. Uranium is associated with MnO_2 in the UO_2/MnO_2 system as U^{6+} and that it has a coordination environment similar or identical to U^{6+} adsorbed on MnO_2. The oxygen atoms in U^{6+}, i.e., UO_2^{2+}, are derived from the MnO_2 (Gordon and Taube, 1962) indicating that this reaction is a heterogeneous process involving innersphere electron transfer. This means that the fate of soluble U^{4+} on the solid MnO_2 is adsorption, subsequent oxidation, and finally U^{6+} adsorption to MnO_2. The subsequent detachment and solubilzation of the sorbed U^{6+} can depend on the parameters of water chemistry that impact U^{6+}. MnO_2 adsorption equilibrium.

Radium also was solubilized as complexes of the same types as these of the uranium. The non-ferrous metals present in the soil were solubilized as complexes with the organic acids generated *in situ* by some of the microorganisms inhabiting the soil.

The dissolved radionuclides and heavy metals were transported by the drainage solutions into the more deeply located soil layers (in the horizon B). The lower part of this horizon (located more deeply than 50 cm from the soil surface) was densely populated by sulfate-reducing bacteria. The growth and activity of these bacteria were enhanced by injecting dissolved organic compounds (lactate and acetate) to this part of the soil profile.

Considerable portions of the non-ferrous metals were precipitated in this anaerobic zone of the soil as the relevant insoluble sulfides as a result of the sulfate-reducing bacteria. Most of the uranium was precipitated as uraninite also by these bacteria. Small portions of uranium and non-ferrous metals were removed from the pregnant leach solutions initially by means of sorption on the soil particles and by the hydroxides and oxides of iron and manganese present in the soil.

The initially adsorbed uranium and non-ferrous metals were then subjected to reduction to the tetravalent solid forms, i.e., to U^{4+} and sulfide respectively. However, the sorption was an essential mechanism for the removal of radium and lead from these waters.

4. Conclusion

The monitored natural attenuation and liming of the acidic soils are still the most largely applied *in situ* methods for treatment of soils contaminated with radionuclides and heavy metals. However, the application of *in situ* bioremediation based on the activity of the indigenous soil microorganisms is steadily increasing under the form of several variants. The treatment connected with solubilization of contaminants and flushing the soil profile usually increases the toxicity of the soil during the operation. It must be noted, however, that the duration of such operations is much shorter than the long periods of spontaneous natural leaching of the soils. Furthermore, regardless of the fact that the natural leaching generates pregnant solutions with lower concentrations, it is difficult to control the distribution of these solutions and to avoid the contamination of other ecosystems.

The efficient immobilization of contaminants in the subhorizon B_2 as a result of the enhanced dissimilatory sulfate reduction produces non-toxic water effluents relatively rich in biodegradable organic compounds. The cleanup of these soil effluents is not a serious problem and can be done by different active or passive systems for wastewater treatment. It is also possible that clean-up of such effluents is done in microbial fuel cells, in which the removal of organics is connected with electricity generation.

It is difficult to predict the future role of the microbial fuel cells as systems of electricity generation connected with the wastewater treatment. At present, it is clear that such systems are promising since a considerable part of the chemical energy of organic contaminants is connected to electricity, reducing in this way the generation of excess amounts of sludge. In any case, after the end of operation for removal or detoxification of the contaminants, the treated soils are subjected to some conventional melioration procedures such as liming (if necessary), grassing, moulching, addition of fertilizers, and animal manure as well as periodic ploughing and irrigation.

References

Anderson K.T., Vrionis H.A., Ortiz-Bernad I., Resch C.T., Long P.E., Dayvault K., Karp K., Marutzky S., Metzler D.K., Peacock A., White D.C., Lowe M., Lovley D.R. Stimulating the *in situ* activity of Geobacter species to remove uranium from the ground water of uranium-contaminated aquifer. Appl. Environ. Microbiol. 69, 5884–5891, 2003.

Bernhard G., Geipel G., Brendler V., Amayri S., Nitsche H. Uranyl (VI) carbonate complex formation: Validation of the $Ca_2UO_2(CO_3)_3$ and $Ca_2UO_2(CO_3)$ (aq.) species. Radiochimica Acta 89, 511–518, 2001.

Clean D.E., Livens F.R., Stennett M.C., Grolimund D., Borca C.N., Hyatt N.C. Remediation of soils contaminated with particulate depleted uranium by multistorage chemical extraction. J. Hazard. Mat. 263, 382–390, 2013.

Elias D.A., Senko J.M., Krumholz L.R. A procedure for quantitation of total oxidized uranium for bioremediation studies. J. Microbiol. Meth. 53, 343–353, 2003.

Finch R.J., Murakami T. Systematics and paragenesis of uranium minerals. *In*: Burns P.C., Finch A.J. (eds.). Uranium: Mineralogy, Geochemistry and the Environment, Review in Mineralogy 38. Mineralogical Society of America: Washington D.C., pp. 91–178, 1999.

Groudev S.N., Georgiev P.S., Spasova I.I., Komnitsas K. Bioremediation of soil contaminated with radioactive elements. Hydrometallurgy 59, 311–318, 2001.

Groudev S.N., Spasova I.I., Nicolova M.V., Georgiev P.S. Bioremediation *in situ* of polluted soil in uranium deposit. *In*: Annable M.D., Teodorescu M., Hlavinek P., Diels L. (eds.). Methods and Techniques for Clean-up Contaminated Sites. NATO Science for Peace and Security Series—C: Environmental Security, Springer, Dordrecht, pp. 25–34, 2008.

Groudev S.N., Spasova I.I., Nicolova M.V., Georgiev P.S. *In situ* bioremediation of contaminated soils in uranium deposit. Hydrometallurgy 104, 518–523, 2010.

Groudev S.N., Georgiev P.S., Spasova I.I., Nicolova M.V. Decreasing the contamination and toxicity of a heavily contaminated soil by *in situ* bioremediation. J. Geochem. Explor. 144, 374–379, 2014.

Groudev S.N., Georgiev P.S., Spasova I.I., Nicolova M.V. Bioremediation of an alkaline soil heavily polluted with radionuclides and heavy metals. *In*: Proceedings of XVI Balkan Mineral Processing Congress. Belgrade, Serbia, June 17–19, vol. II, pp. 1003–1006, 2015.

Hinchee R.E., Means J.J., Burris D.R. Bioremediation of Inorganics. Battelle Press, Columbus, Ohio, 1995.

Knox A.S., Paller M.H., Reible D.D., Ma X., Petrisor I.G. Sequestering agents for active caps—remediation of metals and organics. Soil Sediment Contam. 17, 516–532, 2008.

Malavija P., Singh A. Phytoremediation strategies for remediation of uranium-contaminated environments: a review. Crit. Rev. Environ. Sci. Technol. 42, 2575–2647, 2012.

Mulligan C.N., Yong R., Gibbs B. An evaluation of technologies for the heavy metals evaluation of degraded sediments. J. Hazard. Mat. 85, 145–163, 2001.

Park H.-M., Kim G.-N., Shon D.-B., Lee K.-W., Chung U.-S., Moon J.-K. Remediation of soil contaminated with uranium using a biological method. *In*: Trans Korean Nuclear Society Spring Meeting. Taebaek, Korea, May 26–27, pp. 199–200, 2011.

Romera-Freize A., Garcia Fernandez I., Simon Torres M., Martinez Garzon F.J., Reinado M. Long-term toxicity assessment of soils in a recovered area affected by a mining spill. Environ. Pol. 208, 553–561, 2016.

Senko J.M., Istok J.D., Sutlita J.M., Krumholz L.R. *In-situ* evidence for uranium immobilization and remobilization. Environ. Sci. Technol. 26, 1491–1496, 2002.

Sing K.L., Rao C.M., Sudhakar G. Evaluation of uranium mine tailing remediation by amending land soil and invading native plant species. J. Environ. Sci. Toxicol. Food Technol. 8, 64–81, 2014.

Vrionis H.A., Anderson R.T., Ortiz-Bernad I., O'Neill K.R., Resch C.T., Peacock A.D., Dayvault R., White D.C., Long P.E., Lovley D.R. Microbiological and geochemical heterogeneity in an *in situ* uranium bioremediation field site. Appl. Environ. Microbiol. 71, 6308–6318, 2005.

CHAPTER 17

Macrophyte Role and Metal Removal in Constructed Wetlands for the Treatment of Effluents from Metallurgical Industries

María Alejandra Maine,[1,*] *Hernán R. Hadad,*[1] *Gisela A. Di Luca,*[1]
María de las Mercedes Mufarrege[1] and *Gabriela C. Sánchez*[2]

1. Introduction

Wastewater containing contaminants that are discharged into the water bodies result in negative environmental consequences. Ecological technologies for wastewater treatment have attracted attention as alternative non-conventional solutions (Vymazal, 2011; Wu et al., 2015). Constructed wetlands (CWs) are one of the most widely used ecological technologies due to their high contaminant removal efficiencies and reduced costs for maintenance and operation. However, CWs application is limited for land area requirement.

(CWs), also known as treatment wetlands or wetland systems, are engineered systems designed and constructed to utilize natural processes to remove contaminants from water. They are designed to take advantage of the same processes that occur in natural wetlands but do so within a more controlled environment (Kadlec and Wallace, 2009; Vymazal, 2011). The removal of contaminants in CWs is complex and depends on a variety of removal mechanisms, including sedimentation, filtration, precipitation, volatilization, adsorption, plant uptake, microbial processes, etc.

---

[1] Química Analítica, Facultad de Ingeniería Química, Universidad Nacional del Litoral, Santiago del Estero 2829, Santa Fe (3000), Argentina. Consejo Nacional de Investigaciones Científicas y Técnicas (CONICET).
[2] Química Analítica, Facultad de Ingeniería Química, Universidad Nacional del Litoral, Santiago del Estero 2829, Santa Fe (3000), Argentina.
* Corresponding author: amaine@fiq.unl.edu.ar

CWs can be classified in a variety of ways. The most used is according to the flow regime: Free Water Surface (FWS) wetlands, composed by sediment and floating, submerged and/or emergent macrophytes, similar in appearance to natural marshes; Horizontal subsurface flow (HSSF) wetlands that employ a gravel bed planted with emergent macrophytes, where the water flows horizontally from the inlet to the outlet; and Vertical Flow (VF) wetlands, where water is discharged as rain on all the wetland surface and flows vertically across different layers of sand and/or gravel bed planted with emergent vegetation. The most suitable CW type, substrate, plant species, etc. to be used are chosen according the volume and chemical composition of the effluent to be treated. If an equal treatment performance is targeted, FWSs need the least costs for operation and maintenance but require the largest land area, and VFs need the least area but the costs for maintenance and operation are the highest (Wu et al., 2015).

CWs have been widely studied for the treatment of various types of wastewater such as domestic sewage, agricultural, industrial, mine drainage, leachate, urban runoff, etc. (Brix, 1993; Brix and Arias, 2005; Kadlec and Wallace, 2009; Maine et al., 2006, 2007, 2009, 2013, 2017; Shubiao et al., 2014; Vymazal, 2011; Wu et al., 2015; Zhang et al., 2014). There are hundreds of CWs operating in Europe, Asia, United States, and Australia. In Latin America, in countries such as Mexico, Colombia, Peru, and Bolivia, this technology has been widely used for the depuration of sanitary effluents of small villages, tourist resorts, university campus, etc. where the contaminants to remove are N and P (Maine et al., 2017; Rodriguez et al., 2016; Zurita et al., 2011). In Argentina, despite the environmental conditions are favorable with a great land availability, CWs are not widely implemented for sanitary effluents; however, two CWs have been designed for the treatment in metallurgical industries, where the critical contaminants are metals.

The removal of metals using CWs has been applied worldwide to mine effluents, the pH of which is acid. In the case of effluents from metallurgical industries, the pH is alkaline and the salinity is high. Such is the case of the two mentioned CWs for final effluent treatment at metallurgical industries (CW1 and CW2). Free-water surface CWs were used because they are the most convenient CW type for metal removal. Sediment is majorly responsible for contaminant removal from waters in free water surface wetlands (Di Luca et al., 2011a, 2011b, 2016; Jacob and Otte, 2003; Maine et al., 2009; Marchand et al., 2010; Weis and Weis, 2004; Ye et al., 2001). However, sediments can release or retain contaminants according to environmental conditions such as redox potential, pH, temperature, etc. (Boström et al., 1985; Maine et al., 1992). Contaminant dynamics also depend on the chemical forms in which they are retained by sediment which are studied by sequential extraction schemes (Di Luca et al., 2011a,b).

Despite the importance of sediment, macrophytes are the main biological component of CWs. In the case of nitrogen or phosphorous, main contaminants of domestic sewage, agricultural, industrial, urban runoff, etc., plants assimilate them as nutrients. However, most metals are toxic for plants. It has been suggested that plants uptake high concentrations of metals as a self-defense mechanism against pathogens and herbivores (Poschenrieder et al., 2006). High metal concentrations uptake by some emergent macrophytes have been reported (Arduini et al., 2006; Hechmi et al., 2014; Mangabeira et al., 2011; Mufarrege et al., 2014, 2015, 2016) indicating plants' tolerance. Plants not only retain metals in their tissues and renew the sources of carbon for degrading bacteria but also contribute to wastewater treatment processes in a number of ways such as favoring the settlement of suspended solids, providing surface area for microorganisms, carrying oxygen from the aerial parts to the roots, creating the proper environment in the rhizosphere for the proliferation of microorganisms, and promoting a variety of chemical and biochemical reactions which enhance metal retention by the sediment (Brix, 1994, 1997; Kadlec et al., 2000).

CW1 and CW2 have been operating for 14 and 7 years respectively. As the chemical composition of the wastewaters and the volumes to be treated are different, CWs have different design characteristics. Wastewater from the industrial processes and sewage from the factory are treated together after a primary treatment. The idea of treating sewage was based on the fact that sewage composition is rich in organic matter and nutrients. High nutrient concentrations could improve macrophyte tolerance to metals. This hypothesis was corroborated by our research group at greenhouse scale experiments (Hadad et al., 2007; Mufarrege et al., 2016).

The studied cases of the CWs for the treatment of metallurgical effluents are explained below.

2. Case 1: Wetland for effluent treatment at a tool-manufacturing factory (CW1)

An experimental pilot scale wetland was constructed to assess the feasibility of using this technology for the final treatment of effluents from a metal tool factory. The effluent to be treated presented high salinity and high pH, and contained Cr, Ni, and Zn. Its composition showed a high variability on a daily basis due to different industrial processes. This experimental pilot scale wetland was 6 m long, 3 m wide, and 0.4 m deep. A polyethylene impermeable film (200 μm) was placed at the bottom and a soil layer of 50 cm was added. The incoming wastewater entered the wetland through a 63 mm-diameter PVC pipe with a perpendicular drip dispersion tube. The drip dispersion tube was provided with aligned holes to produce a laminar flow. The outcoming water left the wetland through a "V" notch weir and a pipe carried it into a pond located next to the wetland. Immediately after plant transplantation, the wetland was filled with tapwater up to a depth of approximately 0.40 m. Wastewater was mixed with tap water at gradually increased wastewater/ tap water ratio until only wastewater was added after two weeks. The influent discharge was 1000 L.d⁻¹ and water residence time was approximately 7 d. Previous experiments were conducted to assess the tolerance and efficiency of accumulation of nutrients and pollutants of different plant species (Maine et al., 2001, 2004). Plant species were chosen taking into account these results and literature (Ellis et al., 1994; Gersberg et al., 1986; Jenssen et al., 1993). The plants used were emergent and free floating native species, transplanted from natural environments of the same area. Among free floating species *Salvinia herzogii, Salvinia rotundifolia, Pistia stratiotes* and *Eichhornia crassipes* were selected. Emergent species such as *Cyperusal ternifolius, Typha domingensis, Pontederia cordata* and *Schoenoplectus californicus* were planted. Emergent species were pruned before transplantation. The wetland was monitored for a year. Removal efficiency and macrophyte tolerance to the effluent and its efficiency in contaminants uptake were evaluated. In a first phase, *E. crassipes* and *P. stratiotes* were the species that showed the highest cover (Hadad et al., 2007), but after 6 months of operation, the wetland became a monospecific stand of *T. domingensis*, which attained a high biomass (1.9 kg.m⁻²) (Maine et al., 2005) (Fig. 1).

The wetland removed efficiently 81, 66, and 59% of Cr, Ni, and Zn respectively (Maine et al., 2005). The pilot scale wetland efficiently removed nutrients (74 and 88% of Tot-P and NO_3^- respectively, and metals from the inlet wastewater) (Maine et al., 2005). Cr, Ni, and Zn showed large increments in the inlet area sediment. Metal concentrations in *T. domingensis* root tissues were strongly increased.

Due to the satisfactory performance of the small-scale experimental wetland, a large-scale wetland was constructed for the factory wastewater treatment. The final large-scale is a free water surface wetland of 50 m long,

Figure 1. Pilot scale wetland: (a) initially; (b) after 1 month; (c) after 3 months; (d) after 9 months from the beginning of operation.

40 m wide and 0.5–0.6 m deep. A central baffle was constructed, parallel to the flow direction, dividing the wetland into two sections of an equal area and forcing the effluent to flow in "U" form, resulting in a 5:1 length-wide ratio. The wetland was rendered impermeable with bentonite (6 layers of compacted bentonite, in order to achieve a hydraulic conductivity of 10^{-7} m.s^{-1}) (Fig. 2). A layer of 1 m of soil was placed on top of the bentonite layer. Phreatic water meters were placed around the wetland to monitor groundwater quality as a security measure.

Wastewater discharge is approximately 100 m^3.d^{-1} and the hydraulic residence time ranges from 7 to 12 days. Industrial wastewater and sewage were treated together (25 m^3.d^{-1} of sewage + 75 m^3.d^{-1} of industrial wastewater) after a primary treatment. Wastewater reaches the CW through a PVC pipe provided with a perpendicular distribution pipe with holes at regular distances in order to allow uniform flow distribution. After crossing the wetland, the effluent flows along an excavated channel to a 1.5 ha pond (Fig. 3). CW1 has been in operation since 2003.

Since the wetland became operational, chemical composition of the effluent before and after the treatment was monitored. Measured parameters at the inlet and outlet and estimated removal efficiencies obtained in CW1 are shown in Table 1. The wetland showed high removal efficiencies of Cr, Ni, and Zn. COD and BOD showed good removal efficiencies pointing out high organic matter mineralization in the wetland. Nitrate and nitrite concentrations decreased successfully, while soluble reactive phosphorus (SRP), total phosphorus (TP) and ammonium were not efficiently removed, probably due to the low DO concentration in the outlet zone. An aerator was placed to oxygenate the effluent before discharge.

Figure 2. Waterproofing of the wetland area using bentonite.

Taking into account the results obtained in the pilot scale wetland, several locally available macrophytes were transplanted into the CW1. The use of a high diversity of plants has advantages, such as a higher efficiency in contaminant removal, disturbance resilience, better habitat, etc. (Brisson, 2013). Initially, *E. crassipes, T. domingensis,* and *P. cordata* showed the highest

Figure 3. Aerial view of the treatment wetland and discharge pond.

Table 1. Ranges of measured parameters at the inlet and outlet and estimated removal efficiencies obtained in CW1.

Parameter	Inlet	Outlet	% Removal
pH	10.4–11.8	7.9–9.1	-
DO (mg.L^{-1})	0–6.2	0.3–5.2	-
Conductivity (umho.cm^{-1})	3890–8700	1400–2500	-
Ca^{2+} (mg.L^{-1})	32.3–120.7	11.1–41.2	59.3
Alkalinity	194.6–750.4	136.8–332.3	46.5
NO$_3^-$ (mg.L^{-1})	15.4–98.2	3.6–24.2	80.4
NO$_2^-$ (mg.L^{-1})	0.258–6.22	0.017–0.766	84.1
NH$_4^+$ (mg.L^{-1})	0.154–2.67	0.05–2.14	11.8
SRP (mg.L^{-1})	0.005–0.079	0.005–0.334	13.3
TP (mg.L^{-1})	0.064–1.38	0.129–0.696	22.0
Fe (mg.L^{-1})	0.05–8.54	0.05–0.430	93.4
Cr (mg.L^{-1})	0.023–0.204	0.002–0.033	84.7
Zn (mg.L^{-1})	0.022–0.070	0.015–0.039	61.2
Ni (mg.L^{-1})	0.004–0.101	0.004–0.082	69.5
COD (mg.L^{-1})	57.9–154.0	13.9–39.9	79.5
BOD (mg L^{-1})	9.8–30.9	3.0–14.1	83.2

cover. The development of the vegetation in both, the previous experimental-scale wetland and the CW1, showed the same pattern but occurred at a different time scale. In the first months of operation, *E. crassipes* showed fast growth becoming the dominant macrophyte and covering 80% of CW1 for 2 years. However, its cover decreased after this period. New specimens were transplanted but *E. crassipes* did not tolerate the effluent (Maine et al., 2007). High pH and salinity were the cause of disappearance of floating macrophytes (Hadad et al., 2007), but when *E. crassipes* cover decreased, *T. domingensis* cover began to increase. To enhance *T. domingensis* growth, CW1 water level was decreased and 0.50 m wide baffles were constructed transversally to the effluent circulation. *T. domingensis* became the dominant species, showing the highest tolerance and competitive hierarchy until it became the only species which covered almost all the surface attaining high biomass (Fig. 4). *T. domingensis* development was favored by water level regulation, achieving the best plant growth at 30 cm depth. *T. domingensis* showed optimal growth reaching a higher biomass than that recorded for natural wetlands (8 kg.m^{-2}) (Hadad et al., 2010). In a green house experiment, growth responses to pH and salinity treatments of *T. domingensis* plants sampled from an uncontaminated natural wetland (NW) and from CW were studied. The treatments of salinity (mg.L^{-1})/pH were: 8,000/10 (values found in the CW); 8,000/7; 200/10 and 200/7 (characteristic values found in the NW). *T. domingensis* plants from NW showed growth inhibition and senescence when exposed to the conditions of

Figure 4. Stages of vegetation development in the constructed wetland: (a) during transplantation, where the baffle is observed; (b) 4 months after transplantation; (c) one year of operation (dominance of floating species); (d) after three years, where the dominance of emergent species and the income of the effluent pipe is observed.

pH and salinity commonly found in the CW. Plants from the CW tolerated high pH and salinity treatments. Contrarily, they showed stress when they were exposed to the conditions of pH and salinity commonly found in the NW (Mufarrege et al., 2011). *T. domingensis* is a good choice to treat wastewater of high pH and salinity and common characteristics of many industrial effluents but a suitable plant acclimation is necessary in CWs.

Contaminant removal efficiency of the CW1 was satisfactory in all stages of vegetation dominance (Maine et al., 2009). Metal accumulation occurred mainly in the roots of macrophytes. Metal concentrations in the above ground biomass did not show a significant increase compared to the values found in natural environments. This is an important issue because contaminants did not enter the food chain. During *E. crassipes* dominance, contaminants were retained in the macrophyte biomass; during *T. domingensis* dominance stage, metals were retained mainly in sediment, while P was retained in sediment and macrophyte biomass (Maine et al., 2007). Harvest of floating macrophytes would allow metal and P removal from the system. But, plants have to be safely disposed. Emergent macrophytes favor metal accumulation in the sediment, phytostabilizating them (Maine et al., 2009).

T. domingensis maintained a cover of 60–90% during the past 10 years. Some cover decreases were registered because plants were pruned regularly to enhance growth. But, the most important cover decrease occurred in 2011, when a population of capybaras (*Hydrochoerus hydrochaeris*) caused the depredation of the plant aerial parts. Capybaras, the largest amphibious rodent known in the world, weigh 80 Kg on average. The CW looked like a pond without macrophytes. However, the roots and rhizomes of *T. domingensis* were not damaged and CW1 continued retaining contaminants during the predation period (Maine et al., 2013). The wetland was fenced with wire to stop animals from approaching, which allowed the recovery of the vegetation. Subsequently, *T. domingensis* reached a cover of 80% after 90 days. This luxuriant growth was enhanced by the growth season. It is important to highlight that during this predation event which lasted a few months, the wetland did not decrease its efficiency, retaining the metals in sediment and into the root system of plants, demonstrating the robustness of these systems (Maine et al., 2013).

Regarding sediments, those of the inlet area of the CW showed significantly higher metal concentrations than those registered in the outlet zone, indicating contaminant retention in this area. Using metal fractionation techniques, it was found that metals were bound to sediment fractions that would not release them into water while the chemical and environmental conditions of the system were maintained (Di Luca et al., 2011b; Maine et al., 2013, 2017). Metal concentrations in sediment of the outlet area were not significantly different from the initial values at the beginning of the operation period, suggesting that the wetland only used the sorption capacity of the sediment from the inlet zone. The sorption sites of the outlet sediment remained available for contaminants (Di Luca et al., 2011b; Maine et al., 2013).

Figure 5. (a, b) Greenhouse for ornamental plants cultivated with compost produced at CW1; (c, d) compost processing to be used for plant culture.

Both, bottom sediment and macrophytes were responsible for the removal of contaminants. Emergent macrophytes, as *T. domingensis*, influenced the biogeochemical cycle of sediments through changes in redox potential due to their ability to supply oxygen to the rhizospheric zone. A complete root-rhizome development for a constructed wetland may require 3–5 years. Constructed wetland performance improved with wetland maturity (Kadlec and Wallace, 2009). Vymazal and Krópfelová (2005) reported that for the emergent macrophyte *Phragmites* spp., three to four seasons were usually needed to reach maximum standing crop but in some systems it may take even longer. CW1 became a monoculture of *T. domingensis* and its efficiency has improved over time. When the capacity of a wetland to retain contaminants depends mainly on its sediment sorption ability, it can have a limited lifetime. However, since the conditions for metal removal (high pH, alkalinity, Fe, Ca, and ionic concentrations) are largely provided by the effluents, the sediment may be expected to continue retaining metals as far as the composition of the influent remains the same.

CW1 maintenance was performed yearly after the winter season. If necessary, the solids accumulated in the inlet zone were removed and the macrophyte dead leaves were pruned. These residues were used to process compost for ornamental plants cultivated in a greenhouse constructed at the same factory (Fig. 5). Metal concentration in compost and tissues of the

Figure 6. Wildlife observed in CW1 and the receiving effluent pond.

ornamental species were analyzed on a regular basis. Concentrations were below the levels permitted by national regulations.

The CW1 and the receiving effluent pond provided habitat for many animals in the area such as ducks, geese, coots, coypos, capybaras, turtles, etc. (Fig. 6). Similar results were reported by Mitsch and Gosselink (2000), Vymazal and Kröpfelová (2008), and Kadlec and Wallace (2009).

3. Case 2: Wetland for the effluent treatment of large piece chrome-plating processes (CW2)

CW2 is 20 m long, 7 m wide and 0.5 m deep, resulting in a 3:1 length-wide ratio adequate to achieve a good flow distribution. For its construction, soil movement works were done: excavation, profiling and compacting of side slopes (Fig. 7a).

The wetland was waterproofed with a geomembrane of high density polyethylene, 1.5 mm thick (Fig. 7b). The bed has a slope of 1°. A layer of 1.5 m of soil was placed on top of the geomembrane, acting as the substrate for emergent plants. To enhance plant growth in deeper zones and to increase the effluent circulation through the wetland, baffles of 0.50 m width with plants were constructed transversally to the effluent circulation. The water level on the baffles was 0.30–0.40 m. In the other zones, it was 0.5–0.7 m. This wetland treated all the factory effluents: the chrome plating bath effluents and sewage, the storm water and the cooling circuit water. The first two effluents received a previous primary treatment. The primary treatment of

Figure 7. Construction of CW2: (a) excavation and compacting of side slopes; (b) geomembrane waterproofing; (c) outlet pool; (d) *T. domingensis* recently transplanted.

the industrial effluent consisted of the reduction of Cr(VI) to Cr(III) and the subsequent oxy-hydroxides precipitation. The effluents reached an equalizing chamber and then entered the wetland. During the first year of operation, the CW treated sanitary effluents (with previous primary treatment), pluvial and cooling circuit effluents. After this period, the industrial effluent began to be treated. Mean wastewater discharge was approximately 10 m³.d⁻¹. Minimum water residence time was 7 days. After crossing the wetland, wastewater was discharged by a channel simulating a waterfall, to reach a concrete pool of 4 m x 2 m and a depth of 40 cm. To monitor the system, samples were taken from the concrete pool (Fig. 7c). The treated effluent was left the pool by a channel reaching a nearby pond.

Considering the experience with CW1, the species selected for CW2 was *T. domingensis*. Specimens growing in the factory facilities were transplanted, ensuring that plants were adapted. The plants were pruned to a height of approximately 30 cm keeping their rhizomes. Three plants per square meter were disposed and were irrigated until vigorous growth (Fig. 7d). Then, CW2 was filled with pond water.

This wetland has been in operation since 2009. The wetland demonstrated a good retention efficiency of contaminants. Concentration ranges of the parameters measured and estimated removal efficiencies obtained in CW2 are shown in Table 2. Phosphorous, COD, BOD and metals showed good removal efficiencies. Regarding nitrogen species, removal efficiencies

Table 2. Ranges of measured parameters at the inlet and outlet and estimated removal efficiencies in CW2.

Parameter	Inlet	Outlet	% Removal
pH	7.4–8.3	8.0–8.1	-
DO (mg.L^{-1})	3.2–5.4	4.2–5.8	-
Conductivity (umho.cm^{-1})	975–10060	1058–1358	-
Ca^{2+} (mg.L^{-1})	76.8–120	48.0–88.8	36.9
Alkalinity	101.7–1647.0	167.9–378.2	63.2
NO$_3^-$ (mg.L^{-1})	0.271–1.28	0.158–0.484	74.4
NO$_2^-$ (mg.L^{-1})	0.004–0.223	0.030–0.053	71.2
NH$_4^+$ (mg.L^{-1})	0.957–15.6	0.722–3.89	66.1
SRP (mg.L^{-1})	0.247–0.903	0.291–0.350	58.1
TP (mg.L^{-1})	0.642–1.322	0.398–0.442	52.8
Fe (mg.L^{-1})	0,15–0,56	0,06–0,17	70,4
Cr (mg.L^{-1})	0.012–1.45	0.019–0.025	92.9
Zn (mg.L^{-1})	0.006–0.145	0.003–0.067	51.7
Ni (mg.L^{-1})	0.003–0.082	0.004–0.004	77.5
COD (mg.L^{-1})	21.3–160	< 6–27	78.2
BOD (mg.L^{-1})	10.2–55.5	3.2–17.6	82.5

were satisfactory. Ammonium concentration presented a high variability. However, it is important to highlight that at the outlet, concentrations were below the law regulatory limits.

T. domingensis cover reached 90% in a few months, demonstrating its high productivity (Fig. 8). Macrophyte cover was stable along time probably because *T. domingensis* was planted in a suitable way initially and could develop their root-rhizome system properly.

A decrease in plant cover occurred during the first months of 2012 due to an accidental dump of untreated effluent. The wetland was closed to avoid an effluent outflow. Cr concentration in water decreased after 30 days. The wetland was emptied, plant detritus containing high Cr concentrations was removed and new specimens of *T. domingensis* were planted in the inlet area. Four months after the accidental dump the wetland was operating normally. It is important to highlight that the environment was preserved since the wetland acted as a cushion, retaining the contaminants. Although the high Cr concentrations in the effluent that reached the wetland, Cr concentration did not increase in the sediment inlet as expected. In the inlet of the CW2, detritus from *T. domingensis* were accumulated, in which high metal concentrations were found. Detritus are conformed by death leaves that remain in decomposition after the winter, which is part of the annual cycle of the plant growth. Metals are not only sorbed by live plants. Schneider and

Figure 8. *T. domingensis* cover reached 90% in a few months, demonstrating its high productivity.

Rubio (1999) demonstrated at laboratory scale that the dry biomass of three floating macrophytes (*Potamogeton lucens, Salvinia herzogii* and *Eichhornia crassipes*) was an excellent metal biosorbent. Miretzky et al. (2006) reported similar results when they worked with death biomass of *Spirodela intermedia, Lemna minor* and *Pistia stratiotes* using a multimetal solution. Due to the fact that *T. domingensis* have a significantly high biomass, it shows a higher metal biosorption capacity than free-floating species. When plants die, due to their degradation is slow (Hammerly et al., 1989), they follow retaining metals in CWs, as it was experimentally determined in our work. Detritus is mineralized and becomes part of the sediment. If necessary, detritus can be easily removed from the CW for final disposal. This can be an important advantage for the CW management.

4. Comparison between CW1 and CW2

CW1 is in operation for the last 15 years and CW2 for the last 8 years. Both wetlands performed satisfactory removal efficiencies during normal operation as can be seen in Tables 1 and 2. Effluents to be treated in both CWs presented high pH and conductivity. These parameters were higher in CW1 than in CW2, but in CW2 some conductivity peaks were registered. These are hard conditions for macrophyte growth. However, *T. domingensis* can survive to these conditions. The chemical composition of the effluent to be treated presented high variability, a common characteristic of industrial effluents. However, after flowing through the CWs, the parameters measured in the outlet effluent presented not only lower concentrations but also lower variability than those of the inlet effluent, proving the buffer capacity of the CWs.

   Metals were efficiently removed from the effluent. It is important to highlight that these CWs were final treatments and metal concentrations were low. At the inlet, Cr and Zn concentrations in water were significantly higher in CW2 than in CW1, while Ni exhibited the highest concentrations in the CW1. Contaminant removal efficiencies were satisfactory, except for SRP and NH_4^+ in CW1, probably due to low DO concentration. In CW2, DO concentration was higher and SRP and NH_4^+ removal efficiencies were higher. An important decrease of sulfate was also observed. In both CWs, effluents showed high sulfate concentrations due to its use in the primary treatments. CWs performances were steady over the operation periods, allowing concentrations to meet the regulatory limits set by law.

In both CWs, *T. domingensis* is the dominant macrophyte, covering 60–90% of the surface of the CWs during all the year. *Typha* spp. is one of the most tolerant, invading, and productive macrophyte in treatment wetlands of the entire world (Calheiros et al., 2009; Carranza-Álvarez et al., 2008; Hadad et al., 2006, 2010; Juwarker et al., 1995; Kadlec and Wallace, 2009; Maddison et al., 2005; Maine et al., 2007, 2009, 2013; Manios et al., 2003; Vymazal, 2013; Vymazal and Kröpfelová, 2008). Due to their characteristics, *Typha* spp. creates monospecific communities in the treatment wetlands where they are used.

In the studied CWs, *T. domingensis* showed high metal retention capacity in roots demonstrating its ability of phytostabilization. In a greenhouse experiment, this species was exposed to high concentrations of Cr, Ni, and Zn in combined treatments (Mufarrege et al., 2014). *T. domingensis* decreased the root morphology parameters due to extremely high metal concentrations. However, the metaxylematic vessel's cross-sectional area (CSA) did not decrease to enhance metal transport to aerial parts (Fig. 9). A higher metaxylematic vessel CSA represents a higher efficiency in the uptake and accumulation of contaminants in roots, which could increase the efficiency of a CW in the retention of contaminants. Metals caused growth inhibition and affected anatomical parameters. Despite the sub-lethal effects registered, *T. domingensis* demonstrated that it could uptake Cr, Ni, and Zn efficiently and survive in polluted water bodies due to the morphological plasticity of the root system (Mufarrege et al., 2014). Due to the fact that the metal concentrations were remarkably higher than the concentrations found in natural and constructed wetlands, their results demonstrate the ability of *T. domingensis* to survive an accidental dump of high concentrations of contaminants in an aquatic system.

Table 3 shows Cr, Ni, and Zn concentrations in *T. domingensis* tissues of plants sampled in the inlet and outlet areas of both CWs. Metals concentrations in plant tissues taken at the inlet area were significantly higher than those of the outlet area in both CWs. Metals are retained in roots. Low translocation from roots to aerial parts is an advantage because metals are not available for herbivorous animals. Metals remain immobilized in the CW sediment. Despite the longer operation time of CW1, Cr, and Zn concentrations in tissues were significantly lower than in CW2. This is probably due to the higher Cr and Zn concentrations in CW2 influent than in CW1 influent. Ni concentrations in tissues were significantly higher in CW1 than in CW2 due to this metal low concentration in the influent of CW2 (Tables 1 and 2). It is important to highlight that the emergent macrophytes also have an influence on the biogeochemical cycles of the sediment affecting the redox potential because of their ability to transport oxygen from the root to the rhizosphere zone (Barko et al., 1991; Sorrel and Boon, 1992).

Qualitatively, this oxygenated layer can be visualized by the red color associated with the iron oxidized forms on the root surface and the surrounding sediment.

In both CWs, the inlet sediment showed significantly higher Cr, Ni, and Zn concentrations than those registered in the outlet zone (Table 4). The outlet sediment concentrations were not significantly different from the initial concentrations (measured at the beginning of the study), showing that contaminants are retained in the inlet area. Metal concentrations were higher in the sediment of CW1 probably due to the longer time of operation.

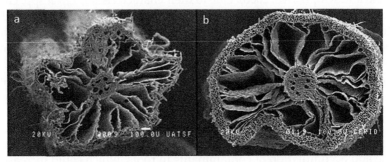

Figure 9. *T. domingensis* scanning electronic microscopy (SEM) images of cross-sectional roots obtained at the end of the experiment in the metal treatment (a) and in the control (b) (Bar = 350 μm).

Table 3. Cr, Ni and Zn concentrations (mg.g^{-1}) in tissues of *T. domingensis* grown in CW1 and CW2.

	CW1					
	Cr		Ni		Zn	
	Leaf	Root	Leaf	Root	Leaf	Root
Inlet Area	0.023	0.356	0.014	0.199	0.034	0.090
Outlet Area	0.010	0.034	0.006	0.030	0.035	0.086
	CW2					
Inlet Area	0.053	0.764	0.009	0.019	0.034	0.199
Outlet Area	0.033	0.195	0.007	0.013	0.014	0.054

Table 4. Cr, Ni and Zn concentrations (mg.g^{-1}) in CW1 and CW2 sediments.

		CW1			CW2	
Sample	Cr	Ni	Zn	Cr	Ni	Zn
Inlet Area	1.582	0.960	0.146	0.865	0.017	0.056
Outlet Area	0.047	0.039	0.063	0.016	0.011	0.024
Initial	0.038	0.028	0.060	0.016	0.011	0.024

Ni concentration was not significantly accumulated at the inlet of the CW2 because the treated effluent contained low concentrations of this metal.

Taking into account not only the concentration but also the mass of each compartment, a mass balance was carried out. Sediment was the main accumulation compartment of metals in the studied free water surface CWs with emergent macrophytes (Maine et al., 2017). This is an advantage since metals were phytostabilized within the treatment system. Sediment sorption is considered to be the main long-term contaminant accumulation mechanism (Machemer et al., 1993; Maine et al., 2009; Wood and Shelley, 1999). However, the sediment could release the contaminants if the environmental conditions changed (Boström et al., 1985). In order to determine the perdurability of the contaminant retention in the sediment of both CWs, a fractionation of sediment was carried out to assess which compounds have retained metals in the sediment.

Figure 10 shows Cr, Ni, and Zn fractionation in sediments at the end of the studied period in the inlet areas of CW1 and CW2.

The extraction sequence can be seen as an inverse scale of relative availability of metals, being the exchangeable fraction the most labile and bioavailable. It can be seen that this fraction is significantly lower than the other fractions in all cases. In both CWs, Cr, and Ni are mainly bound to Fe-Mn oxides in CW1. This fraction is known to be "the sink of metals" since they could be sorbed onto or co-precipitated with Fe-Mn oxides (Marchand et al., 2010). In CW2, Cr was mostly bound to organic matter. Organic matter has the ability to form complexes and adsorb cations due to the presence

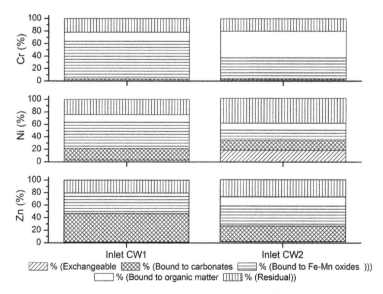

Figure 10. Cr, Ni and Zn fractionation in sediments at the end of the studied period in the inlet areas of CW1 and CW2.

of negatively charged groups (Laveuf and Cornu, 2009). Ni was found in the residual fraction in CW2. The low Ni concentration in the different fractions of CW2 sediment confirmed the low concentrations of this metal in the effluent, indicating that there was no accumulation in sediment. Zn was bound predominantly to the Fe-Mn oxides in CW2 and to carbonates in CW1. Calcium carbonate precipitation is thermodynamically favored in CW1 and Zn can co-precipitate with it. Zn retention as Zn bound to carbonates has repeatedly been reported in the literature (Banerjee, 2003; Lee et al., 2005; Stone and Marsalek, 1996). These fractions can be considered steady under the chemical and environmental conditions of the wetlands.

5. Conclusions

- In both CWs, removal efficiencies were satisfactory allowing effluents to comply with regulations. CWs decreased not only the mean value but also the variability in the concentration of contaminants in the inlet effluents.

- Metals were efficiently removed in both CWs, being accumulated mainly in sediment of the inlet area. Metals were bound to sediment fractions that would not release them to water while the chemical and environmental conditions of the systems remained the same.

- CWs outlet area sediments have not begun to accumulate contaminants, suggesting wetland potential for long term performance.

- According to the conditions considered in this study, *T. domingensis* is a suitable species for the treatment of metallurgical effluents. *T. domingensis* was efficient in metal retention, especially in roots, demonstrating its phytostabilization capacity. This is desirable because metals remain immobilized in the sediment of the CWs.

- *T. domingensis* detritus accumulated high metal concentrations. These detritus can be easily removed for their final disposition.

- In FWS wetlands, metal concentration in macrophyte tissues is related mainly to influent concentration, while metal concentration in sediment depends not only on influent concentration but also on time of operation of the CWs.

- The CWs faced with accidental events were capable of recovering their performance, demonstrating their robustness.

References

Arduini I., Masoni A., Ercoli L. Effects of high chromium applications on miscanthus during the period of maximum growth. Environ. Exp. Bot. 58, 234–243, 2006.

Banerjee A.D.K. Heavy metal levels and solid phase speciation in street dusts of Delhi, India. Environ. Pollut. 123, 95–105, 2003.

Barko J.W., Gunnison D., Carpenter S.R. Sediment interactions with submersed macrophyte growth and community dynamics. Aquat. Bot. 41, 41–65, 1991.

Boström B., Ahlgren L., Bell R. Internal nutrient loading in a eutrophic lake, reflected in seasonal variations of some sediment parameters. Verhandlungen IVL. 22, 3335–3339, 1985.

Brisson J. Ecosystem services of wetlands: does plant diversity really matter? *In*: Chazarenc, F., Gagnon, V., Méchineau, M. (eds.). Proceedings of 5th International Symposium on Wetland Pollutant Dynamics and Control, WETPOL 2013. Ecole des Mines de Nantes-GEPEA Nantes (France) 13–17 october, pp. 10–11, 2013.

Brix H. Wastewater treatment in constructed wetlands: system design, removal processes and treatment performance. USA, Lewis Pub. 1993.

Brix H. Functions of macrophytes in constructed wetlands. Water Sci. Technol. 29, 71–78, 1994.

Brix H. Do macrophytes play a role in constructed treatment wetlands? Water Sci. Technol. 35, 11–17, 1997.

Brix H., Arias C.A. The use of vertical flow constructed wetlands for onsite treatment of domestic wastewater: new danish guidelines. Ecol. Eng. 25, 491–500, 2005.

Calheiros C.S., Rangel A.O., Castro P.M. Treatment of industrial wastewater with two-stage constructed wetlands planted with *Typha latifolia* and *Phragmites australis*. Bioresour. Technol. 100, 3205–3213, 2009.

Carranza-Álvarez C., Josabad Alonso-Castro A., Alfaro de la Torre M.C., García de la Cruz R.F. Accumulation and distribution of heavy metals in *Scirpus americanus* and *Typha latifolia* from an artificial lagoon in San Luis Potosí, México. Water Air Soil Pollut. 188, 297–309, 2008.

Di Luca G.A., Mufarrege M.M., Sánchez G.C., Hadad H.R., Maine M.A. P distribution in different sediment fraction of a constructed wetland. Water Sci. Technol. 63, 2374–2380, 2011a.

Di Luca G.A., Maine M.A., Mufarrege M.M., Hadad H.R., Sánchez G.C., Bonetto C.A. Metal retention and distribution in the sediment of a constructed wetland for industrial wastewater treatment. Ecol. Eng. 37, 1267–1275, 2011b.

Di Luca G.A., Mufarrege M.M., Hadad H.R., Maine M.A. Distribution of high Zn concentrations in unvegetated and *Typha domingensis* Pers. vegetated sediments. Environ. Earth Sci. 75, 773–782, 2016.

Ellis J., Shutes R., Revitt D., Zhang T. Use of macrophytes for pollution treatment in urban wetlands. Conserv. Recycl. 11, 1–12, 1994.

Gersberg R.M., Elkins B.V., Lyon S.R., Goldman, C.R. Role of aquatic plants in wastewater treatment by artificial wetlands. Water Res. 20, 363–368, 1986.

Hadad H.R., Maine M.A., Bonetto C.A. Macrophyte growth in a pilot-scale constructed wetland for industrial wastewater treatment. Chemosphere 63, 1744–1753, 2006.

Hadad H.R., Maine M.A., Natale G.S., Bonetto C. The effect of nutrient addition on metal tolerance in *Salvinia herzogii*. Ecol. Eng. 31, 122–131, 2007.

Hadad H.R., Mufarrege M.M., Pinciroli M., Di Luca G.A., Maine M.A. Morphological response of *Typha domingensis* to an industrial effluent containing heavy metals in a constructed wetland. Arch. Environ. Contam. Toxicol. 58, 666–675, 2010.

Hammerly J., Leguizamon M., Maine M.A., Schiver D. Decomposition rate of plant material in the Parana Medio River (Argentina). Hydrobiologia 183, 179–184, 1989.

Hechmi N., Aissa N.B., Abdenaceur H., Jedidi N. Evaluating the phytoremediation potential of *Phragmites australis* grown in penta-chlorophenol and cadmium co-contaminated soils. Environ. Sci. Pollut. Res. 21, 1304–1313, 2014.

Jacob D., Otte M. Conflicting processes in the wetland plant rhizosphere: metal retention or mobilization? Water Air Soil Pollut. 3, 91–104, 2003.

Jenssen P.D., Mahlum T., Krogstad T. Potential use of constructed wetlands for wastewater treatment in northern environments. Water Sci. Technol. 28, 149–157, 1993.

Juwarker A.S., Oke B., Patnaik S.M. Domestic wastewater treatment through constructed wetland in India. Water Sci. Technol. 32, 291–294, 1995.

Kadlec R., Knight R., Vymazal J., Brix H., Cooper P., Haberl R. Constructed Wetlands for Pollutant Control. Processes, Performance, Design and Operation. IWA 2000.

Kadlec R.H. , Wallace S.D. Treatment Wetlands (2nd ed.). CRC Press, Boca Raton, Fl, 2009.

Laveuf C., Cornu S. A review on potentiality of rare earth elements to trace pedogenetic processes. Geoderma. 154, 1–12, 2009.

Lee P.K., Yu Y.H., Yun S.T., Mayer B. Metal contamination and solid phase partitioning of metals in urban roadside sediments. Chemosphere 60, 672–689, 2005.

Machemer S., Reynolds J., Laudon L., Wildeman T. Balance of S in a constructed wetland built to treat acid mine drainage. Idaho Springs, Colorado, USA. Appl. Geochem. 8, 587–603, 1993.

Maddison M., Soosaar K., Lôhmus K., Mander Ü. Cattail population in wastewater treatment wetlands in Estonia: biomass production, retention of nutrients, and heavy metals in phytomass. J. Environ. Sci. Health. 40, 1157–1166, 2005.

Maine M.A., Leguizamon M., Hammerly J., Pizarro M. Influence of the pH and redox potential on phosphorus activity in the Parana Medio system. Hydrobiologia 228, 83–90, 1992.

Maine M., Duarte M., Suñe N. Cadmium uptake by floating macrophytes. Water Res. 35, 2629–2634, 2001.

Maine M.A., Suñe N.L., Lagger, S.C. Chromium bioaccumulation: Comparison of the capacity of two floating aquatic macrophytes. Water Res. 38, 1494–1501, 2004.

Maine M.A., Suñe N., Hadad H., Sánchez G., Bonetto C. Phosphate and metal retention in a small-scale constructed wetland for waste-water treatment. *In*: Golterman H.L., Serrano L. (eds.). Phosphates in Sediments. Backhuys Publishers, Leiden, The Netherlands, pp. 21–31, 2005.

Maine M.A., Suñé N., Hadad H.R., Sánchez G., Bonetto C. Nutrient and metal removal in a constructed wetland for waste-water treatment from a metallurgic industry. Ecol. Eng. 26, 341–347, 2006.

Maine M.A., Suñé N., Hadad H.R., Sánchez G., Bonetto C. Removal efficiency of a constructed wetland for wastewater treatment according to vegetation dominance. Chemosphere 68, 1105–1113, 2007.

Maine M.A., Hadad H.R., Sánchez G., Caffaratti S., Bonetto C. Influence of vegetation on the removal of heavy metals and nutrients in a constructed wetland. J. Environ. Manage. 90, 355–363, 2009.

Maine M.A., Hadad H.R., Sánchez G.C., Mufarrege M.M., Di Luca G.A., Caffaratti S.E., Pedro M.C. Sustainability of a constructed wetland faced with a depredation event. J. Environ. Manage. 128, 1–6, 2013.

Maine M.A., Hadad H.R., Sánchez G.C., Di Luca G.A., Mufarrege M.M., Caffaratti S.E., Pedro M.C. Long-term performance of two free-water surface wetlands for metallurgical effluent treatment. Ecol. Eng. 98, 372–377, 2017.

Mangabeira P.A., Ferreira A.S., de Almeida A.A.F., Fernandes V.F., Lucena E., Souza V.L., dos Santos Júnior A.J., Oliveira A.H., Grenier-Loustalot M.F., Barbier F., Silva D.C. Compartmentalization and ultra-structural alterations induced by chromium in aquatic macrophytes. Biometals 24, 1017–1026, 2011.

Manios T., Stentiford E., Millner P. The effect of heavy metals accumulation on the chlorophyll concentration of *Typha latifolia* plants, growing in a substrate containing sewage sludge compost and watered with metalliferous water. Ecol. Eng. 20, 65–74, 2003.

Marchand L., Mench M., Jacob D.L., Otte M.L. Metal and metalloid removal in constructed wetlands, with emphasis on the importance of plants and standardized measurements: a review. Environ. Pollut. 158, 3447–3461, 2010.

Miretzky P., Saralegui A., Fernandez-Cirelli A. Simultaneous heavy metal removal mechanism by dead macrophytes. Chemosphere 66, 247–254, 2006.

Mitsch W.J., Gosselink J.G. Wetlands (3rd ed.). John Wiley and Sons, NY, 2000.

Mufarrege M.M., Di Luca G.A., Hadad H.R., Maine, M.A. Adaptability of *Typha domingensis* to high pH and salinity. Ecotoxicology 20, 457–465, 2011.

Mufarrege M.M., Hadad H.R., Di Luca G.A., Maine, M.A. Metal dynamics and tolerance of *Typha domingensis* exposed to high concentrations of Cr, Ni and Zn. Ecotoxicol. Environ. Saf. 105, 90–96, 2014.

Mufarrege M.M., Hadad H.R., Di Luca G.A., Maine M.A. The ability of *Typha domingensis* to accumulate and tolerate high concentrations of Cr, Ni, and Zn. Environ. Sci. Pollut. Res. 22, 286–292, 2015.

Mufarrege M.M., Di Luca G.A., Hadad H.R., Sanchez G.C., Pedro M.C., Maine M.A. Effects of the presence of nutrients in the removal of high concentrations of Cr(III) by *Typha domingensis*. Environ. Earth Sci. 75, 887–895, 2016.

Poschenrieder C., Tolrá R., Barceló J. Can metal defend plants against biotic stress? Trends Plant Sci. 11, 288–295, 2006.

Rodríguez L.C., Corrales A., Lara-Borrero J. Pathogen removal in constructed wetlands for stormwater harvesting at Pontificia Universidad Javeriana. *In*: Hadad H.R., Maine M.A. (eds.). Proceedings III Conferencia Panamericana de Sistemas de Humedales para el Tratamiento y Mejoramiento de la Calidad del Agua. Ediciones UNL, Santa Fe, Argentina, pp. 145–146, 2016.

Schneider I., Rubio J. Sorption of heavy metal ions by the nonliving biomass of freshwater macrophytes. Environ. Sci. Technol. 33, 2213–2217, 1999.

Shubiao W., Kuschk P., Brix H., Vymazal J., Dong R. Development of constructed wetlands in performance intensifications for wastewater treatment: A nitrogen and organic matter targeted review. Water Res. 57, 40–55, 2014.

Sorrell B.K., Boon P.L. Biogeochemistry of billabong sediments II. Seasonal variations in methane production. Freshwater Biol. 27, 435–445, 1992.

Stone M., Marsalek J. Trace metal composition and speciation in street sediment: sault Ste. Marie, Canada. Water Air Soil Pollut. 87, 149–169, 1996.

Vymazal, J. Constructed wetlands for wastewater treatment: five decades of experience. Environ. Sci. Technol. 45, 61–69, 2011.

Vymazal J. Emergent plants used in free water surface constructed wetlands: A review. Ecol. Eng., 61, 582–592, 2013.

Vymazal J., Krópfelová L. Growth of *Phragmites australis* and *Phalaris arundinacea* in constructed wetlands for wastewater treatment in the Czech Republic. Ecol. Eng. 25, 606–621, 2005.

Vymazal J., Kröpfelová L. Nitrogen and phosphorus standing stocks in *Phalaris arundinacea* and *Phragmites australis* in a constructed wetland: 3-year study. Arch. Agron. Soil Sci. 54, 297–308, 2008.

Weis J.S., Weis P. Metal uptake, transport and release by wetland plants: Implications for phytoremediation and restoration. Review. Environ. Int. 30, 685–700, 2004.

Wood T., Shelley M.A. Dynamic model of bioavailability of metals in constructed wetland sediments. Ecol. Eng. 12, 231–252, 1999.

Wu S., Wallace S., Brix H., Kuschk P., Kipkemoi Kirui W., Masi F., Dong R. Treatment of industrial effluents in constructed wetlands: Challenges, operational strategies and overall performance. Environ. Pollut. 201, 107–120, 2015.

Ye Z.H., Whiting S.N., Qian J.H., Lytle C.M., Lin Z.Q., Terry, N. Trace element removal from coal ash leachate by a 10-year-old constructed wetland. J. Environ. Qual. 30, 1710–1719, 2001.

Zhang D.Q., Jinadasa K.B., Richard M.G., Liu Y., Ng W.J., Tan S.K. Application of constructed wetlands for wastewater treatment in developing countries: A review of recent developments (2000–2013). J. Environ. Manage. 141, 116–131, 2014.

Zurita F., Belmont M.A., De Anda J., White J.R. Seeking away to promote the use of constructed wetlands for domestic wastewater treatment in developing countries. Water Sci. Technol. 63, 654–659, 2011.

Index

Printed and bound by CPI Group (UK) Ltd, Croydon, CR0 4YY

01/11/2024

01782623-0008